# NCCER's Instructor Resource Center

**This website is passcode-protected and is only meant for instructor use. The access code below provides you with access to the following:**

- Test Generator software and module exams
- Lesson plans
- PowerPoints®
- Performance Profile sheets

AF683849

**Your access code is good for all levels of the specified craft and edition it accompanies.**

**To get started, follow the steps below:**

1. **Visit www.nccerirc.com.**
2. **Select the appropriate craft area.**
3. **Locate your book cover/edition and click "Login"**
4. **If you are a first-time user, select "Register" and agree to the privacy statement. Select "Yes" if you have ever created a Pearson account. Select "No" if you need to set up a new one.**
5. **Enter your access code provided below and follow the prompts.**

The Access Code is valid for one-time redemption upon intial registration. You may then log-in using your username and password every time you return to this craft area.

**Note:** Additional resources may be added to this site as they become available.

**Pearson Customer Technical Support**

**For assistance with the NCCER IRC, you can access our 24-hour Tech Support Site at http://247pearsoned.custhelp.com**

Access Code Millwright

PEEL HERE →

**STORE IN A SECURE AREA!**

PEARSON

nccer

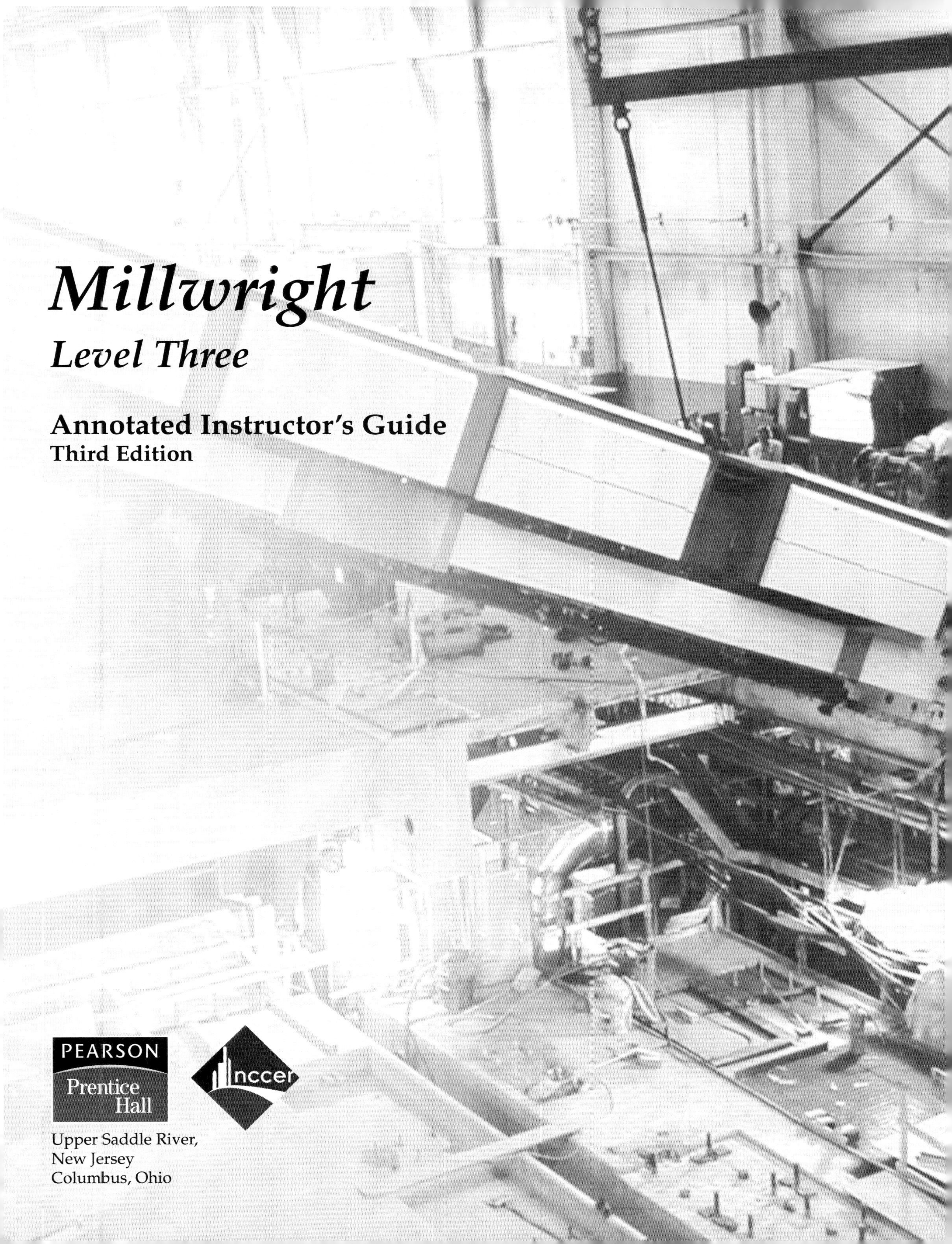

# *Millwright*

## *Level Three*

**Annotated Instructor's Guide**
**Third Edition**

PEARSON
Prentice Hall

nccer

Upper Saddle River,
New Jersey
Columbus, Ohio

**National Center for Construction Education and Research**

*President:* Don Whyte
*Director of Curriculum Revision and Development:* Daniele Stacey
*Millwright Project Manager:* Carla Sly
*Production Manager:* Tim Davis
*Quality Assurance Coordinator:* Debie Ness
*Editors:* Rob Richardson and Matt Tischler
*Desktop Publishing Coordinator:* James McKay
*Desktop Publisher:* Brittany Ferguson

NCCER would like to acknowledge the contract service provider for this curriculum:
Topaz Publications, Liverpool, New York.



PEARSON Prentice Hall

nccer

10 9 8 7 6 5 4
ISBN 0-13-614432-2
ISBN 13 978-0-13-614432-8

# PREFACE

## TO THE TRAINEE

In choosing to begin millwright training, you are seizing the opportunity to enjoy a career that demands complex mechanical knowledge and sharp analytical ability in a variety of industrial work environments. Since its humble beginnings in the construction of wood mills, the millwright trade has expanded to include work in metal and machinery of ever-increasing technology and precision. Millwrights install, align, and troubleshoot machinery in factories, power plants (particularly the precision machinery required in nuclear power plants), and other industrial sites. They install conveyor systems, connect machinery to power supplies and piping, direct hoisting and setting of machines, and adjust the moving and stationary parts of machines to certain specifications. Millwrights are extremely skilled at mathematics and interpreting blueprints and specs to set machines at perfect measurements, sometimes working with clearances no bigger than thousandths of an inch.

Millwrights are a specialized and elite group, as there are only about 60,000 millwrights employed in the United States today. However, over the next decade, there will be a demand for a third more (U.S. Bureau of Labor Statistics). As the population grows, especially in developing countries, increased demands for energy and travel in particular will require more millwrights working in power plants, refineries, and factories. Trained and experienced millwrights enjoy a comfortable salary and the chance for different avenues of professional development. Millwrights can progress upwards in their trade, undergoing full apprenticeship, becoming supervisors, and/or obtaining higher education; millwrights may also opt for careers in related professions, such as machinist, equipment engineer, or aircraft assembler.

If you have a talent for detail and problem-solving, and if you enjoy working independently in a variety of work environments, you will excel in this training program. NCCER wishes you luck as you embark on your chosen career path. With solid millwright training and experience, there are countless avenues open to you in an industrial sector that shows no signs of slowing down.

We invite you to visit the NCCER website at www.nccer.org for the latest releases, training information, newsletter, and much more. You can also reference the Contren® product catalog online at www.crafttraining.com. Your feedback is welcome. You may email your comments to curriculum@nccer.org or send general comments and inquiries to info@nccer.org.

## CONTREN® LEARNING SERIES

The National Center for Construction Education and Research (NCCER) is a not-for-profit 501(c)(3) education foundation established in 1995 by the world's largest and most progressive construction companies and national construction associations. It was founded to address the severe workforce shortage facing the industry and to develop a standardized training process and curricula. Today, NCCER is supported by hundreds of leading construction and maintenance companies, manufacturers, and national associations. The Contren® Learning Series was developed by NCCER in partnership with Pearson Education, Inc., the world's largest educational publisher.

Some features of NCCER's Contren® Learning Series are as follows:

- An industry-proven record of success
- Curricula developed by the industry for the industry
- National standardization, providing portability of learned job skills and educational credits
- Compliance with the Office of Apprenticeship requirements for related classroom training (*CFR 29:29*)
- Well-illustrated, up-to-date, and practical information

NCCER also maintains a National Registry that provides transcripts, certificates, and wallet cards to individuals who have successfully completed modules of NCCER's Contren® Learning Series. *Training programs must be delivered by an NCCER Accredited Training Sponsor in order to receive these credentials.*

# Contents

# Contren® Curricula

NCCER's training programs comprise more over 50 construction, maintenance, and pipeline areas and include skills assessments, safety training, and management education.

Boilermaking
Cabinetmaking
Carpentry
Concrete Finishing
Construction Craft Laborer
Construction Technology
Core Curriculum:
  Introductory Craft Skills
Drywall
Electrical
Electronic Systems Technician
Heating, Ventilating, and
  Air Conditioning
Heavy Equipment Operations
Highway/Heavy Construction
Hydroblasting
Industrial Coating and Lining
  Application Specialist
Industrial Maintenance
  Electrical and Instrumentation
  Technician
Industrial Maintenance Mechanic
Instrumentation
Insulating
Ironworking
Masonry
Millwright
Mobile Crane Operations
Painting
Painting, Industrial
Pipefitting
Pipelayer
Plumbing
Reinforcing Ironwork
Rigging
Scaffolding
Sheet Metal
Site Layout
Sprinkler Fitting
Welding

**Pipeline**
Control Center Operations,
  Liquid
Corrosion Control
Electrical and Instrumentation
Field Operations, Liquid
Field Operations, Gas
Maintenance
Mechanical

**Safety**
Field Safety
Safety Orientation
Safety Technology

**Management**
Introductory Skills for the
  Crew Leader
Project Management
Project Supervision

**Spanish Translations**
Andamios
Currículo Básico
  Habilidades Introductorias
  del Oficio
Instalación de Rociadores
  Nivel Uno
Introducción a la Carpintería
  Nivel Uno
Orientación de Seguridad
Principios Básicos de Maniobras
Seguridad de Campo

**Supplemental Titles**
Applied Construction Math
Careers in Construction

# Acknowledgments

This curriculum was revised as a result of the farsightedness and leadership of the following sponsors:

ABC SW Pelican Training Center and Performance Contractors, Inc.
Bierlein Companies
Cianbro Institute
Constellation Energy
QCI/GE Energy Services
Shaw Group
Sunoco, Inc.
Timec Company, Inc./Frontier Oil
Turner Industries
Zachry Construction Corporation

This curriculum would not exist were it not for the dedication and unselfish energy of those volunteers who served on the Authoring Team. A sincere thanks is extended to the following:

Terry Auger
Barry Daigle
Mark Farrar
Ed Jamison
Gerald Kenyon
Ed LePage
Larry Mustin
Richard Platt
Manuel Ramos
Bill Wall
John Ziegler

## NCCER PARTNERING ASSOCIATIONS

American Fire Sprinkler Association
Associated Builders and Contractors, Inc.
Associated General Contractors of America
Association for Career and Technical Education
Association for Skilled and Technical Sciences
Carolinas AGC, Inc.
Carolinas Electrical Contractors Association
Center for the Improvement of Construction Management and Processes
Construction Industry Institute
Construction Users Roundtable
Design Build Institute of America
Electronic Systems Industry Consortium
Merit Contractors Association of Canada
Metal Building Manufacturers Association
NACE International
National Association of Minority Contractors
National Association of Women in Construction
National Insulation Association
National Ready Mixed Concrete Association
National Systems Contractors Association
National Technical Honor Society
National Utility Contractors Association
NAWIC Education Foundation
North American Crane Bureau
North American Technician Excellence
Painting & Decorating Contractors of America
Portland Cement Association
SkillsUSA
Steel Erectors Association of America
Texas Gulf Coast Chapter, ABC
U.S. Army Corps of Engineers
University of Florida
Women Construction Owners & Executives, USA

# Product Supplements

## Windows/Macintosh-Based TestGen

ISBN 13 978-0-13-614783-1
ISBN 0-13-614783-6 $30

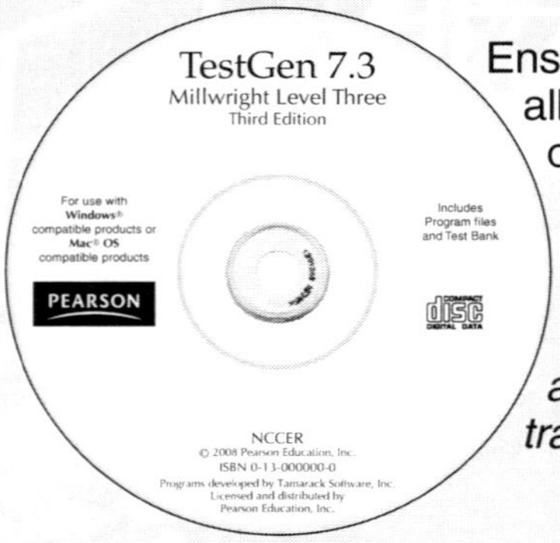


Ensure test security with NCCER's computerized testing software. This software allows instructors to scramble the module exam questions and answer keys in order to print multiple versions of the same test, customize tests to suit the needs of their training units*, add questions, or easily create a final exam.

**Due to NCCER's Accreditation Guidelines, instructors may not delete existing questions from exams. Doing so may seriously jeopardize either the accreditation status or the training program sponsor or any recognition of trainees, instructors, and trainers through the NCCER National Registry.*

## Transparency Masters

ISBN 13 978-0-13-6147992
ISBN 0-13-614799-2 $25

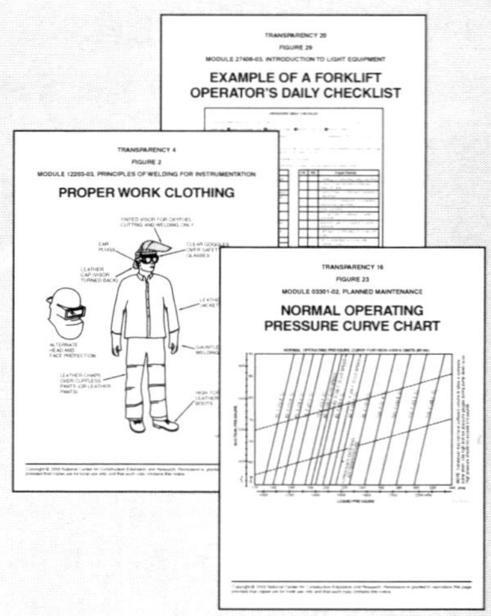


Spend more time training and less time at the copier. In response to instructor feedback NCCER offers loose, reproducible copies of the overhead transparencies referenced in the Instructor's Guides. The transparency masters package includes most of the Trainee Guide graphics, enlarged for projection and printed on loose sheets* for easy copying onto transparency film using your photocopier.

* *Transparency masters are provided on regular, loose sheets of paper, not acetates.*

To order any of these supplements, contact Pearson Education, Inc. at

**800-922-0579**

Module 15301-08

# *Advanced Trade Math*

## NCCER STANDARDIZED CRAFT TRAINING PROGRAM

The National Center for Construction Education and Research (NCCER) provides a standardized national program of accredited craft training. Key features of the program include instructor certification, competency-based training, and performance testing. The program provides trainees, instructors, and companies with a standard form of recognition through a National Craft Training Registry. The program is described in full in the *Guidelines for Accreditation*, published by NCCER. For more information on standardized craft training, contact the NCCER by writing us at 3600 NW 43rd St., Bldg. G, Gainesville, FL 32606; calling 352-334-0911; or emailing info@nccer.org. More information may be found at our website, www.nccer.org.

## HOW TO USE THIS ANNOTATED INSTRUCTOR'S GUIDE

Each page presents two sections of information. The larger section displays each page exactly as it appears in the Trainee Module. The narrow column ties suggested trainee and instructor actions to each page and provides icons (detailed below) to call your attention to material, safety, audiovisual, or testing requirements. The bottom of each page includes space for your notes.

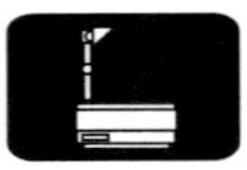
The **Audiovisual** icon indicates an appropriate time to show a transparency or other audiovisual aid.

The **Classroom** icon prompts you to define a term, stress a point, ask trainees to explain a concept, or give examples.

The **Demonstration** icon directs you to show trainees how to perform tasks.

The **Examination** icon tells you to administer the written module examination.

The **Homework** icon is placed where you may wish to assign reading for the next class, assign a project, or advise trainees to prepare for an examination.

The **Laboratory** icon is used when trainees are to practice performing tasks.

The **Materials** icon is a reminder for you to gather materials needed for classes, labs, and testing.

The **Performance Testing** icon tells you to administer a performance test or a portion thereof.

The **Safety** icon is used to emphasize safety issues. It is often keyed to *Caution* and *Warning!* statements in the Trainee Module.

The **Teaching Tip** icon indicates additional guidance is available, such as how to conduct an exercise, get the most educational value from a field trip, or encourage class participation. Teaching Tips may expand on a feature (*Think About It*, *Did You Know?*) or provide *Quick Quizzes* or similar exercises. You will be referred to the Teaching Tips section at the back of the module if there is additional material.

The **Combination** icon indicates that the laboratory listed corresponds with a performance task. If desired, you can note the proficiency of the trainees during the laboratory, and use it to satisfy performance testing requirements.

## PREPARATION

Before teaching this module, you should review the Objectives, Performance Tasks, Materials and Equipment List, and Module Outline. Be sure to allow ample time to prepare your own training or lesson plan and gather all required materials and equipment.

## MODULE OVERVIEW

This module offers advanced practice in geometry, ratios, trigonometry, and algebra.

## PREREQUISITES

Please refer to the Course Map in the Trainee Module. Prior to training with this module, it is recommended that the trainee shall have successfully completed the following: *Core Curriculum; Millwright Level One;* and *Millwright Level Two.*

## OBJECTIVES

Upon completion of this module, the trainee will be able to do the following:

1. Use tables of equivalents.
2. Use unit conversion tables.
3. Perform right angle trigonometry.
4. Calculate weights of objects.

## PERFORMANCE TASKS

This is a knowledge-based module. There are no Performance Tasks.

## MATERIALS AND EQUIPMENT LIST

Overhead projector and screen
Transparencies
Blank acetate sheets
Transparency pens
Whiteboard/chalkboard
Markers/chalk
Pencils and scratch paper
Scientific calculator
Ruler
Framing square
Quick Quiz*
Module Examination**

* Located in the back of this module
**Located in the Test Booklet

## ADDITIONAL RESOURCES

This module is intended to present thorough resources for task training. The following reference work is suggested for both instructors and motivated trainees interested in further study. This is optional material for continued education rather than for task training.

*Applied Construction Math,* Latest Edition. Upper Saddle River, NJ: Prentice Hall Publishing.

## TEACHING TIME FOR THIS MODULE

An outline for use in developing your lesson plan is presented below. Note that each Roman numeral in the outline equates to one session of instruction. Each session has a suggested time period of 2½ hours. This includes 10 minutes at the beginning of each session for administrative tasks and one 10-minute break during the session. Approximately 20 hours are suggested to cover *Advanced Trade Math.* You will need to adjust the time required for hands-on activity and testing based on your class size and resources.

| Topic | Planned Time |
|---|---|
| **Session I. Introduction; Tables of Equivalents; Unit Conversion Tables** | |
| A. Introduction | ____________ |
| B. Tables of Equivalents | ____________ |
| C. Unit Conversion Tables | ____________ |
| **Sessions II through VI. Trigonometry** | |
| A. Trigonometry | ____________ |
| B. Pythagorean Theorem | ____________ |
| C. Trigonometric Functions | ____________ |
| D. Triangle Calculation | ____________ |
| E. Determining the Angles When Side Lengths Are Known | ____________ |
| F. Interpolation | ____________ |
| G. Law of Sines | ____________ |
| **Session VII. Calculating the Weight of an Object** | |
| A. Calculating the Weight of an Object | ____________ |
| **Session VIII. Review and Testing** | |
| A. Module Review | ____________ |
| B. Module Examination | ____________ |

1. Trainees must score 70% or higher to receive recognition from NCCER.
2. Record the testing results on Craft Training Report Form 200, and submit the results to the Training Program Sponsor.

Millwright Level Three

15301-08

# Advanced Trade Math

Assign reading of Module 15301-08.

# 15301-08
# *Advanced Trade Math*

*Topics to be presented in this module include:*



## Overview

Millwrights use mathematics to calculate lengths, volumes, and weights. This module offers advanced practice in geometry, ratios, trigonometry, and algebra. You will learn to lay out special angle cuts, determine the different sides of an offset, and calculate the weight of an object.

*Instructor's Notes:*

## Objectives

When you have completed this module, you will be able to do the following:

1. Use tables of equivalents.
2. Use unit conversion tables.
3. Perform right angle trigonometry.
4. Calculate weights of objects.

## Trade Terms

| | |
|---|---|
| Adjacent side | Ratio |
| Cosine | Reference angle |
| Hypotenuse | Sine |
| Opposite side | Tangent |

## Required Trainee Materials

1. Pencil and paper
2. Scientific calculator

## Prerequisites

Before you begin this module, it is recommended that you successfully complete *Core Curriculum; Millwright Level One;* and *Millwright Level Two.*

This course map shows all of the modules in the third level of the *Millwright* curriculum. The suggested training order begins at the bottom and proceeds up. Skill levels increase as you advance on the course map. The local Training Program Sponsor may adjust the training order.

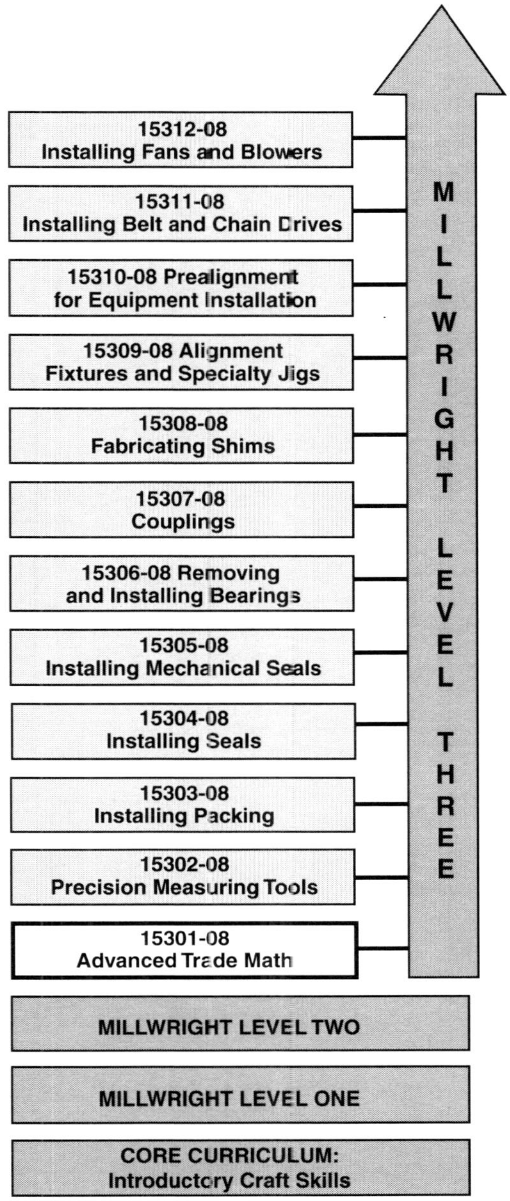


301CMAP.EPS



Ensure that you have everything required to teach the course. Check the Materials and Equipment list at the front of this module.

See the general Teaching Tip at the end of this module.

Explain that terms shown in bold are defined in the Glossary at the back of this module.

Show Transparency 1, Objectives. Review the goals of the module, and explain what will be expected of the trainee.

Review the modules covered in Level Three and explain how this module fits in.

See the Teaching Tip for Sections 1.0.0–5.0.0 at the end of this module.

**Discuss the math skills typically required of millwrights.**

**Explain how tables of equivalents are used to convert different values and measures.**

**Show Transparency 2 (Table 1). Demonstrate how to use Table 1 to determine decimal equivalents of fractions.**

**Show Transparency 3 (Table 2) and explain the equivalent units of measure.**

**Explain how and why unit conversion tables are used.**

**Show Transparency 4 (Table 3). Demonstrate how to use Table 3 to perform unit conversions.**

**Assign reading of Sections 4.0.0–4.6.0.**

**Ensure that you have everything required to teach this session.**

## 1.0.0 ◆ INTRODUCTION

In addition to arithmetic, a millwright needs advanced mathematical skills. This module explains tables of equivalents and unit conversions, how to perform right angle trigonometry, and how to calculate the weight of an object. All these skills can be applied to your everyday duties as a millwright, and the use of mathematics can become a valuable tool to make your job easier.

## 2.0.0 ◆ TABLES OF EQUIVALENTS

A measurement that is expressed in one type of unit can be quickly read as another type of unit by using a table of equivalents. These tables can be used to convert different types of numeric values and measures. Tables of equivalents are usually written so that the equivalent numbers or measurements are listed across from each other.

To use a table of equivalents, locate the number that you wish to convert in the table. Follow that line across the table until you locate the equivalent measurement. *Table 1* lists decimal equivalents of common fractions. *Table 2* lists equivalent units of measure.

You will find that some equivalencies will be so useful that you will memorize them over time. Fraction equivalents in decimal form are easiest to remember, especially quarters, halves, eighths, and sixteenths, because those are the fractions shown on the common tape measure. Remember that one sixteenth is 0.0625 and one eighth is 0.125. You have probably already memorized one quarter (0.25), one half (0.50), and three quarters (0.75), because of the coinage system, where a 25-cent coin is called a quarter.

## 3.0.0 ◆ UNIT CONVERSION TABLES

A number expressed in one unit of measurement can be converted to another unit of measurement using multiplication or division. When a number is converted its value is not changed, it is simply expressed in different units. *Table 3* lists standard conversions obtained by multiplication.

Using *Table 3*, follow these steps to perform conversions:

*Step 1* Locate the unit of measurement to be converted in the To Change column.

*Step 2* Locate the desired unit of measurement in the second (To) column.

*Step 3* Multiply the corresponding number in the Multiply By column by the original quantity.

> **NOTE**
> If you are converting a quantity to smaller units, your answer will contain more of the smaller units. If you are converting a quantity to larger units, your answer will contain fewer of the larger units.

## 4.0.0 ◆ TRIGONOMETRY

Trigonometry is a branch of mathematics that deals with the study of angles, triangles, and distances. Every triangle has six components: three sides and three angles. If the length of one side is known and any two of the other components are known, the remaining angles and sides can be calculated using trigonometry. These calculations can be useful for the following tasks:

- Laying out areas
- Determining machine and vessel placement
- Determining stress loads
- Laying out piping offsets

Right angle trigonometry is about the relationships between the angles and sides of a right triangle. If the triangle from which you are obtaining dimensions and angles is not a right triangle, you can make two right triangles out of it as shown in *Figure 1*. Determine the angles and sides that you need for the original triangle, and then use that information to obtain the other angles and sides.

### 4.1.0 Pythagorean Theorem

The simplest triangles are right triangles. A right triangle has one 90-degree, or right, angle. The angles of any triangle add up to 180 degrees, so if you know any two angles, subtract the sum of the known angles from 180, and you will know the third angle. A right angle is usually indicated with a small box drawn in the angle. The right triangle is very important to millwrights because it is used to determine the components of a piping offset.

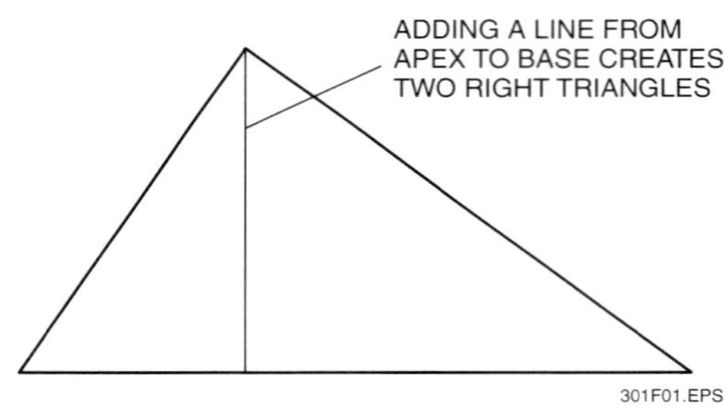


*Figure 1* ◆ Making two right triangles.



Instructor's Notes:

1.2 MILLWRIGHT ◆ LEVEL THREE

**Table 1** Decimal Equivalents of Common Fractions

| Fraction | Decimal Equivalent | | Fraction | Decimal Equivalent | |
|---|---|---|---|---|---|
| | English (in) | Metric (mm) | | English (in) | Metric (mm) |
| 1/64 | 0.015625 | 0.3969 | 33/64 | 0.515625 | 13.0969 |
| 1/32 | 0.03125 | 0.7938 | 17/64 | 0.53125 | 13.4938 |
| 3/64 | 0.046875 | 1.1906 | 35/64 | 0.546875 | 13.8906 |
| 1/16 | 0.0625 | 1.5875 | 9/16 | 0.5625 | 14.2875 |
| 5/64 | 0.078125 | 1.9844 | 37/64 | 0.578125 | 14.6844 |
| 3/32 | 0.09375 | 2.3813 | 19/32 | 0.59375 | 15.0813 |
| 7/64 | 0.109375 | 2.7781 | 39/64 | 0.609375 | 15.4781 |
| 1/8 | 0.1250 | 3.1750 | 5/8 | 0.6250 | 15.8750 |
| 9/64 | 0.140625 | 3.5719 | 41/64 | 0.640625 | 16.2719 |
| 5/32 | 0.15625 | 3.9688 | 21/32 | 0.65625 | 16.6688 |
| 11/64 | 0.171875 | 4.3656 | 43/64 | 0.671875 | 17.0656 |
| 3/16 | 0.1875 | 4.7625 | 11/16 | 0.6875 | 17.4625 |
| 13/64 | 0.203125 | 5.1594 | 45/64 | 0.703125 | 17.8594 |
| 7/32 | 0.21875 | 5.5563 | 23/32 | 0.71875 | 18.2563 |
| 15/64 | 0.234375 | 5.9531 | 47/64 | 0.734375 | 18.6531 |
| 1/4 | 0.250 | 6.3500 | 3/4 | 0.750 | 19.0500 |
| 17/64 | 0.265625 | 6.7469 | 49/64 | 0.765625 | 19.4469 |
| 9/32 | 0.28125 | 7.1438 | 25/32 | 0.78125 | 19.8438 |
| 19/64 | 0.296875 | 7.5406 | 51/64 | 0.796875 | 20.2406 |
| 5/16 | 0.3125 | 7.9375 | 13/16 | 0.8125 | 20.6375 |
| 21/64 | 0.328125 | 8.3384 | 53/64 | 0.828125 | 21.0344 |
| 11/32 | 0.34375 | 8.7313 | 27/32 | 0.84375 | 21.4313 |
| 23/64 | 0.359375 | 9.1281 | 55/64 | 0.859375 | 21.8281 |
| 3/8 | 0.3750 | 9.5250 | 7/8 | 0.8750 | 22.2250 |
| 25/64 | 0.390625 | 9.9219 | 57/64 | 0.890625 | 22.6219 |
| 13/32 | 0.40625 | 10.3188 | 29/32 | 0.90625 | 23.0188 |
| 27/64 | 0.421875 | 10.7156 | 59/64 | 0.921875 | 23.4156 |
| 7/16 | 0.4375 | 11.1125 | 15/16 | 0.9375 | 23.8125 |
| 29/64 | 0.453125 | 11.5094 | 61/64 | 0.953125 | 24.2094 |
| 15/32 | 0.46875 | 11.9063 | 31/32 | 0.96875 | 24.6063 |
| 31/32 | 0.484375 | 12.3031 | 63/64 | 0.984375 | 25.0031 |
| 1/2 | 0.500 | 12.7000 | 1 | 1.000 | 25.4000 |

301T01.EPS



**Define trigonometry and explain how it is used in millwright work.**

**Discuss right angle trigonometry.**

**Show Transparency 5 (Figure 1). Explain how right triangles can be made from any triangle. Draw other examples.**

**Discuss different types of triangles.**

**Explain that the third angle of a triangle can be found if two angles are known.**

**Explain that millwrights use right triangles to determine the components of piping offsets.**

MODULE 15301-08 ◆ ADVANCED TRADE MATH 1.3

**Table 2** Equivalent Units of Measure

**Linear Measure**

12 inches = 1 foot    3 feet = 1 yard    5½ yards = 1 rod
40 rods = 1 furlong    8 furlongs = 1 mile

| | | |
|---|---|---|
| 10 millimeters | = | 1 centimeter |
| 100 centimeters | = | 1 meter |
| 1,000 millimeters | = | 1 meter |
| 1,000 meters | = | 1 kilometer |

**Equivalent Values**

| Inches | | Feet | | Yards | | Rods | | Furlongs | | Miles |
|---|---|---|---|---|---|---|---|---|---|---|
| 36 | = | 3 | = | 1 | | | | | | |
| 198 | = | 16.5 | = | 5.5 | = | 1 | | | | |
| 7,920 | = | 660 | = | 220 | = | 40 | = | 1 | | |
| 63,360 | = | 5,280 | = | 1,760 | = | 320 | = | 8 | = | 1 |

**Square Measure**

| | | |
|---|---|---|
| 144 square inches | = | 1 square foot |
| 9 square feet | = | 1 square yard |
| 30¼ square yards | = | 1 square rod |
| 160 square rods | = | 1 acre |
| 640 acres | = | 1 square mile |

1 square mile = 640 acres = 102,400 square rods = 3,097,600 square yards

1 square mile = 27,878,400 square feet = 4,014,489,600 square inches

**Cubic Measure**

| | | |
|---|---|---|
| 1,728 cubic inches | = | 1 cubic foot |
| 27 cubic feet | = | 1 cubic yard |
| 128 cubic feet | = | 1 cord |
| 24¾ cubic feet | = | 1 perch |
| 1 cubic yard | = | 27 cubic feet = 46,656 cubic inches |

**Liquid Measure**

1 gallon occupies 231 cubic inches
7.48 gallons occupy 1 cubic foot
1 liter occupies 1,000 cubic centimeters
1,000 liters occupy 1 cubic meter

301T02.EPS



Instructor's Notes:

**Table 3** Standard Conversions

| To Change | To | Multiply By |
|---|---|---|
| Inches | Feet | 0.0833 |
| Inches | Millimeters | 25.4 |
| Feet | Inches | 12.0 |
| Feet | Yards | 0.333 |
| Yards | Feet | 3.0 |
| Square inches | Square feet | 0.00694 |
| Square feet | Square inches | 144.0 |
| Square feet | Square yards | 0.11111 |
| Square yards | Square feet | 9.0 |
| Cubic inches | Cubic feet | 0.00058 |
| Cubic feet | Cubic inches | 1,728.0 |
| Cubic feet | Cubic yards | 0.03703 |
| Cubic yards | Cubic feet | 27.0 |
| Cubic inches | Gallons | 0.00433 |
| Cubic feet | Gallons | 7.48 |
| Gallons | Cubic inches | 231.0 |
| Gallons | Cubic feet | 0.1337 |
| Gallons | Pounds of water | 8.33 |
| Pounds of water | Gallons | 0.12004 |
| Ounces | Pounds | 0.0625 |
| Pounds | Ounces | 16.0 |
| Inches of water | Pounds per square inch | 0.0361 |
| Inches of water | Inches of mercury | 0.0735 |
| Inches of water | Ounces per square inch | 0.578 |
| Inches of water | Pounds per square foot | 5.2 |
| Inches of mercury | Inches of water | 13.6 |
| Inches of mercury | Feet of water | 1.1333 |
| Inches of mercury | Pounds per square inch | 0.4914 |
| Ounces per square inch | Inches of mercury | 0.127 |
| Ounces per square inch | Inches of water | 1.733 |
| Pounds per square inch | Inches of water | 27.72 |
| Pounds per square inch | Feet of water | 2.310 |
| Pounds per square inch | Inches of mercury | 2.04 |
| Pounds per square inch | Atmospheres | 0.0681 |
| Feet of water | Pounds per square inch | 0.434 |
| Feet of water | Pounds per square foot | 62.5 |
| Feet of water | Inches of mercury | 0.8824 |
| Atmospheres | Pounds per square inch | 14.696 |
| Atmospheres | Inches of mercury | 29.92 |
| Atmospheres | Feet of water | 34.0 |
| Long tons | Pounds | 2,240.0 |
| Short tons | Pounds | 2,000.0 |
| Short tons | Long tons | 0.89285 |

301T03.EPS



MODULE 15301-08 ◆ ADVANCED TRADE MATH 1.5

**Show Transparencies 6 and 7 (Figures 2 and 3). Point out and define the hypotenuse, opposite side, adjacent side, and reference angle.**

The sides of a right triangle have been named for reference. The side opposite the right angle is always called the **hypotenuse**, and the two sides adjacent to, or connected to, the right angle are called the legs. If one of the other angles is labeled angle A, the leg of the triangle that is not connected to angle A is called its **opposite side**. The remaining leg that is connected to angle A is called the **adjacent side**. The reference angle is the angle to which a given set of trigonometric relationships refer. *Figure 2* shows a right triangle.

**Explain how the set, run, and travel of a piping offset can be compared to a triangle.**

The sides of the piping offset have also been named for reference. These sides are called the set, run, and travel. The set is the distance, measured center to center, that the pipeline is to be offset from the previous line of pipe. The run is the total linear distance required for the offset. The travel is the center-to-center measurement of the offset piping. The angle of the fittings is the number of degrees the piping changes direction.

**Present the Pythagorean theorem and explain its practical application.**

The Pythagorean theorem states that the square of the hypotenuse is equal to the sum of the squares of the other two sides. For example, in triangle abc with c being the hypotenuse and a and b being the two legs, the Pythagorean theorem states that $a^2 + b^2 = c^2$.

**Define the trig functions of sine, cosine, and tangent. Explain how these functions are used to find unknown sides and angles.**

## 4.2.0 Trigonometric Functions

Trigonometry calculations are made by using **ratios** of two of the sides of a right triangle. The ratio of the two sides is directly related to the size of one of the angles in the triangle, known as the reference angle. *Figure 3* shows the sides and reference angle. These ratios are known as the trigonometric, or trig, functions of the reference angle. The major trig functions are sine, cosine, and tangent.

The ratios for sine, cosine, and tangent can easily be remembered by the following memory device: SOH, CAH, TOA. The first acronym means: Sine equals Opposite side divided by the Hypotenuse. The second says: Cosine equals the Adjacent side divided by the Hypotenuse. The third acronym stands for: Tangent equals the Opposite side divided by the Adjacent side. Another mnemonic (memory trick) for the three ratios is: Some Old Horse Caught Another Horse Taking Oats Away.

**Show Transparency 8 (Figure 4). Explain the relationship between trig functions and piping offsets.**

These trig functions are used to find unknown sides and angles. *Figure 4* shows the relationship between trig functions. Use these functions to find another side when you have one side and the reference angle:

- Hypotenuse × sine = opposite side
- Hypotenuse × cosine = adjacent side
- Adjacent side × tangent = opposite side
- Adjacent side × cosine = hypotenuse
- Opposite side / sine = hypotenuse
- Opposite side / tangent = adjacent side

To find the angle from any two sides:

- Opposite side divided by the hypotenuse = sine
- Adjacent side divided by the hypotenuse = cosine
- Opposite side divide by the adjacent side = tangent

**Show Transparency 9 (Figure 5). Explain how scientific calculators are used to simplify trig functions.**

A scientific calculator (*Figure 5*) simplifies trigonometric functions. If a calculator is not available, however, it is possible to look up the appropriate value for the sine, cosine, or tangent on a table of trigonometric functions and perform the multiplication or division on paper.

On the calculator, once you have set up the formula as 30 × sin 30 degrees, pressing the = button will get 15 inches as the answer. If you are using a table, looking up sin 30 will give you 0.5000 as the value of the function. Insert that value in the formula, and the adjacent side will again be 15 inches.

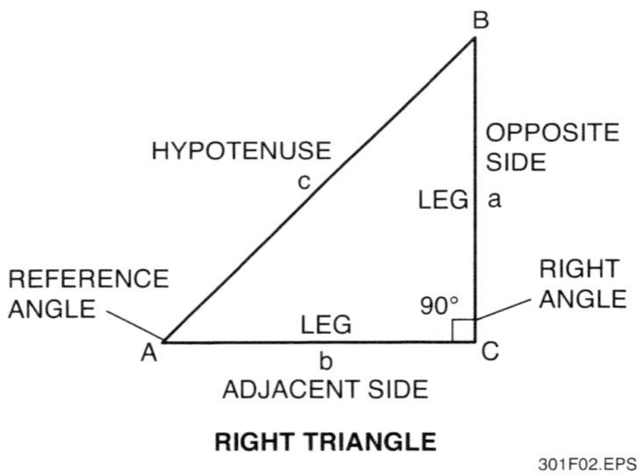


*Figure 2* ◆ Right triangle.

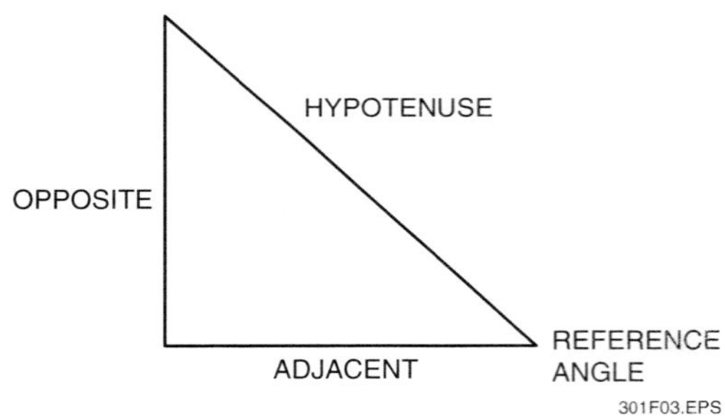


*Figure 3* ◆ Triangle terms.

1.6 MILLWRIGHT ◆ LEVEL THREE

Instructor's Notes:

### 4.2.1 Sine

The sine of the reference angle is the ratio formed by dividing the length of the side opposite the angle by the length of the hypotenuse. In triangle ABC (*Figure 6*), if the side opposite angle A is divided by the hypotenuse, the ratio is called the sine of angle A. It is usually written as sin A and read as the sine of A. The formula for sin A is written as follows:

sin A = side opposite A/hypotenuse

**Show Transparency 10 (Figure 6). Discuss the sine function and demonstrate how it applies to triangle ABC.**

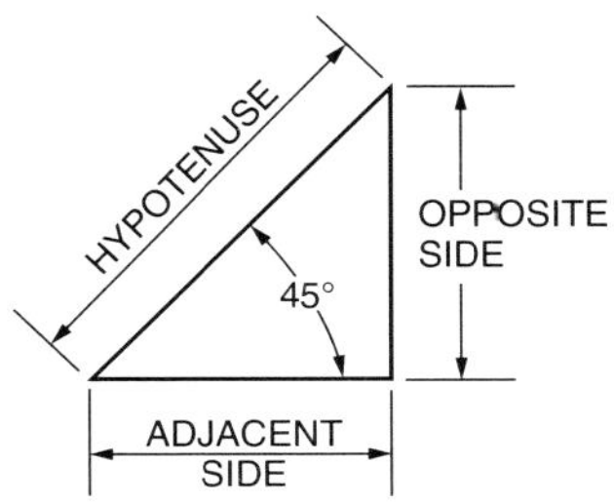


**TO FIND THE ANGLE FROM ANY TWO SIDES:**

OPPOSITE SIDE DIVIDED BY HYPOTENUSE = SINE

ADJACENT SIDE DIVIDED BY HYPOTENUSE = COSINE

OPPOSITE SIDE DIVIDED BY ADJACENT SIDE = TANGENT

| Length of sides when angle and one side are known | Angle of Offset | | | | |
|---|---|---|---|---|---|
| | 60° | 45° | 30° | 22.5° | 15° |
| Opposite = Hypotenuse × Sine | 0.866 | 0.707 | 0.5 | 0.383 | 0.259 |
| Adjacent = Hypotenuse × Cosine | 0.5 | 0.707 | 0.866 | 0.924 | 0.966 |
| Opposite = Adjacent × Tangent | 1.732 | 1.0 | 0.577 | 0.414 | 0.268 |
| Hypotenuse = Adjacent / Cosine | 0.5 | 0.707 | 0.866 | 0.924 | 0.966 |
| Hypotenuse = Opposite / Sine | 0.866 | 0.707 | 0.5 | 0.383 | 0.259 |
| Adjacent = Opposite / Tangent | 1.732 | 1.0 | 0.577 | 0.414 | 0.268 |

301F04.EPS

*Figure 4* ◆ Relationship between trig functions and piping offsets.

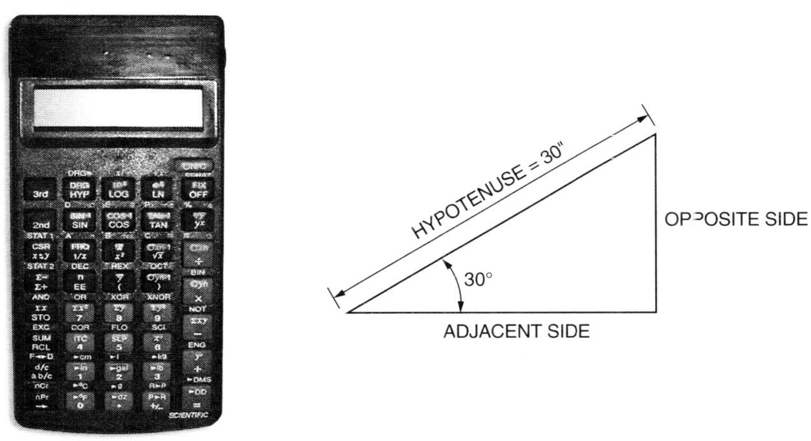


301F05.EPS

*Figure 5* ◆ Scientific calculator with sin, cos, and tan keys.



MODULE 15301-08 ◆ ADVANCED TRADE MATH 1.7

**Discuss the steps to find the sine of an angle for triangle ABC.**

**Show Transparency 11 (Table 4). Demonstrate how to use Table 4 to find the size of the angle for the triangle ABC.**

**Show Transparency 12 (Figure 7). Review the ladder angle example and explain how to find the height at which the ladder touches the wall.**

**Show Transparency 13 (Figure 8) and explain how to find the length of the hypotenuse for piping offset 1.**

Follow these steps to find the sine of an angle:

> **NOTE**
> The following steps refer to triangle ABC in which the side opposite angle A is 3 inches long and the hypotenuse is 5 inches long.

*Step 1* Find the length of the side opposite the angle.

3 inches

*Step 2* Find the length of the hypotenuse.

5 inches

*Step 3* Form a ratio, using the formula for sine.

sin A = 3/5

*Step 4* Divide the numerator of the ratio by the denominator.

3/5 = 0.6

In this example, sin A is equal to 0.600. The size of sin A can be found by using a table of trig functions. This type of table (*Table 4*) can be used to find the size of an angle from its decimal value or the decimal value from the size of the angle.

The size of angle A can be found by locating the closest value to 0.6000 under the sine column. This number is 0.6018, which corresponds to 37 degrees listed in the angle column. Therefore, angle A is approximately 37 degrees.

If a 16-foot ladder is set against a wall at a 61-degree angle, the height at which the ladder touches the wall can be found by using the formula for sine. *Figure 7* shows a representation of the ladder angle.

The height at which the ladder touches the wall is found as follows:

sin angle = side opposite/hypotenuse

sin 61 degrees = X/16 feet

sin 61 degrees = 0.8746

0.8746 × 16 = 13.9936

The height at which the ladder touches the wall is 14 feet.

Using the relationships shown in *Figure 4*, sine can be used to determine the opposite side in any right triangle when the hypotenuse is known. In piping offset 1 (*Figure 8*), the hypotenuse is given as 30 inches and the reference angle as 30 degrees. The length of the opposite side can be found as follows:

Opposite side = hypotenuse × sin 30 degrees

Opposite side = 30 inches × 0.500 = 15 inches

If the side opposite of 15 inches is known and the hypotenuse is unknown, you can use division with the sine function to determine the length of the hypotenuse. For example:

Opposite side = hypotenuse × sin 30 degrees

15 inches = hypotenuse × 0.500

15 inches/0.500 = hypotenuse

30 inches = hypotenuse

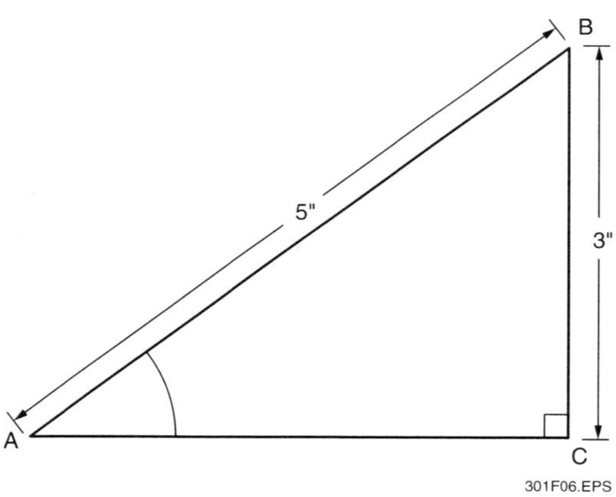


*Figure 6* ◆ Triangle ABC.

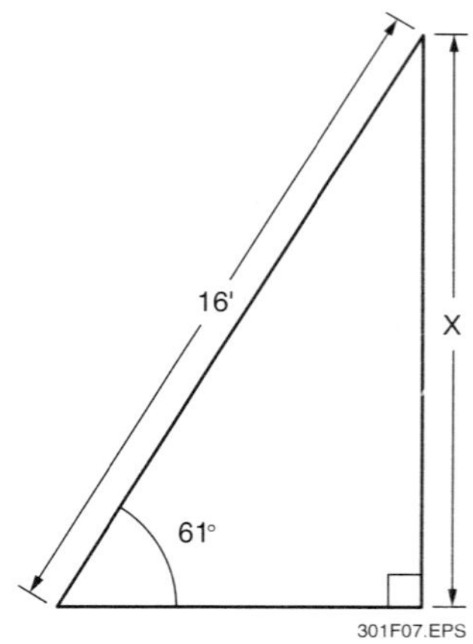


*Figure 7* ◆ Representation of ladder angle.



*Instructor's Notes:*

1.8 MILLWRIGHT ◆ LEVEL THREE

**Table 4** Trig Functions

| Degree | Sin | Cos | Tan | Csc | Sec | Cot | |
|---|---|---|---|---|---|---|---|
| 0.0 | 0.0 | 1.0 | 0.0 | --- | 1.0 | --- | 90.0 |
| 1.0 | 0.0175 | 0.9998 | 0.0175 | 57.2987 | 1.0002 | 57.29 | 89.0 |
| 2.0 | 0.0349 | 0.9994 | 0.0349 | 28.6537 | 1.0006 | 28.6363 | 88.0 |
| 3.0 | 0.0523 | 0.9986 | 0.0524 | 19.1073 | 1.0014 | 19.0811 | 87.0 |
| 4.0 | 0.0698 | 0.9976 | 0.0699 | 14.3356 | 1.0024 | 14.3007 | 86.0 |
| 5.0 | 0.0872 | 0.9962 | 0.0875 | 11.4737 | 1.0038 | 11.4301 | 85.0 |
| 6.0 | 0.1045 | 0.9945 | 0.1051 | 9.5668 | 1.0055 | 9.5144 | 84.0 |
| 7.0 | 0.1219 | 0.9925 | 0.1228 | 8.2055 | 1.0075 | 8.1443 | 83.0 |
| 8.0 | 0.1392 | 0.9903 | 0.1405 | 7.1853 | 1.0098 | 7.1154 | 82.0 |
| 9.0 | 0.1564 | 0.9877 | 0.1584 | 6.3925 | 1.0125 | 6.3138 | 81.0 |
| 10.0 | 0.1736 | 0.9848 | 0.1763 | 5.7588 | 1.0154 | 5.6713 | 80.0 |
| 11.0 | 0.1908 | 0.9816 | 0.1944 | 5.2408 | 1.0187 | 5.1446 | 79.0 |
| 12.0 | 0.2079 | 0.9781 | 0.2126 | 4.8097 | 1.0223 | 4.7046 | 78.0 |
| 13.0 | 0.225 | 0.9744 | 0.2309 | 4.4454 | 1.0263 | 4.3315 | 77.0 |
| 14.0 | 0.2419 | 0.9703 | 0.2493 | 4.1336 | 1.0306 | 4.0108 | 76.0 |
| 15.0 | 0.2588 | 0.9659 | 0.2679 | 3.8637 | 1.0353 | 3.7321 | 75.0 |
| 16.0 | 0.2756 | 0.9613 | 0.2867 | 3.28 | 1.0403 | 3.4874 | 74.0 |
| 17.0 | 0.2924 | 0.9563 | 0.3057 | 3.4203 | 1.0457 | 3.2709 | 73.0 |
| 18.0 | 0.309 | 0.9511 | 0.3249 | 3.2361 | 1.0515 | 3.0777 | 72.0 |
| 19.0 | 0.3256 | 0.9455 | 0.3443 | 3.0716 | 1.0576 | 2.9042 | 71.0 |
| 20.0 | 0.342 | 0.9397 | 0.364 | 2.9238 | 1.0642 | 2.7475 | 70.0 |
| 21.0 | 0.3584 | 0.9336 | 0.3839 | 2.7904 | 1.0711 | 2.6051 | 69.0 |
| 22.0 | 0.3746 | 0.9272 | 0.404 | 2.6695 | 1.0785 | 2.4751 | 68.0 |
| 23.0 | 0.3907 | 0.9205 | 0.4245 | 2.5593 | 1.0864 | 2.3559 | 67.0 |
| 24.0 | 0.4067 | 0.9135 | 0.4452 | 2.4586 | 1.0946 | 2.246 | 66.0 |
| 25.0 | 0.4226 | 0.9063 | 0.4663 | 2.3662 | 1.1034 | 2.1445 | 65.0 |
| 26.0 | 0.4384 | 0.8988 | 0.4877 | 2.2812 | 1.1126 | 2.0503 | 64.0 |
| 27.0 | 0.454 | 0.891 | 0.5095 | 2.2027 | 1.1223 | 1.9626 | 63.0 |
| 28.0 | 0.4695 | 0.8829 | 0.5317 | 2.1301 | 1.1326 | 1.8807 | 62.0 |
| 29.0 | 0.4848 | 0.8746 | 0.5543 | 2.0627 | 1.144 | 1.804 | 61.0 |
| 30.0 | 0.5 | 0.866 | 0.5773 | 2.0 | 1.1547 | 1.7321 | 60.0 |
| 31.0 | 0.515 | 0.8571 | 0.6009 | 1.9416 | 1.1666 | 1.6643 | 59.0 |
| 32.0 | 0.5299 | 0.848 | 0.6249 | 1.8871 | 1.1792 | 1.6003 | 58.0 |
| 33.0 | 0.5446 | 0.8387 | 0.6494 | 1.8361 | 1.1924 | 1.5399 | 57.0 |
| 34.0 | 0.5592 | 0.829 | 0.6745 | 1.7883 | 1.2062 | 1.4826 | 56.0 |
| 35.0 | 0.5736 | 0.8191 | 0.7002 | 1.7434 | 1.2208 | 1.4281 | 55.0 |
| 36.0 | 0.5878 | 0.809 | 0.7265 | 1.7013 | 1.2361 | 1.3764 | 54.0 |
| 37.0 | 0.6018 | 0.7986 | 0.7535 | 1.6616 | 1.2521 | 1.327 | 53.0 |
| 38.0 | 0.6157 | 0.788 | 0.7813 | 1.6243 | 1.269 | 1.2799 | 52.0 |
| 39.0 | 0.6293 | 0.7772 | 0.8098 | 1.589 | 1.2868 | 1.2349 | 51.0 |
| 40.0 | 0.6428 | 0.766 | 0.8391 | 1.5557 | 1.3054 | 1.1918 | 50.0 |
| 41.0 | 0.6561 | 0.7547 | 0.8693 | 1.5243 | 1.325 | 1.1504 | 49.0 |
| 42.0 | 0.6691 | 0.7431 | 0.9004 | 1.4945 | 1.3456 | 1.1106 | 48.0 |
| 43.0 | 0.682 | 0.7314 | 0.9325 | 1.4663 | 1.3673 | 1.0724 | 47.0 |
| 44.0 | 0.6947 | 0.7193 | 0.9657 | 1.4396 | 1.3902 | 1.0355 | 46.0 |
| 45.0 | 0.7071 | 0.7071 | 1.000 | 1.4142 | 1.4142 | 1.0 | 45.0 |
| | Cos | Sin | Cot | Sec | Csc | Tan | Degree |

301T04.EPS



MODULE 15301-08 ◆ ADVANCED TRADE MATH 1.9

**Explain the cosine function and how it is used.**

**Show Transparency 14 (Figure 9). Explain how to find the cosine of angle A for triangle ABC.**

**Show Transparency 15 (Figure 10). Explain how to find the hypotenuse for piping offset 2.**

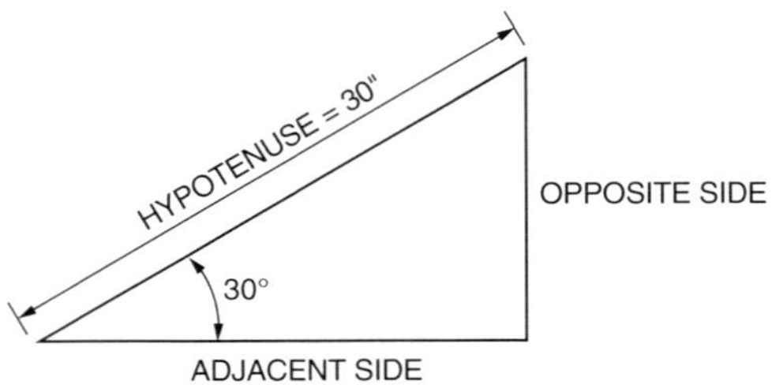


*Figure 8* ◆ Offset 1.

### 4.2.2 Cosine

The ratio formed by dividing the length of the side adjacent to an angle by the length of the hypotenuse is called the cosine of that angle. If the side adjacent to angle A is divided by the hypotenuse, the ratio is called the cosine of angle A. It is usually written as cos A. The formula for cos A is written as follows:

cos A = side adjacent to A/hypotenuse

This formula can be used to find the cosine of angle A in triangle ABC when the side adjacent to A is 39 inches and the hypotenuse is 45 inches (*Figure 9*). The cosine of angle A is found as follows:

cos A = side adjacent to A/hypotenuse

cos A = 39/45 = 13/15

cos A = 0.8667

Angle A is approximately 30 degrees (from *Table 4* or from calculator)

If the problem is to find the length of the hypotenuse of a simple offset when the angles of the fittings are 45 degrees and the adjacent side is 12 inches, the length of the hypotenuse can be found using the formula for cosine. See piping offset 2 (*Figure 10*).

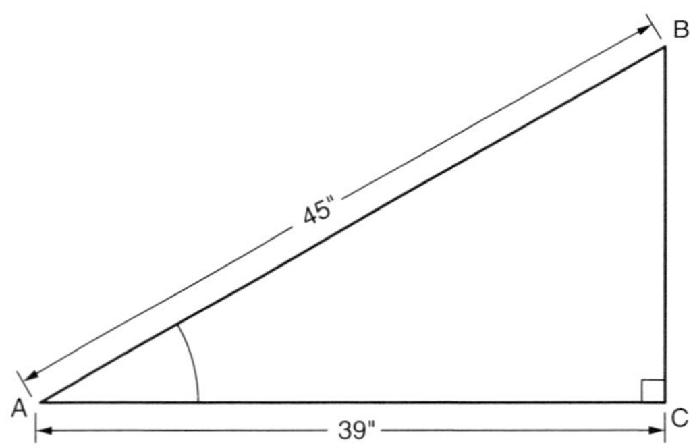


*Figure 9* ◆ Triangle ABC.

The length of the hypotenuse is found as follows:

Adjacent = hypotenuse × cos 45 degrees

12 inches = hypotenuse × 0.7071

12 inches/0.7071 = hypotenuse

16.97 inches = hypotenuse

The length of hypotenuse is 17 inches.

**Explain the tangent function and how it is used.**

**Show Transparency 16 (Figure 11). Explain how to find the tangent for triangle ABC.**

### 4.2.3 Tangent

The ratio formed by dividing the length of the side opposite an angle by the length of the side adjacent to the angle is called the tangent of that angle. If the side opposite angle A is divided by the side adjacent to angle A, the ratio is called the tangent of angle A. It is usually written as tan A. The formula for tan A is written as follows:

tan A = side opposite A/side adjacent to A

This formula can be used to find the tangent of angle A in triangle ABC when the side opposite angle A is 17 inches and the side adjacent to angle A is 21 inches (*Figure 11*).

The tangent of angle A is found as follows:

tan A = side opposite A/side adjacent to A

tan A = 17/21

tan A = 0.8095

Angle A is approximately 39 degrees (from *Table 4*).

The tangent trig function can be used to determine the opposite side of an offset when the angles of the corners and the length of the adjacent side are known.

In piping offset 3 (*Figure 12*), the adjacent side is given as 14 inches and the angle of the elbows is 30 degrees. The length of the opposite is found as follows:

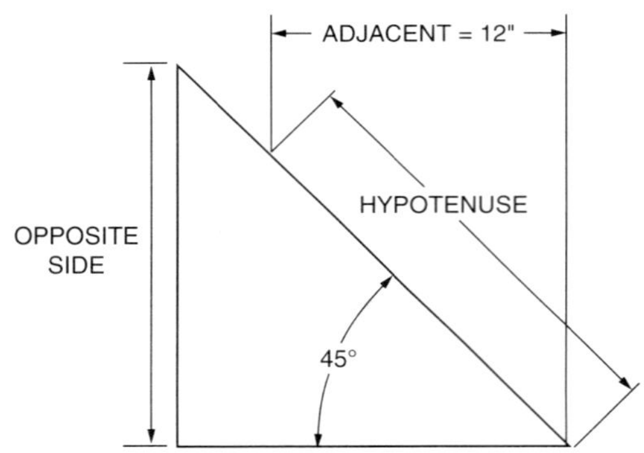


*Figure 10* ◆ Offset 2.



Instructor's Notes:

1.10 MILLWRIGHT ◆ LEVEL THREE

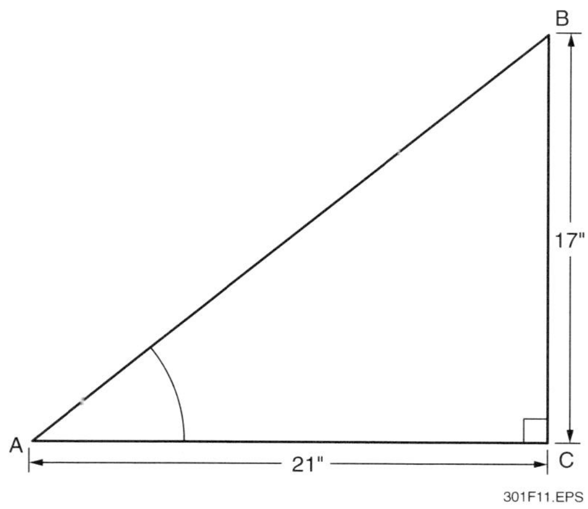


*Figure 11* ◆ Triangle ABC.

Opposite = adjacent × tan 30 degrees

Opposite = 14 inches × 0.577

Opposite = 8.078 inches

Refer back to *Table 1*, which shows that 8.078 converts to 8 5/64 inches. Rounded to the nearest 1/16-inch, it equals approximately 8 1/16 inches.

Suppose that for piping offset 3, the opposite side of 8 1/16 inches and the adjacent side of 14 inches are known, but you need to determine the angle of the elbows. The tangent is the opposite divided by the adjacent. The angle of the elbows is found as follows:

opposite/adjacent = tan

8 1/16 inches/14 inches = tan

8.0625/14 = tan (from *Table 1*)

0.576 = tan

Angle = 30 degrees (from *Table 4*)

### 4.3.0 Triangle Calculation

To calculate the sine, cosine, and tangent, of *Figure 13*, state the formula for each, substitute the line segment lengths, and use either a table or calculator to find the value of the function. For example:

tangent = opposite side over adjacent side

tangent angle A = 6/8 = 0.75

Having determined the tangent of angle A, you can find the angle to determine what angle miter you need to use. Either the calculator or the table will tell you that the angle corresponding to tan = 0.75 is 36.87 degrees. Therefore, since there are 180 degrees in any triangle, angle B is 90 – 36.87 = 53.13 degrees. Given the sides, you can know all the angles of any right triangle.

Now try the same procedure with only one side and a reference angle known, using *Figure 14* as an example. The angle is 60 degrees, and the side opposite is 48 inches long. To calculate the adjacent side (side b), use the tangent ratio, because it states the relationship between the opposite side and the adjacent side. The tangent is given as equal to the opposite divided by the adjacent.

tan A = opposite/adjacent

Multiply both sides by the unknown length b:

adjacent × tan 60 = 48 inches

Divide both sides by tan 60:

adjacent = 48/tan 60 = 27.7 inches

The Pythagorean theorem can now be used to find the length of the hypotenuse, or use the sine or cosine to find it. Since you started with side a, the opposite side, use the sine.

**Show Transparency 17 (Figure 12). Explain how to find the length of the opposite and the angle of the elbows for piping offset 3.**

**Show Transparency 18 (Figure 13). Explain how to calculate the sine, cosine, and tangent for the triangle shown in Figure 13.**

**Show Transparency 19 (Figure 14). Explain how to calculate the sine, cosine, and tangent for the triangle shown in Figure 14.**

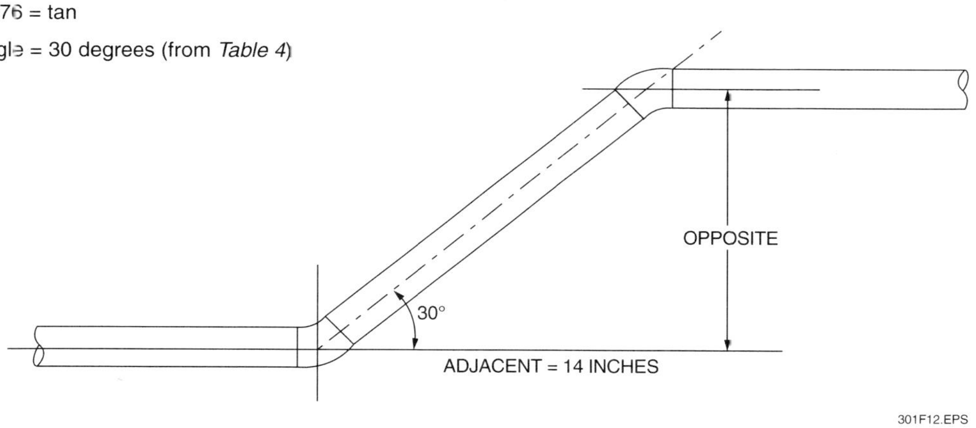


*Figure 12* ◆ Offset 3.



MODULE 15301-08 ◆ ADVANCED TRADE MATH 1.11

**Define and explain the inverse functions.**

**Describe and/or demonstrate how to use a scientific calculator to convert sine, cosine, and tangent values to angles.**

**Show Transparency 20 (Figure 15). Follow the steps for using a scientific calculator to solve the example formula.**

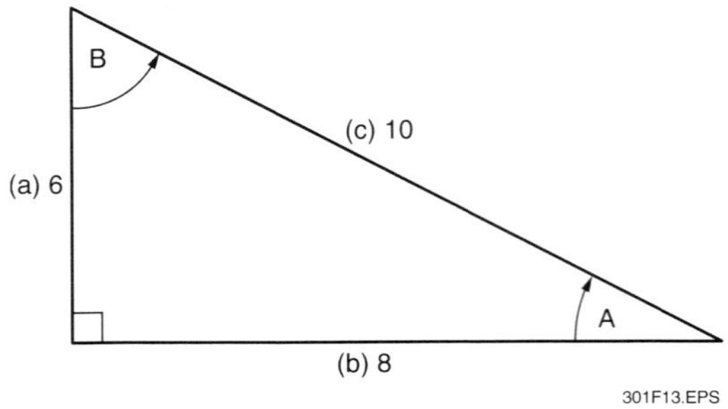


*Figure 13* ◆ Calculating the sine, cosine, and tangent of a right triangle.

sin 60 = side a/hypotenuse

sin 60 × hypotenuse = side a = 48 inches

hypotenuse = 48/sin 60 = 55.43 inches

The angles are easy; you know that the right angle is 90 degrees and the other angle is 60 degrees, so the third angle is 90 – 60 = 30 degrees.

The sine, cosine, and tangent are the most commonly used trigonometric functions, and tables showing their values are readily available. These functions are also available on the scientific calculator. However, each function also has an inverse, a ratio that is inverted.

The sine's opposite is the cosecant, the formula being cosecant = hypotenuse divided by the opposite side. The cosine's inverse is the secant, using the equation cosecant = hypotenuse divided by the adjacent side. The tangent's inverse is the cotangent, the adjacent side divided by the opposite side.

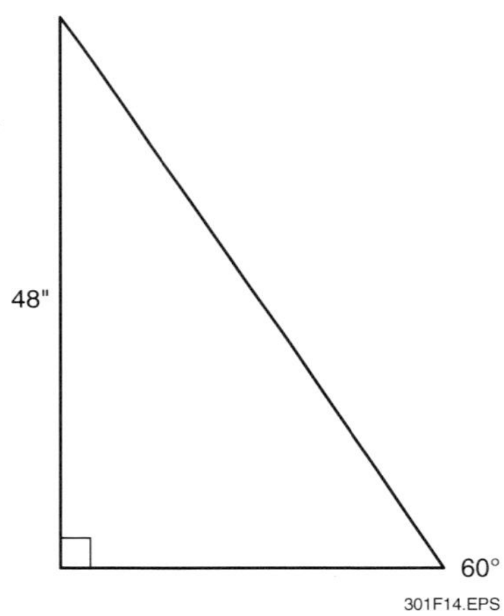


*Figure 14* ◆ Calculating the sides of a right triangle.

The inverse functions were used for many years because calculations were performed with a slide rule, and it was easier to multiply than to divide. The millwright could obtain the desired solutions by using whichever of the six formulas could be solved by multiplication. Inverse functions are performed less frequently today because of the convenience of the scientific calculator.

### *4.3.1 Using a Scientific Calculator to Convert Sine, Cosine, or Tangent Values to Angles*

To convert sine, cosine, or tangent values to angles using a scientific calculator, the calculator must have the SIN, COS, TAN, $SIN^{-1}$, $COS^{-1}$, and $TAN^{-1}$ keys. Decide which angle in the right triangle you are trying to find, then calculate the ratios of the line segments associated with that angle. Remember that the sine is the length of the line segment directly across from the angle divided by the hypotenuse; the cosine is the length of the line segment adjacent (connected) to the angle divided by the hypotenuse; and the tangent is the length of the line segment directly across from the angle divided by the length of the line segment adjacent to the angle.

As long as you know any two of the three lengths, you can calculate the angle with one of the three ratio methods.

**NOTE**

You only need one of the three ratios to find the angle.

Using the right triangle shown in *Figure 15*, you can find the length of line segment b using the Pythagorean theorem:

*Step 1* Write the formula in the form that solves for line segment b:

$$b^2 = c^2 - a^2$$

*Step 2* Replace the letters with the known numbers:

$$b^2 = 26.8^2 - 17.4^2$$

*Step 3* Use the calculator to solve the formula as follows:

- Enter 26.8 on the calculator keyboard; the calculator displays 26.8.
- Press the $x^2$ key; the calculator displays 718.24.
- Press the minus (–) key; the calculator displays 718.24.



Instructor's Notes:

1.12 MILLWRIGHT ◆ LEVEL THREE

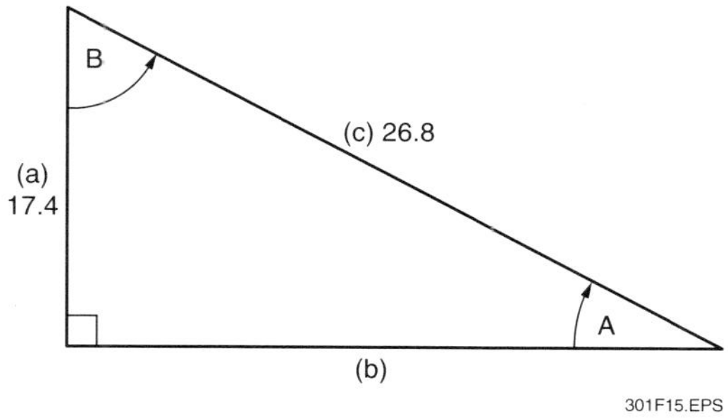


*Figure 15* ◆ Example of solving ratio problems in right triangles.

- Enter 17.4 on the calculator keyboard; the calculator displays 17.4.
- Press the $x^2$ key; the calculator displays 302.76.
- Press the equal (=) key; the calculator displays 415.48 ($b^2$ = 415.48).
- Press the $\sqrt{\ }$ key; the calculator displays 20.38332652

Rounding off to one decimal place like the other two line segments, b = 20.4.

Find the degrees in angle A of *Figure 15* by applying the appropriate line segment ratio (sine, cosine, or tangent) and using the calculator. Because all three line segment lengths are now known, you can use any one of the three ratios. For this example, use sine:

sin A = side opposite divided by the hypotenuse

sin A = a/c

***Step 1*** Enter 17.4; the calculator displays 17.4.

***Step 2*** Press the divide key (÷); the calculator displays 17.4.

***Step 3*** Enter 26.8; the calculator displays 26.8.

***Step 4*** Press the equal (=) key; the calculator displays 0.649253731.

When rounded off to five decimal places, sin A = 0.64925

This decimal answer is the ratio or sine between the two line segments. It is not the angle. In other words, the ratio between the two line segments is approximately $\frac{6}{10}$, 0.6, or $\frac{3}{5}$. To figure out the angle formed by the two line segments that have the calculated sine ratio, use the $SIN^{-1}$ key on the scientific calculator. Use the SIN key to determine the sine ratio when the angle is known. When you know the sine but not the angle, reverse the process. The small minus one ($^{-1}$) means that the process is reversed or inverted. The technical term for this reversal is **reciprocal**. $SIN^{-1}$ is the reciprocal of SIN. The calculator does the math as long as you press the correct keys.

From the sine value of angle A in *Figure 15*, which is 0.64925, use your calculator's inverse or $SIN^{-1}$ key to determine the degrees in angle A. The calculator illustrations are shown in *Figure 16*.

sin A = 0.64925

***Step 1*** Enter 0.64925; the calculator displays 0.64925 (*Figure 16A*).

> **NOTE**
>
> If your calculator has the $SIN/SIN^{-1}$, $COS/COS^{-1}$, or $TAN/TAN^{-1}$ combined on one key, you may have to press a key such as the 2nd or 3rd key to switch between the two functions.

***Step 2*** On this particular calculator, the SIN and the $SIN^{-1}$ function are located on the same key. To switch from the SIN to the $SIN^{-1}$ function, the key marked 2nd must be pressed before pressing the $SIN^{-1}$ key (*Figure 16B*). By pressing the 2nd key first, the functions of all the calculator keys are changed to the functions shown on the upper part of the keys. If you wish to change back to the functions on the lower half of the keys, you press the key marked 3rd first.

***Step 3*** Press the $SIN^{-1}$; the calculator displays 40.48507893, which is the number of degrees in angle A (*Figure 16C*).

***Step 4*** Round off answer to one decimal place—angle A = 40.5 degrees.

Use the same process to determine the angle if you know the cosine or tangent ratio, except use the $COS^{-1}$ key when you know the cosine ratio, and use the $TAN^{-1}$ key when you know the tangent ratio.

To determine the sine, cosine, or tangent ratio when the angle is known:

***Step 1*** Enter the value of the angle in degrees.

***Step 2*** Press the SIN, COS, or TAN key to display the ratio in decimal form.

### 4.3.2 Obtuse Triangle

An obtuse triangle (*Figure 17*) is not a right triangle. The largest angle of this triangle is more than 90 degrees. Since this is not a right triangle, make two right triangles out of it to use trig functions. This can be done with two sides and one angle, or



**Review how obtuse triangles are broken into two right triangles to perform trig functions.**

MODULE 15301-08 ◆ ADVANCED TRADE MATH 1.13

Show Transparency 21 (Figure 17). Explain how the Pythagorean theorem is used to determine the length of the segments of an obtuse triangle.

Show Transparency 22 (Figure 18). Explain angle-side relationships.

(A) (B) (C)

301F16.EPS

*Figure 16* ◆ Steps for using a calculator to convert a sine value to an angle.

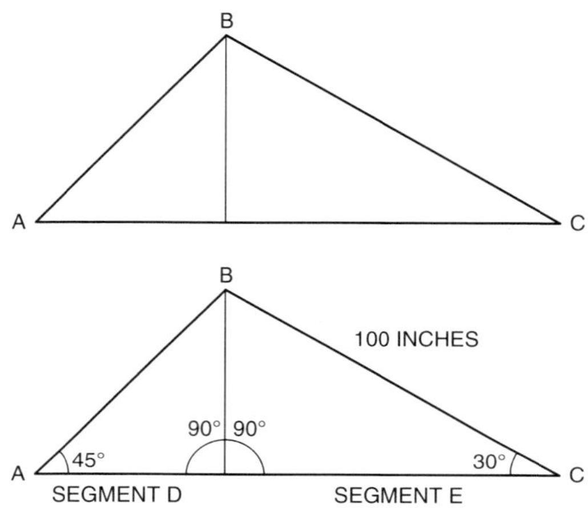


301F17.EPS

*Figure 17* ◆ Solving an obtuse triangle.

with two angles. (You know the third angle of the triangle, because the total of the three angles must be 180 degrees, so 30 + 45 = 75; 180 – 75 = 105.)

Take the length of the side you have as the hypotenuse of the triangle made by drawing a line from the apex at a right angle to the base of the triangle. To determine the length of the height line:

sin 30 = opposite divided by 100 inches

opposite = 100 × sin 30 = 50 inches

You have the height, and that gives the length of segment d, using the tangent of 45 degrees:

tan 45 = 50 inches / adjacent side (segment d)

segment d × tan 45 = 50 inches

segment d = 50 inches / tan 45 = 50 inches

To determine the line segment from angle B to angle C, use the Pythagorean theorem:

$(50^2 + 50^2) = BC^2$

$\sqrt{5{,}000} = BC = 70.71$ inches

Only one more length is needed, that of the base line. You already know the length of segment d. In order to get the length of the other segment, use either the Pythagorean theorem or the cosine of angle A and the length of the line AC to find the answer.

cos 30 degrees = segment e / 100 inches

segment e = 100 inches × cos 30 = 86.6 inches

The overall length for the base is segment e + segment d, 86.6 + 50 = 136.6 inches.

## 4.4.0 Determining the Angles When Side Lengths Are Known

Determining the angles to be used in an offset bend is not difficult. Both angles must be the same number of degrees so that the piping run remains parallel. Use the ratios between the sides to determine the value of the trig functions, then use the function to determine the angle. (See *Figure 18*.)

Refer to *Figure 19*. The sine of angle A is found as follows:



*Instructor's Notes:*

1.14 MILLWRIGHT ◆ LEVEL THREE

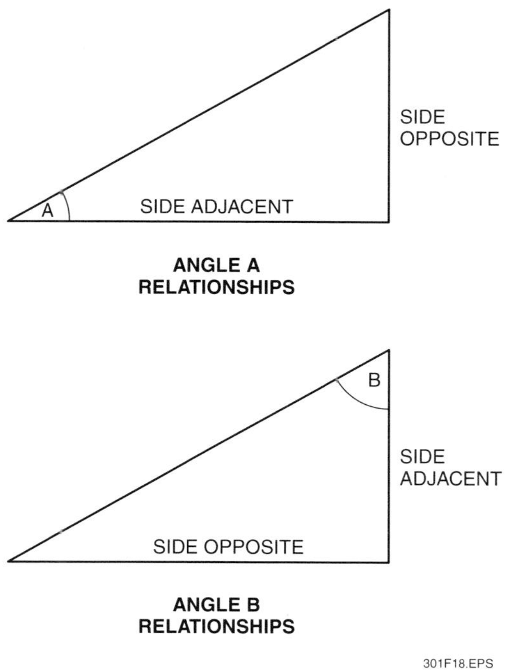


*Figure 18* ◆ Angle-side relationships.

sin A = side opposite A/hypotenuse

sin A = 5/7.5

sin A = 0.667

Angle A is approximately 42 degrees.

Refer to piping offset 4 (*Figure 20*). If the length of the opposite side is 17 inches and the hypotenuse is 32 inches, the offset angle can be determined using the formula for sine.

The angle of the offset is found as follows:

opposite/hypotenuse = sine

17/32 = sine

0.53125 = sine

Angle X is approximately 32 degrees.

The sine can also be used to calculate the hypotenuse of a rolling offset. The rolling offset is similar to the simple offset except there is one more factor to calculate: the roll of the offset. The roll is the distance that one end of the hypotenuse is displaced laterally. While a triangle best represents a simple offset, a rectangular box best represents the rolling offset, with the hypotenuse moving diagonally across the box from corner to corner (see *Figure 21*).

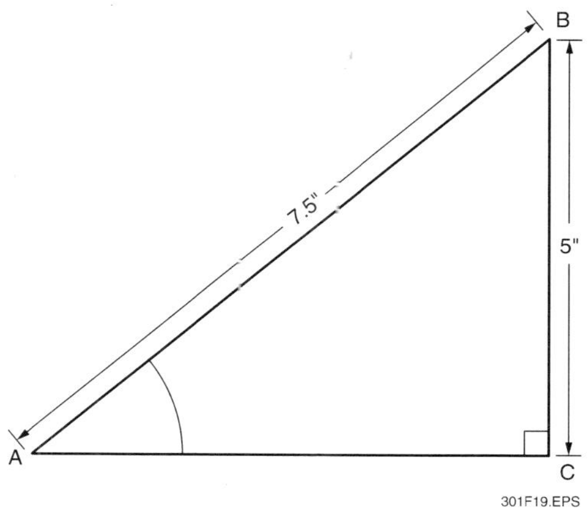


*Figure 19* ◆ Triangle ABC.

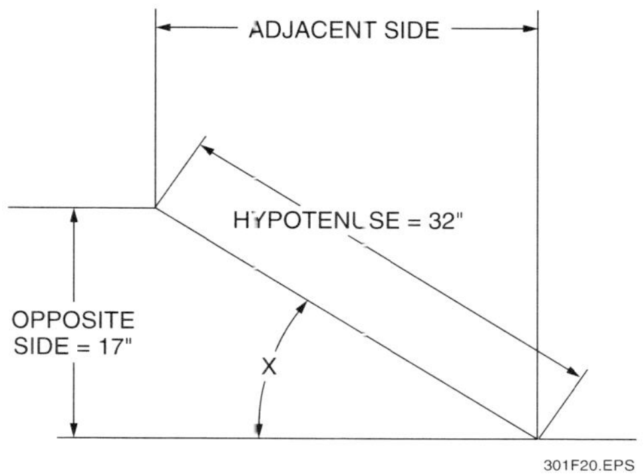


*Figure 20* ◆ Offset 4.

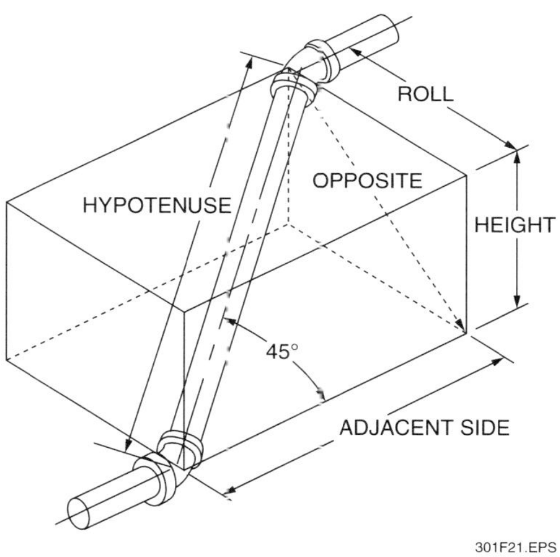


*Figure 21* ◆ Rolling offset.

**Show Transparency 23 (Figure 19). Explain how to find the sine of angle A.**

**Show Transparency 24 (Figure 20). Explain how to determine the offset angle.**

**Show Transparency 25 (Figure 21). Explain how to calculate the hypotenuse of a rolling offset.**



MODULE 15301-08 ◆ ADVANCED TRADE MATH 1.15

**Explain how to use a framing square to calculate the hypotenuse of a rolling offset.**

**Show Transparency 26 (Figure 22).**

**Define interpolation and discuss increasing and decreasing functions.**

**Review the example for interpolating numerical values from angles.**

Assume that the roll of a 60-degree offset is 8 inches and the height is 15 inches. Follow these steps to calculate the hypotenuse of a rolling offset, using a framing square.

*Step 1* Lay out the roll on one side of the framing square and the height on the other side of the square as shown in *Figure 22*.

*Step 2* Measure the distance between these two points on the framing square. In this example, the measurement is 17 inches. That is the opposite side.

*Step 3* Divide the measurement by the sine for the angle of the fitting.

Instead of the framing square, you can easily calculate the opposite side of a rolling offset using the Pythagorean theorem if you know the height and the roll. If the height is 15 inches and the roll is 8 inches, the opposite side is:

$$\sqrt{(8^2 + 15^2)} = \sqrt{289} = 17 \text{ inches}$$

To figure the length of the pipe, calculate: 17/sin60 = 19.629, or $19\frac{5}{8}$".

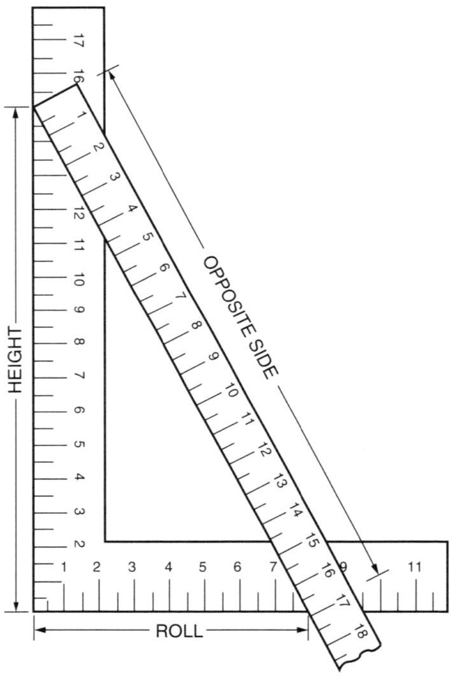


301F22.EPS

*Figure 22* ◆ Laying out a rolling offset using a framing square.

## 4.5.0 Interpolation

When precise measurements involving trigonometry are made, it is sometimes necessary to find an unknown value between two given values on a trig functions table. Calculating an unknown value in this way is called interpolation. It is used when a trig functions calculator is unavailable.

When a value is interpolated, increasing and decreasing functions are calculated differently. The functions of sines and tangents are called increasing functions because they increase when the size of the angle increases. The functions of cosines are called decreasing functions because they decrease when the size of the angle decreases.

These calculations can be used for interpolating numerical values from angles and interpolating angles from numerical values.

### *4.5.1 Interpolating Numerical Values from Angles*

Follow these steps to calculate the numerical values of the functions of angles.

**NOTE**

For the following steps, assume that the angle to be interpolated is sin 37.4 degrees. Use *Table 4* to find the numerical values for the calculations.

*Step 1* Find the numerical value of the next larger angle in the proper column.

sin 38 degrees = 0.6157

*Step 2* Find the numerical value of the next smaller angle in the proper column.

sin 37 degrees = 0.6018

*Step 3* Subtract the smaller value from the larger value.

0.6157 – 0.6018 = 0.0139

*Step 4* Multiply the difference by the fraction of the degree to be interpolated.

$0.0139 \times 0.4 = 0.00556$

**NOTE**

If calculating an increasing function, go to Step 5. If calculating a decreasing function, go to Step 6.



Instructor's Notes:

1.16 MILLWRIGHT ◆ LEVEL THREE

*Step 5* Add the answer from Step 4 to the value of the smaller angle found in Step 2.

0.00556 + 0.6018 = 0.60736,
rounded to 0.6074

*Step 6* Subtract the answer from Step 4 from the value of the smaller angle found in Step 2.

### *4.5.2 Interpolating Angles from Numerical Values*

Follow these steps to calculate an angle when the numerical value is known:

> **NOTE**
> For the following steps, assume that the numerical value of the function of the unknown angle is 0.8942. Use *Table 4* to find the numerical values for the calculations.

*Step 1* Find the next larger numerical value in the proper column.

cos 26 degrees = 0.8988

*Step 2* Find the next smaller numerical value in the proper column.

cos 27 degrees = 0.8910

*Step 3* Subtract the smaller value from the larger value.

0.8988 – 0.8910 = 0.0078

*Step 4* Subtract the value of the unknown angle from the value of the smaller angle.

0.8988 – 0.8942 = 0.0046

*Step 5* Divide the number found in Step 4 by the number found in Step 3.

0.0046/0.0078 = 0.5897,
rounded to 0.6

*Step 6* Add the answer from Step 5 to the value of the smaller angle found in Step 1.

26 degrees + 0.6 = 26.6 degrees

## 4.6.0 Law of Sines

In a triangle ABC where the angles A, B, and C are opposite the corresponding sides of lengths a, b, and c (see *Figure 23*):

(sin A) / a = (sin B) / b = (sin C) / c

The other form of the law is that:

a / (sin A) = b / (sin B) = c / (sin C)

This means that if you have either one angle and two sides, or two angles and one side, you can calculate the other parts of the triangle. For example, if you have a value for angles A and B, and a length for any side, first calculate the other angle. Because the angles of a triangle add up to 180 degrees, if A is 30 degrees and B is 75 degrees, then C = 180 – (30 + 75) = 180 – 105 = 75.

Now that the angles are known, set up the equation using the known angles and side (assume that you know the length of side c):

a / (sin 30) = b / (sin 75) = c / (sin 75)

Since b and c are the same, plug a value for c into the equation. If c is 25, then b is 25 also, since sin 75 is the same. Now, you have:

a / (sin 30) = 25 / (sin 75)

a / 0.5 = 25 / 0.9659... = 25.88...

Multiply both sides of the equation by 0.5, and you find that:

a = 12.94

Suppose that you have two sides and one angle that is opposite that angle. Now place the sides in the equation. If a is 15 inches, and b is 17 inches, and angle A is 35 degrees, then the equation becomes:

a / (sin A) = 15 / (sin 35) = 26.15 = 17 / (sin B)

(sin B) = 17 / 26.15... = 0.65

Angle B = 40.5456 degrees

Angle C = 180 – (35 + 40.5456) = 104.4544

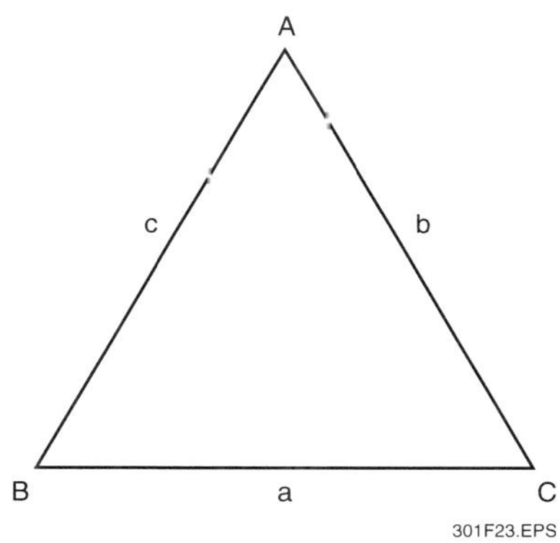


*Figure 23* ◆ Law of sines.

**Review the example for interpolating angles from numerical values.**

**Show Transparency 27 (Figure 23). Explain the law of sines and review the examples.**



MODULE 15301-08 ◆ ADVANCED TRADE MATH 1.17

**Assign reading of Section 5.0.0.**

**Ensure that you have everything required to teach this session.**

**Discuss situations in millwright work where it might be necessary to determine the weight of an object.**

**Show Transparency 28 (Table 5).**

**Explain the calculations for determining the weight and volume of the steel turbine shaft.**

**Explain the calculations for determining the volume of the rectangular housing and shaft.**

**Explain the calculations for determining the weight of the rectangular tank.**

Having already determined that a / (sin A) = 26.15…, it is clear that:

c / (sin 104.4544) = 26.15

c = 26.15 × (sin 104.45) = 25.32 inches

There are special cases where the law will produce either two possible answers, or one impossible answer. These special cases are unlikely to appear in the real world, and will be obviously incorrect if they do. In the case of the two possible answers, it will be evident which of the two is correct, and in the other case, the answer will be obviously impossible (as when the solution turns out to be a negative answer).

## 5.0.0 ◆ CALCULATING THE WEIGHT OF AN OBJECT

It is frequently necessary for workers to calculate the weight of an object, especially when the object must be rigged for transport. To do so, you will calculate the volume of the object, in cubic feet, and multiply the volume by the weight per cubic foot shown for that material in *Table 5*.

For example, the volume and weight of a steel turbine shaft 12 inches in diameter and 15 feet long can be found by calculating the volume of the cylinder, using the formula:

$V = \pi r^2 h$

Substitute values into the equation:

$V = \pi r^2 h$

V = 3.1416(0.5*0.5)*15 =11.781 cu ft

Now that you have the volume of the shaft in cubic feet, determine the weight of the shaft in pounds by multiplying the volume by the weight of a cubic foot of rolled steel:

W = 11.78 * 495 = 5,831.1 pounds

By this calculation, you will be able to determine what equipment you might use to transport the object to its destination. You would not, for example, use a one- or two-ton jib crane to move this shaft. Even a three-ton crane would be above 97 percent of rated capacity, and would require a lift plan, detailing the weights of all the components including the rigging itself, quite possibly exceeding the capacity of the crane.

If a solid form has a large hollow area, first calculate the volume of the entire object, then subtract the volume of the open area from the total volume. For example, a housing that is rectangular might have a 24-inch OD shaft passing through the middle. The housing is 4 feet wide, 4 feet deep, and 10 feet long. The calculation of the volume of the rectangular solid is:

4 feet × 4 feet × 10 feet (4*4*10) = 160 cubic feet

The shaft is 24 inches in diameter; the calculation is:

π*1*1*10 = 31.416 cubic feet

Subtract the shaft from the housing volume and you find the housing volume is 128.584 cubic feet. If the housing is cast iron, it weighs 128.584*450 = 57,863 pounds (rounded to the nearest pound).

A particular issue in calculating weight is the weight of a vessel or tank. If the weight is available from the manufacturer, that is well and good, but it is often necessary for the millwright to calculate the weight of an empty tank and/or the weight of the contents of the tank or vessel. The way to calculate the weight of the empty tank is to determine the wall thickness of the tank and determine the volume of the wall, base, and top (if there is a top). For a rectangular tank, the base is calculated in the normal manner: length × width × thickness.

For a rectangular tank 25 feet by 15 feet made out of $\frac{3}{4}$-inch steel, the calculation is easiest in feet. First, find the decimal dimension for $\frac{3}{4}$ inch in feet, 0.75 divided by 12 (because an inch is $\frac{1}{12}$ of a foot).

Assume that the tank is made out of mild steel, at approximately 491 pounds per cubic foot. The solution for the volume of the base is:

25 × 15 × 0.0625 = 23.4375 cubic feet

This assumes that the base is measured on the inside of the tank. That means that when the tank walls are calculated, the walls will extend $\frac{3}{4}$ of an inch beyond the bottom of the inside. This assumption requires adding $\frac{3}{4}$ of an inch vertically at each corner. Assuming the depth of the tank is 10 feet, the formula for the volume of the walls is:

2(25.0625 × 10.0625 × 0.0625) +
2(15.0625 × 10.0625 × 0.0625) = 50.47

**Table 5** Weight of Common Materials

| Material | Weight in lbs/cu ft |
|---|---|
| Aluminum | 165.0 |
| Cast Brass | 134.0 |
| Cast Copper | 542.0 |
| Cast Iron | 450.0 |
| Concrete | 150.0 |
| Dense Earth | 125.0 |
| Lead | 708.0 |
| Rolled Nickel | 541.0 |
| Rolled Steel | 495.0 |
| Mild Steel | 491.0 |
| Tungsten | 1,224.0 |
| Water | 62.4 |

301T05.EPS



Instructor's Notes:

1.18 MILLWRIGHT ◆ LEVEL THREE

Add the volumes and multiply by 491 to get the weight of the tank:

491(23.4375 + 50.47) = 36,289 (rounded up)

This is the weight of the tank, not including any reinforcement such as beams or angle iron because it is assumed to be in the ground. Had there been angle iron or beams involved, the type and dimension would be specified with the weight per linear foot.

If it is necessary to calculate the pressure on the material under the tank, you now need to calculate the weight of the contents.

Fluids can be calculated if the weight per volume unit is given for the fluid used. Fluids can also be calculated based on the weight of water per cubic foot, multiplied by the specific gravity of the actual fluid, and multiplied by the volume of the space in the tank that is to be filled. Note that water has a specific gravity of 1.0, and ethanol, as used in this example, has a specific gravity of 0.79.

First, calculate the volume:

$25 \times 15 \times 10 = 3{,}750$ cubic feet

Calculate the weight of a cubic foot of the fluid (in this case ethanol), starting with the standard weight of water per cubic foot:

62.425 lbs / cu ft $\times$ 0.79 (the specific gravity of ethanol) = 49.31575 lbs / cu ft

Finally, calculate the weight of the fluid in the tank:

$49.31575 \times 3{,}750 = 184{,}934$ pounds

A cylindrical vessel can be treated in the same manner. The diameter and height of the vessel are required, as well as the wall thickness. Assuming a 10-foot inside diameter, a height of 15 feet for the cylindrical part, and a 1-inch wall, the cylindrical volume is easy. However, if the top and bottom are hemispheres, then the volume and weight of those must be calculated also.

The equation for the cylinder is $V = r\pi^2h$. The equation for the volume of a sphere is $(4r\pi^2) / 3$. If the top end is flat and the bottom is a hemisphere, the flat end could be calculated as a cylinder of height equal to the wall thickness, while the bottom would be calculated by using the sphere formula and dividing it by 2.

The quickest way to do the calculation for the vessel cylinder is to figure the plate it would be rolled from. The circumference at the inside wall is close enough for most calculations of the length of the plate:

$C = \pi D = 3.1416 \times 10 = 31.416$ feet

The other dimension is 15 feet, so:

$31.416 \times 15 = 471.24$

The cubic measure for the wall is either divided by 12 or multiplied by 1⁄12; the decimal equivalent is 0.08333. Either way, the result is 39.27 cubic feet.

A flat top or bottom is easy enough if you remember to add the wall thickness to the radius for the cylinder:

$V = \pi r^2h = 3.1416 \times (5.0833)^2 \times 0.0833 = 6.76$ cu ft

The hemisphere is a little more complicated. This requires two calculations, first for the volume of the hemispherical space inside the wall, and then for the hemispherical space outside the wall. The difference is the volume of the wall:

$0.5 \times (4\pi r^2) / 3 =$
$0.5 \times 4 \times 3.1416 \times (5.0833)^2 / 3 = 54.12$

$0.5 \times (4\pi r^2) / 3 =$
$0.5 \times 4 \times 3.1416 \times (5.0)^2 / 3 = 52.36$

$54.12 - 52.36 = 1.76$

Finally, add the volumes together and multiply the total by the weight per cubic foot of whatever material the vessel is made out of, in this case mild steel:

$39.27 + 6.76 + 1.76 = 47.79$

$47.79 \times 491 = 23{,}465$ pounds

To calculate the weight of the fluid in the cylinder, you need the internal volume of the vessel, and the weight per cubic foot of the fluid. The calculation for the cylinder volume uses the formula for cylinders:

$V = \pi r^2h = 3.1416 \times 5^2 \times 15 = 1{,}178.1$ cu ft

This doesn't include the volume of the hemisphere, but it has already been calculated to be 52.36 cubic feet. Therefore, the total volume of fluid inside the vessel, if it was filled completely, is:

$1{,}178.1 + 52.36 = 1{,}230$ cu ft, rounded off

If the fluid is ethanol, the weight per cubic foot is already calculated to be 49.31575 pounds. The calculation for the weight of the fluid in the vessel is:

$1{,}230 \times 49.31575 = 60{,}658$ pounds rounded off

Add some support pieces of ironwork to hold the vessel upright. Assume 92 lineal feet of wide-flange beam, labeled as W 10 $\times$ 45, which means it weighs 45 pounds per lineal foot:

$92 \times 45 = 4{,}140$ pounds

The total weight for the filled vessel comes to:

$60{,}658 + 23{,}465 + 4{,}140 = 88{,}263$ pounds, or 44.1 tons

**Explain the calculations for determining the weight of the fluid in the tank.**

**Explain the calculations for determining the weight and volume of the cylindrical vessel and the weight of the fluid inside.**

**Ask trainees if they have any questions about how to calculate the weight of an object.**



MODULE 15301-08 ◆ ADVANCED TRADE MATH 1.19

Have trainees complete the Review Questions, and go over the answers prior to administering the Module Examination.

## Review Questions

1. The fraction equivalent of 6.125 is _____.
   a. 6⅛
   b. 6¼
   c. 6½
   d. 6⅞

2. The decimal equivalent of $\frac{9}{16}$ inch is _____.
   a. 0.375
   b. 0.4125
   c. 0.5000
   d. 0.5625

3. One atmosphere equals _____ pounds per square inch.
   a. 0.8842
   b. 14.696
   c. 29.92
   d. 34

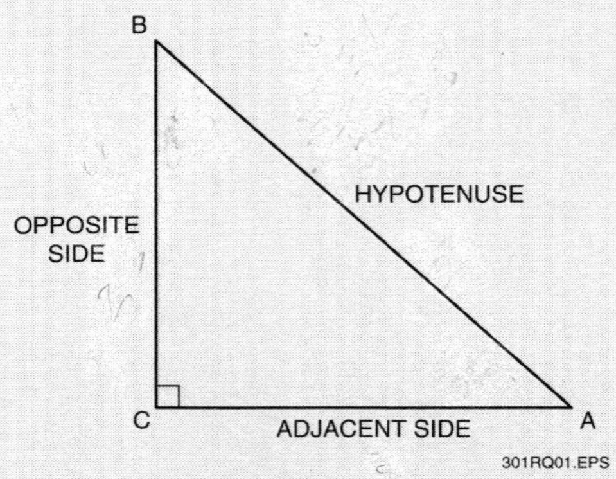


**Figure 1**

4. If the opposite side in *Figure 1* is 4 feet, and the hypotenuse is 35 feet, the adjacent side is _____. (Round off to the nearest tenth.)
   a. 31 feet
   b. 34.8 feet
   c. 39 feet
   d. 51 feet

5. If the opposite side in *Figure 1* is 6 feet and the adjacent side is 8 feet, the hypotenuse is _____. (Round off to the nearest tenth.)
   a. 4 feet
   b. 8 feet
   c. 10 feet
   d. 12 feet

6. The sine of a reference angle is the ratio of the _____.
   a. opposite side to the adjacent
   b. opposite angle to the reference angle
   c. side opposite to the hypotenuse
   d. side adjacent to the hypotenuse

7. The cosine of the reference angle is the ratio of the _____.
   a. opposite side to the adjacent
   b. opposite angle to the reference angle
   c. side opposite to the hypotenuse
   d. side adjacent to the hypotenuse

8. If the opposite side is 4 feet and the hypotenuse is 35 feet, angle A is _____.
   a. 3.6 degrees
   b. 6.6 degrees
   c. 11.2 degrees
   d. 15 degrees

9. If side AC is 14 feet and the hypotenuse is 19 feet, the value of cos angle A, to four places, is _____.
   a. 0.3571
   b. 0.5000
   c. 0.7368
   d. 0.8055

10. The ratio of the two sides of a right triangle is directly related to the size of the _____ in the triangle.
    a. apex
    b. left angle
    c. reference angle
    d. right angle

11. If the opposite is 25 feet, and angle A is 22.5 degrees, the hypotenuse is _____ feet.
    a. 35.2
    b. 45.5
    c. 65.3
    d. 71

12. If angle B is 30 degrees and the hypotenuse is 48 inches, the adjacent side is _____ inches.
    a. 12
    b. 16
    c. 24
    d. 36



Instructor's Notes:

1.20 MILLWRIGHT ◆ LEVEL THREE

# Review Questions

13. If angle A is 45 degrees and the side adjacent is 48 inches, side AB is _____ inches.
    a. 48
    b. 56.6
    c. 67.9
    d. 72

14. If the opposite side is 20 inches and angle B is 75 degrees, the hypotenuse is _____ inches.
    a. 45.2
    b. 62.8
    c. 74.6
    d. 77.3

15. If angle A is 30 degrees, and side BC is 45 inches, side AC is _____ inches.
    a. 48
    b. 77.9
    c. 84.5
    d. 90.1

16. If angle A is 30 degrees, angle B is _____ degrees.
    a. 22.5
    b. 30
    c. 60
    d. 90

17. If angle A is 45 degrees and side AC is 48 inches, side BC is _____ inches.
    a. 45
    b. 48
    c. 52
    d. 56.6

18. If the adjacent side is 24 inches and the hypotenuse is 48 inches, angle A is _____ degrees.
    a. 30
    b. 45
    c. 60
    d. 90

19. If the opposite side is 6 feet and the adjacent side is 8 feet, angle A is _____ degrees.
    a. 10
    b. 22.5
    c. 36.9
    d. 45

20. If angle A is 15 degrees and side BC is 20 inches, side AC is _____ inches.
    a. 35
    b. 48.3
    c. 69.5
    d. 74.6

21. If the opposite side is 25 feet and angle A is 22.5 degrees, the adjacent side is _____ feet.
    a. 16
    b. 25
    c. 45.3
    d. 60.4

22. If the adjacent side is 3 inches and the hypotenuse is 7 inches, angle A is _____ degrees.
    a. 64.6
    b. 70.1
    c. 72
    d. 84.6

23. If the opposite side is 2 feet and the hypotenuse is 11 feet, angle A equals _____ degrees.
    a. 10.5
    b. 15
    c. 22.4
    d. 25.4

24. If the height of a rolling offset is 6 feet, the roll is 8 feet, and the angle is 45 degrees, the hypotenuse is _____ feet.
    a. 10
    b. 11.2
    c. 14.1
    d. 17.2

25. The specific gravity of a fluid is 0.75, and water weighs 62.4 pounds per cubic foot. If the container is a cylinder 3 feet inside diameter by 10 feet high, the fluid weighs _____ pounds. (Round to nearest pound.)
    a. 1,561
    b. 2,104
    c. 3,308
    d. 4,008



MODULE 15301-08 ◆ ADVANCED TRADE MATH 1.21

**Summarize the major concepts presented in the module.**

**Administer the Module Examination. Record the results on Craft Training Report Form 200, and submit the results to the Training Program Sponsor.**

## Summary

Millwrights use many different tools. Mathematical tools will allow you to calculate the shape and dimensions of assemblies before you build them. The use of trigonometric ratios, sine, cosine, and tangents will allow you to determine the opposite side, adjacent side, angles, and hypotenuse of a triangle from incomplete information. Mathematics will also allow you to cut custom fitting angles and determine the weight of an object based on volume.

## Notes



*Instructor's Notes:*

1.22 MILLWRIGHT ◆ LEVEL THREE

## Trade Terms Introduced in This Module

***Adjacent side***: The side of a right triangle that is next to the reference angle.

***Cosine:*** Trigonometric ratio between the adjacent side and the hypotenuse, written as adjacent divided by the hypotenuse.

***Hypotenuse***: The longest side of a right triangle. It is always located opposite the right angle.

***Opposite side***: The side of a right triangle that is located directly across from the reference angle.

***Ratio***: A comparison of one value to another value.

***Reciprocal:*** The inverse of any fraction. Since a whole number is equal to the number divided by one, the reciprocal of the number is one divided by the number.

***Reference angle:*** The angle to which the sides are related as adjacent and opposite.

***Sine:*** Ratio between the opposite side and the hypotenuse, written as the opposite divided by the hypotenuse.

***Tangent:*** Ratio between the opposite side and the adjacent side, written as the opposite divided by the adjacent.



MODULE 15301-08 ◆ ADVANCED TRADE MATH 1.23

# Resources & Acknowledgments

## Additional Resources

This module is intended to be a thorough resource for task training. The following reference work is suggested for further study. This is optional material for continued education rather than for task training.

*Applied Construction Math*, Latest Edition. Upper Saddle River, NJ: Prentice Hall Publishing.

## Figure Credits

Topaz Publications, Inc., 301F05 (photo), 301F16



Instructor's Notes:

1.24 MILLWRIGHT ◆ LEVEL THREE

## MODULE 15301-08 — TEACHING TIPS

The following are suggested activities or instructional methods to help you teach the material in this module.

### General

When you call on someone to answer a question, the rest of the class relaxes or even tunes out because they expect that the question and answer will take place only between you and the trainee you called on. Instead, use this technique to involve more trainees in answering questions and to keep them on their toes.

1. Ask trainees to define a term or explain a concept.
2. After one trainee has answered, ask a trainee seated nearby if the answer is right. Then ask whether a trainee in the back of the room agrees.
3. Ask trainees to explain why they think an answer is right or wrong.
4. Use the session to clear up incorrect ideas and encourage trainees to learn from their mistakes.

### Sections 1.0.0 – 5.0.0

#### *Quick Quiz*

This Quick Quiz will familiarize trainees with advanced math trade terms. You will need photocopies of the quiz provided on the following page. Trainees will need pencils. If you allow trainees to use the Trainee Guide, decrease the amount of time you give them to complete the quiz.

1. Make a photocopy of the quiz for each trainee.
2. Give trainees between 5 and 10 minutes to complete the quiz.
3. Go over the answers to the quiz.
4. Ask trainees if they have questions.

#### *Answers to Quick Quiz*

1. g
2. b
3. e
4. c
5. d
6. f
7. h
8. a
9. i

## Quick Quiz

### *Trade Terms*

For each description listed, identify the term that the text best describes. Write the corresponding letter in the blank provided.

_____ 1. The angle to which the sides are related as adjacent and opposite is the _____.

_____ 2. The trigonometric ratio between the adjacent side and the hypotenuse is the _____.

_____ 3. A comparison of one value to another value is a(n) _____.

_____ 4. The longest side of a right triangle is the _____.

_____ 5. The side of a right triangle that is located directly across from the reference angle is the _____.

_____ 6. The inverse of any fraction is its _____.

_____ 7. The ratio between the opposite side and the hypotenuse is the _____.

_____ 8. The side of a right triangle that is next to the reference angle is the _____.

_____ 9. The ratio between the opposite side and the adjacent side is the _____.

a. adjacent side

b. cosine

c. hypotenuse

d. opposite side

e. ratio

f. reciprocal

g. reference angle

h. sine

i. tangent

# MODULE 15301-08 — ANSWERS TO REVIEW QUESTIONS

| Answer | | Section |
|---|---|---|
| 1. | a | 2.0.0; Table 1 |
| 2. | d | 2.0.0; Table 1 |
| 3. | b | 3.0.0; Table 3 |
| 4. | b* | 4.1.0 |
| 5. | c* | 4.1.0 |
| 6. | c | 4.2.0 |
| 7. | d | 4.2.0 |
| 8. | b* | 4.2.0 |
| 9. | c* | 4.2.0; Table 4 |
| 10. | c | 4.2.0 |
| 11. | c* | 4.2.1 |
| 12. | c* | 4.2.1 |
| 13. | c* | 4.2.2 |
| 14. | d* | 4.2.2 |
| 15. | a* "b" IS CORRECT | 4.2.3 CONFIRMED NEXT PAGE |
| 16. | c* | 4.2.3 |
| 17. | b* | 4.2.3 |
| 18. | c* | 4.2.3 |
| 19. | c* | 4.2.3 |
| 20. | d* | 4.2.3 |
| 21. | d* | 4.2.3 |
| 22. | a* | 4.3.1 |
| 23. | a* | 4.3.1 |
| 24. | c* | 4.4.0 |
| 25. | c* | 5.0.0 |

*Calculations for these answers are provided on the following page.

4. $\sqrt{(35^2 - 4^2)} = \sqrt{1,209}$

5. $h^2 = 6^2 + 8^2 = 100$

   $h = \sqrt{10}$

8. Sin A = 4/35

9. Cos A = 14/19

11. Hypotenuse = 25/sin 22.5 = 65.3

12. b = 48 sin 30 = 24

13. AC = 48/cos 45 = 67.9

14. Hypotenuse = 20/cos 75 = 77.3

15. AC = 45/tan 30 = 77.9

16. B = 180 – (90 + 30) = 60 degrees

17. BC = 48 tan 45 = 48

18. Cos A = 24/48 = 0.5 A = 60 degrees

19. Tan A = 6/8 = 0.75 A = 36.9

20. AC = 20/tan 15 = 74.6

21. Adjacent side = 25/tan 22.5 = 60.4

22. Cos A = 3/7 = 0.42857 $\cos^{-1}$ = 64.67

23. Sin A = 2/11 = 0.181818

    $\sin^{-1}$ = 10.5 degrees

24. $\sqrt{(6^2 + 8^2)} = \sqrt{100}$ = 10/sin 45 = 14.1

25. $1.5^2 \times 3.1416 \times 10 \times 0.75 \times 62.4$ = 3,308 pounds

# CONTREN® LEARNING SERIES — USER UPDATE

The NCCER makes every effort to keep these textbooks up-to-date and free of technical errors. We appreciate your help in this process. If you have an idea for improving this textbook, or if you find an error, a typographical mistake, or an inaccuracy in NCCER's Contren® textbooks, please write us, using this form or a photocopy. Be sure to include the exact module number, page number, a detailed description, and the correction, if applicable. Your input will be brought to the attention of the Technical Review Committee. Thank you for your assistance.

*Instructors* – If you found that additional materials were necessary in order to teach this module effectively, please let us know so that we may include them in the Equipment/Materials list in the Annotated Instructor's Guide.


**Write:** Product Development and Revision
National Center for Construction Education and Research
3600 NW 43rd St., Bldg. G, Gainesville, FL 32606

**Fax:** 352-334-0932

**E-mail:** curriculum@nccer.org


Craft ______ Module Name ______

Copyright Date ______ Module Number ______ Page Number(s) ______

Description ______

______

______

______

(Optional) Correction ______

______

______

(Optional) Your Name and Address ______

______

______

Module 15302-08

# *Precision Measuring Tools*

## NCCER STANDARDIZED CRAFT TRAINING PROGRAM

The National Center for Construction Education and Research (NCCER) provides a standardized national program of accredited craft training. Key features of the program include instructor certification, competency-based training, and performance testing. The program provides trainees, instructors, and companies with a standard form of recognition through a National Craft Training Registry. The program is described in full in the *Guidelines for Accreditation*, published by NCCER. For more information on standardized craft training, contact the NCCER by writing us at 3600 NW 43rd St., Bldg. G, Gainesville, FL 32606; calling 352-334-0911; or emailing info@nccer.org. More information may be found at our website, www.nccer.org.

## HOW TO USE THIS ANNOTATED INSTRUCTOR'S GUIDE

Each page presents two sections of information. The larger section displays each page exactly as it appears in the Trainee Module. The narrow column ties suggested trainee and instructor actions to each page and provides icons (detailed below) to call your attention to material, safety, audiovisual, or testing requirements. The bottom of each page includes space for your notes.

The **Audiovisual** icon indicates an appropriate time to show a transparency or other audiovisual aid.

The **Classroom** icon prompts you to define a term, stress a point, ask trainees to explain a concept, or give examples.

The **Demonstration** icon directs you to show trainees how to perform tasks.

The **Examination** icon tells you to administer the written module examination.

The **Homework** icon is placed where you may wish to assign reading for the next class, assign a project, or advise trainees to prepare for an examination.

The **Laboratory** icon is used when trainees are to practice performing tasks.

The **Materials** icon is a reminder for you to gather materials needed for classes, labs, and testing.

The **Performance Testing** icon tells you to administer a performance test or a portion thereof.

The **Safety** icon is used to emphasize safety issues. It is often keyed to *Caution* and *Warning!* statements in the Trainee Module.

The **Teaching Tip** icon indicates additional guidance is available, such as how to conduct an exercise, get the most educational value from a field trip, or encourage class participation. Teaching Tips may expand on a feature (*Think About It*, *Did You Know?*) or provide *Quick Quizzes* or similar exercises. You will be referred to the Teaching Tips section at the back of the module if there is additional material.

The **Combination** icon indicates that the laboratory listed corresponds with a performance task. If desired, you can note the proficiency of the trainees during the laboratory, and use it to satisfy performance testing requirements.

## PREPARATION

Before teaching this module, you should review the Objectives, Performance Tasks, Materials and Equipment List, and Module Outline. Be sure to allow ample time to prepare your own training or lesson plan and gather all required materials and equipment.

## MODULE OVERVIEW

This module presents information on precision measuring tools used by millwrights, and explains how to select, inspect, use, and care for them.

## PREREQUISITES

Please refer to the Course Map in the Trainee Module. Prior to training with this module, it is recommended that the trainee shall have successfully completed the following: *Core Curriculum; Millwright Level One; Millwright Level Two;* and *Millwright Level Three,* Module 15301-08.

## OBJECTIVES

Upon completion of this module, the trainee will be able to do the following:

1. Use levels.
2. Use calipers.
3. Use micrometers.
4. Use dial indicators.
5. Use universal bevel protractors.
6. Use gauge blocks.
7. Use speed measurement tools.
8. Use pyrometers.

## PERFORMANCE TASKS

Under the supervision of the instructor, the trainee should be able to do the following:

1. Use levels.
2. Use calipers.
3. Use micrometers.
4. Use dial indicators.
5. Use universal bevel protractors.
6. Use gauge blocks.
7. Use speed measurement tools.
8. Use pyrometers.

## MATERIALS AND EQUIPMENT LIST

Overhead projector and screen
Transparencies
Blank acetate sheets
Transparency pens
Whiteboard/chalkboard
Markers/chalk
Pencils and scratch paper
Appropriate personal protective equipment
Master, mechanic's, optical, and electronic levels
Inside, outside, vernier, and dial calipers
Outside, inside, and depth micrometers
Various micrometer tips
Height gauges and surface plates
Various sized dial indicators
Universal bevel protractor
Stroboscopic and mechanical tachometers
Optical, thermocouple, and infrared pyrometers
Module Examination*
Performance Profile Sheets*

* Located in the Test Booklet

## SAFETY CONSIDERATIONS

Ensure that the trainees are equipped with appropriate personal protective equipment and know how to use it properly.

## TEACHING TIME FOR THIS MODULE

An outline for use in developing your lesson plan is presented below. Note that each Roman numeral in the outline equates to one session of instruction. Each session has a suggested time period of 2½ hours. This includes 10 minutes at the beginning of each session for administrative tasks and one 10-minute break during the session. Approximately 20 hours are suggested to cover *Precision Measuring Tools.* You will need to adjust the time required for hands-on activity and testing based on your class size and resources. Because laboratories often correspond to Performance Tasks, the proficiency of the trainees may be noted during these exercises for Performance Testing purposes.

| Topic | Planned Time |
|---|---|
| **Session I. Introduction; Levels** | |
| A. Introduction | ____________ |
| B. Levels | ____________ |
| C. Laboratory | ____________ |
| Trainees practice using levels. This laboratory corresponds to Performance Task 1. | |
| **Session II. Calipers** | |
| A. Calipers | ____________ |
| B. Laboratory | ____________ |
| Trainees practice using calipers. This laboratory corresponds to Performance Task 2. | |
| **Session III. Micrometers** | |
| A. Micrometers | ____________ |
| B. Laboratory | ____________ |
| Trainees practice using micrometers. This laboratory corresponds to Performance Task 3. | |
| **Session IV. Dial Indicators; Universal Bevel Protractors** | |
| A. Dial Indicators | ____________ |
| B. Laboratory | ____________ |
| Trainees practice using dial indicators. This laboratory corresponds to Performance Task 4. | |
| C. Universal Bevel Protractors | ____________ |
| D. Laboratory | ____________ |
| Trainees practice using universal bevel protractors. This laboratory corresponds to Performance Task 5. | |
| **Session V. Gauge Blocks** | |
| A. Gauge Blocks | ____________ |
| B. Laboratory | ____________ |
| Trainees practice using gauge blocks. This laboratory corresponds to Performance Task 6. | |
| **Session VI. Speed Measurement Tools** | |
| A. Speed Measurement Tools | ____________ |
| B. Laboratory | ____________ |
| Trainees practice using speed measurement tools. This laboratory corresponds to Performance Task 7. | |

**Session VII. Pyrometers**

A. Pyrometers ______________

B. Laboratory ______________

Trainees practice using pyrometers. This laboratory corresponds to Performance Task 8.

**Session VIII. Review and Testing**

A. Module Review ______________

B. Module Examination ______________

1. Trainees must score 70% or higher to receive recognition from NCCER.
2. Record the testing results on Craft Training Report Form 200, and submit the results to the Training Program Sponsor.

C. Performance Testing ______________

1. Trainees must perform each task to the satisfaction of the instructor to receive recognition from NCCER. If applicable, proficiency noted during laboratory exercises can be used to satisfy the Performance Testing requirements.
2. Record the testing results on Craft Training Report Form 200, and submit the results to the Training Program Sponsor.

15302-08

# Precision Measuring Tools

Assign reading of Module 15302-08.

# 15302-08
# *Precision Measuring Tools*

*Topics to be presented in this module include:*



## Overview

This module explores the tools required to achieve the close tolerances required for modern machinery. Here you will learn how to use the most precise measuring equipment available, and how to apply some of the values shown on the equipment.

*Instructor's Notes:*

## Objectives

When you have completed this module, you will be able to do the following:

1. Use levels.
2. Use calipers.
3. Use micrometers.
4. Use dial indicators.
5. Use universal bevel protractors.
6. Use gauge blocks.
7. Use speed measurement tools.
8. Use pyrometers.

## Trade Terms

Graduated
Increment
Plumb
Transducer

## Required Trainee Materials

1. Pencil and paper
2. Appropriate personal protective equipment

## Prerequisites

Before you begin this module, it is recommended that you successfully complete *Core Curriculum; Millwright Level One; Millwright Level Two;* and *Millwright Level Three,* Module 15301-08.

This course map shows all of the modules in the third level of the *Millwright* curriculum. The suggested training order begins at the bottom and proceeds up. Skill levels increase as you advance on the course map. The local Training Program Sponsor may adjust the training order.

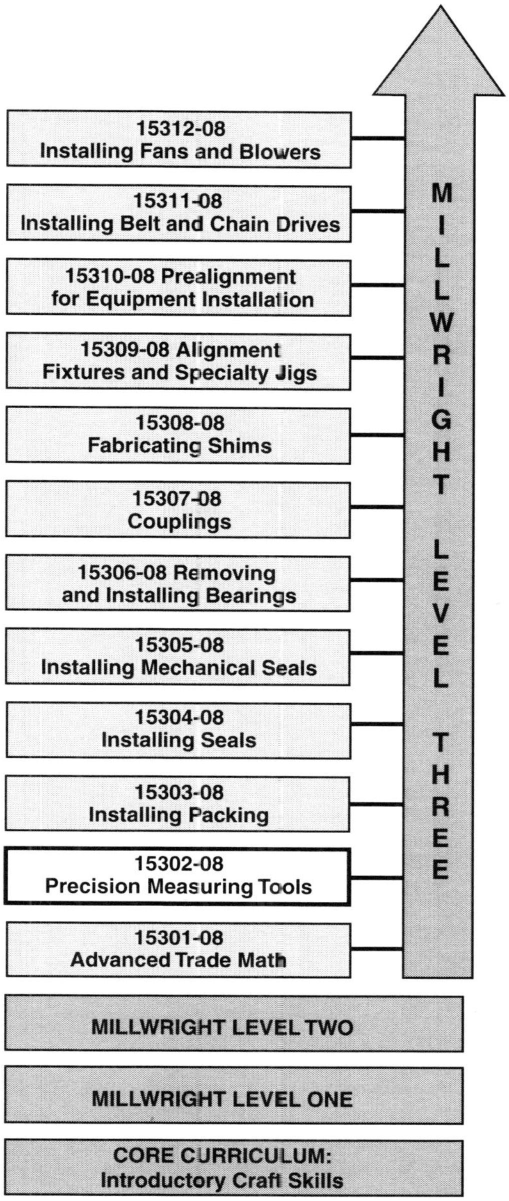


302CMAP.EPS



**Ensure that you have everything required to teach the course. Check the Materials and Equipment list at the front of this module.**

**See the general Teaching Tip at the end of this module.**

**Explain that terms shown in bold are defined in the Glossary at the back of this module.**

**Show Transparency 1, Objectives, and Transparency 2, Performance Tasks. Review the goals of the module, and explain what will be expected of the trainees.**

**Review the modules covered in Level Three and explain how this module fits in.**

Discuss precision measuring tools commonly used by millwrights.

Review the general guidelines for using these tools.

Introduce levels and explain how they are used. Discuss different types of levels.

Provide an assortment of levels for the trainees to examine.

Show Transparency 3 (Figure 1). Explain how master levels are used.

## 1.0.0 ◆ INTRODUCTION

Millwrights use precision measuring tools to perform specialized measuring jobs. As a millwright, you must be able to recognize, select, inspect, maintain, and use precision measuring tools to perform your job efficiently. The following are guidelines to be followed when using any precision instrument or tool:

- Keep the tool clean and store it in a protective case.
- Store tools with the jaws open at least 0.010 inch.
- Check the instrument or tool for accuracy by zeroing it before use.
- Periodically calibrate the tool against a master gauge set.
- Do not drop or lay anything on top of a precision instrument or tool.
- Stabilize the temperature of the instrument to match the ambient temperature of the area where it will be used.
- Wipe the instruments clean after use and coat with high-grade instrument oil.
- Practice using each instrument, mindful that feel and accurately reading the scales will consistently yield good measurements.

## 2.0.0 ◆ LEVELS

Levels are layout or testing tools used to check the levelness of horizontal or perpendicular surfaces. Some precision levels are calibrated to indicate the levelness of a surface in degrees, minutes, and seconds. Millwrights use the following types of levels:

- Master
- Mechanic's
- Optical
- Electronic

### 2.1.0 Master Levels

Master levels (*Figure 1*) are used by millwrights to level machinery during setup. The master level is a type of spirit level that is accurate to 0.0005 inch per foot on its main, or horizontal, vial. Master levels generally have three vials, two auxiliary vials to check cross level and one to measure level. The amount of liquid each vial contains is not enough to fill the vial. This creates a bubble that can be seen when the vial is held in the horizontal position. When the bubble is centered between the lines on the outside of the vial, the surface is **plumb** or level.

The easiest way to check a master level for accuracy is to place the level on a surface, take a reading, then turn the level 180 degrees in the same place, and compare the reading. If the reading is the same, the level is correct. The machinist's level can be adjusted by turning the adjustment screws at the ends of the main vial.

Some master levels are sensitive to small temperature changes. They must be handled carefully to prevent body heat from affecting the readings. Handle the level by the phenolic part to protect against heat transfer.

**CAUTION**

Never use abrasives to clean a master level. Removing any of the material would damage the accuracy of the level.

**NOTE**

Make sure the level is in place long enough to reach the same temperature as the object being measured.

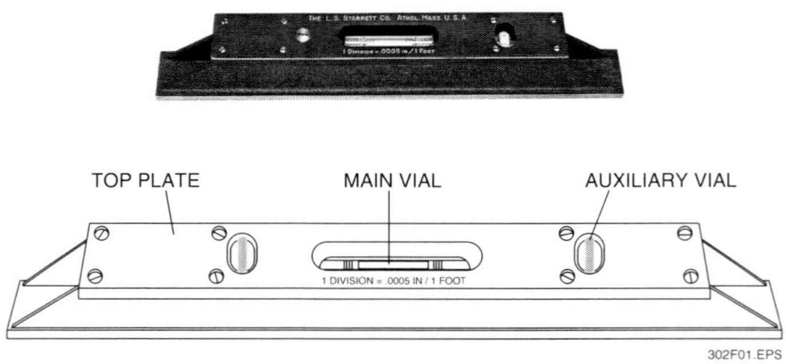


*Figure 1* ◆ Master level.



Instructor's Notes:

2.2 MILLWRIGHT ◆ LEVEL THREE

## 2.2.0 Mechanic's Levels

The 6- to 8-inch mechanic's level is a type of spirit level used for jobs that do not require extreme accuracy. It has an accuracy of 0.005 inch per foot. A V-groove along the flat surface is used for accurate settings on round pieces, such as shafts. *Figure 2* shows a No. 98 mechanic's level.

> **CAUTION**
> Do not use emery cloth or sandpaper on a mechanic's level as the use of these items will affect the accuracy of the instrument.

## 2.3.0 Optical Levels

An optical, or tilting, level consists of a telescope mounted on a base. Optical levels (*Figure 3*) can be used to level any equipment but are needed to level equipment more than 25 feet long. An optical level is a sensitive precision instrument that should be operated by trained personnel only. Always handle and store optical levels carefully, and keep them free of grease or grit.

An optical level has an optical micrometer that can give displacement readings in thousandths of an inch or in millimeters. Optical levels also have a telescopic sight with a sensitive spirit level. When the optical level is adjusted to indicate level, the spirit level is perfectly horizontal.

Follow these steps to use an optical level:

*Step 1* Mount the optical level onto a support, such as a tripod.

*Step 2* Place the optical level in the location for sighting (*Figure 4*).

*Step 3* Adjust the four leveling screws on the foot plate for rough leveling. They should be turned opposite to one another.

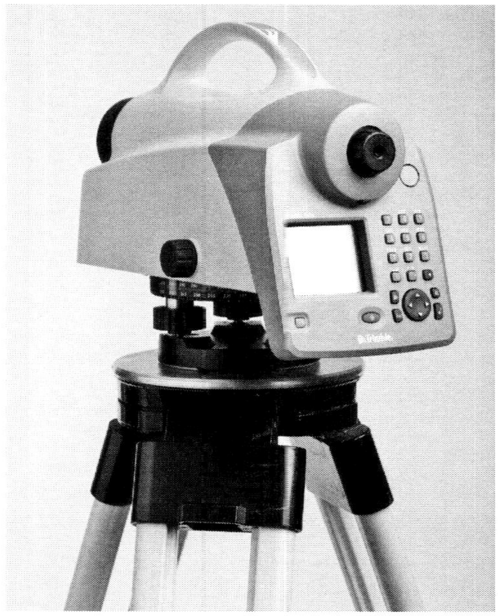

302F03.EPS

*Figure 3* ◆ Optical level.

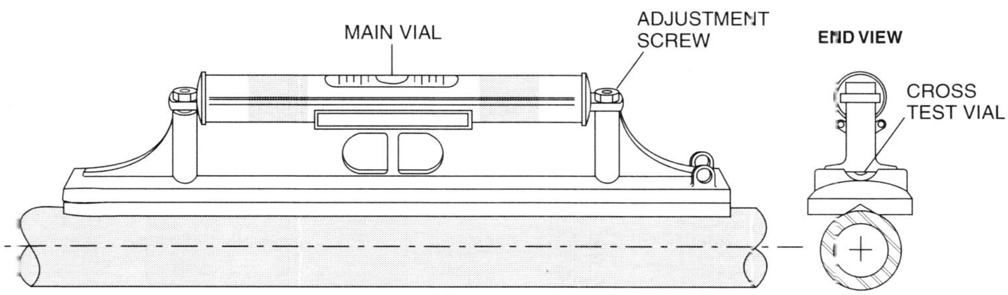


302F02.EPS

*Figure 2* ◆ No. 98 mechanic's level.

**Show Transparencies 4 and 5 (Figures 2 and 4).**

**Discuss mechanic's levels and optical levels and explain how each is used.**

**Describe the steps for positioning an optical level for sighting.**



MODULE 15302-08 ◆ PRECISION MEASURING TOOLS 2.3

> **NOTE**
> Turn two of the leveling screws alternately to center the bubble in the circular level.

*Step 4* Tighten the leveling screws that are not in contact with the foot plate to prevent the level from rocking. When the leveling screws are fingertight, check the level by turning it first to 90 degrees from the first point, and then to 180 degrees from the first direction.

*Step 5* Adjust the tilting wheel located directly below the eyepiece for fine leveling. The instrument is level when the two halves of the bubble, viewed through the eyepiece, appear to be aligned.

*Step 6* Set the micrometer scale to zero.

> **CAUTION**
> Do not overtighten any of the adjusting screws of the optical level. The screws should be fingertight only.

*Step 7* Adjust the eyepiece for initial focusing until the cross hairs appear sharp against a white surface.

*Step 8* Turn the focusing knob with the infinity direction mark to focus on the leveling rod.

*Step 9* Rotate the micrometer drum off zero to raise or lower the line of sight until the nearest paired-line marking is bisected to obtain a reading. Add the drum dimension to the scale reading if the line of sight was lowered, or subtract it from the scale reading if the line of sight was raised.

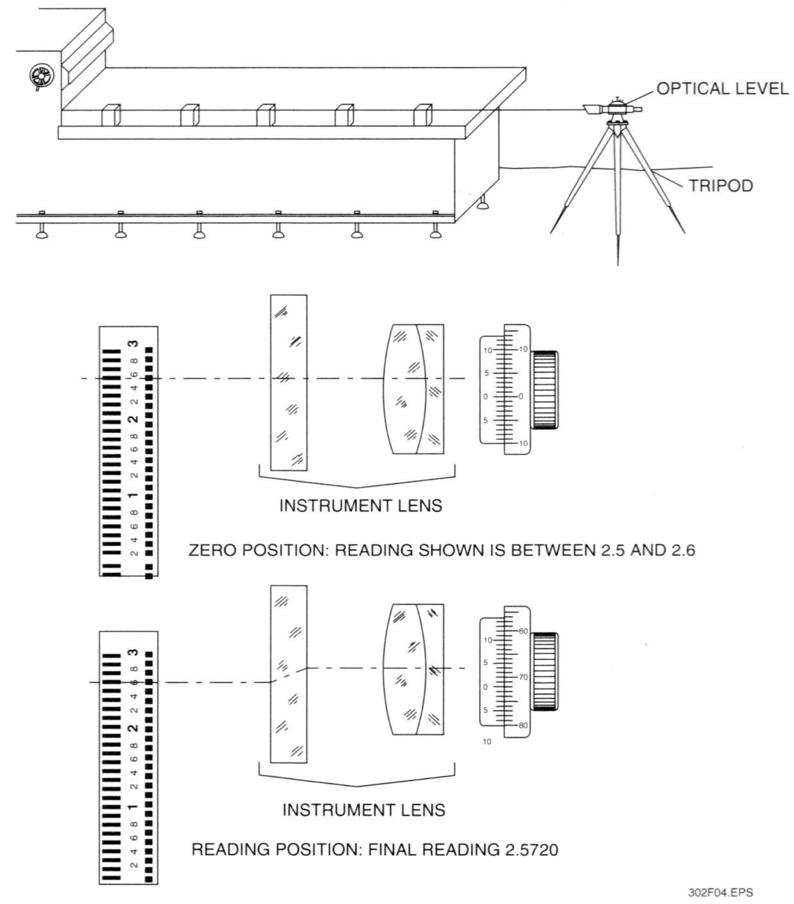


*Figure 4* ◆ Optical level placed for sighting.



Instructor's Notes:

2.4 MILLWRIGHT ◆ LEVEL THREE

Alignment and leveling have both benefited from the introduction of laser equipment. Laser levels are available at many levels of accuracy and price. The simpler versions (*Figure 5*) are easily used, and reasonably accurate at short ranges, while the high precision lasers used for machine alignment and high accuracy (*Figure 6*) are accurate to within a thousandth of an inch or less. The lasers used for site layout are set up on a tripod and leveled, and then can be used to set levels to within their accuracy all over a site. The lasers used for alignments are expensive, and have much higher accuracy levels, accurate to 33 feet.

302F05.EPS

*Figure 5* ◆ Laser level.

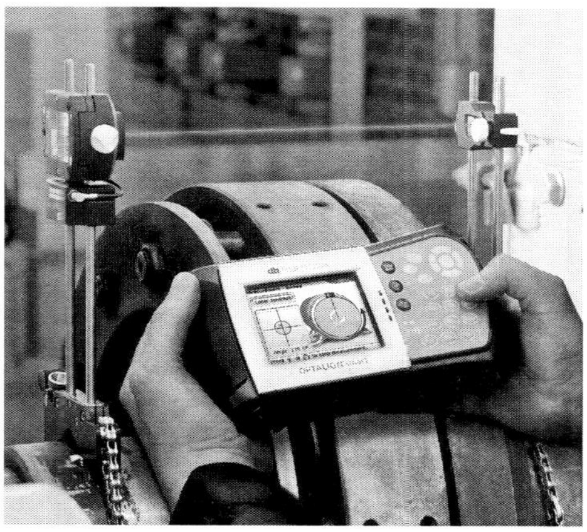

302F06.EPS

*Figure 6* ◆ Alignment laser.

## 2.4.0 Electronic Levels

The most accurate type of level is the electronic level (*Figure 7*). Some models are accurate to 2 millionths of an inch (0.000002 inch). An electronic level consists of a pickup unit and an indicator unit connected by an electrical cable. Levelness is shown on the indicator unit.

Follow these steps to use an electronic level:

*Step 1* Clean the surface to be checked.

*Step 2* Position the pickup unit so that the horizontal center line is parallel to the surface being checked.

*Step 3* Ensure that the indicator on the electronic level points to zero to determine if the surface is level.

*Step 4* Adjust the workpiece until the indicator reads zero, indicating that the surface is level.

Refer to the manufacturer's instructions for operating a specific electronic level.

## 3.0.0 ◆ CALIPERS

Calipers are measuring tools used for layout and for transferring measurements. Contact measurements are made with calipers by two methods. In the first method, the dimensions of the caliper are set using a steel rule, and then the caliper is placed on the workpiece. In the second method, the caliper is placed on the workpiece, and then the distance between the contact points is measured using a steel rule. The four types of calipers are inside, outside, vernier, and dial.

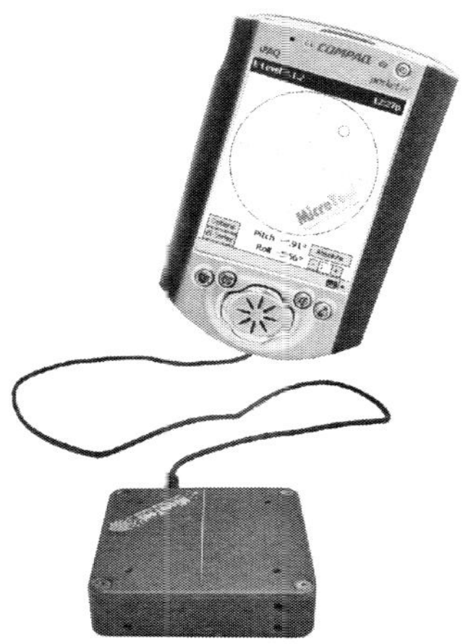


302F07.EPS

*Figure 7* ◆ Electronic level.



**Explain how laser levels and alignment lasers are used.**

**Discuss electronic levels and explain the steps for using them.**

**Demonstrate how to use different types of levels.**

**Have trainees practice using levels. Note the proficiency of each trainee. This laboratory corresponds to Performance Task 1.**

**Have trainees review Sections 3.0.0–3.3.0.**

**Ensure that you have everything required to teach this session.**

**See the Teaching Tip for Section 3.0.0 at the end of this module.**

**Explain the different types of calipers and explain how each is used.**

**Provide calipers for trainees to examine.**

MODULE 15302-08 ◆ PRECISION MEASURING TOOLS 2.5

Describe inside and outside calipers and review the steps for using them.

Discuss vernier calipers and explain how they are used.

Show Transparency 6 (Figure 11). Explain how to read a vernier scale.

### 3.1.0 Inside and Outside Calipers

The two most commonly used general types of calipers are inside and outside calipers (*Figure 8*). Inside calipers are used for measuring inside diameters and the size of inside surfaces, such as slots or holes. Outside calipers are used for measuring outside diameters and distances between outside surfaces.

Follow these steps to use inside or outside calipers.

*Step 1* Set one leg of the caliper firmly on the workpiece to be measured.

*Step 2* Adjust the caliper until the other leg rests on the largest part of the workpiece diameter.

*Step 3* Tighten the adjusting screw.

*Step 4* Remove the caliper, being careful not to change the adjustment.

*Step 5* Measure the space between the caliper legs, using a steel rule, to determine the measurement of the workpiece (*Figure 9*).

### 3.2.0 Vernier Calipers

The most common vernier tool is the vernier caliper (*Figure 10*). It is made in standard lengths of 6, 12, 24, 36, and 48 inches and can measure to an accuracy of 1/1,000 (0.001) inch over a large range of sizes. Vernier calipers are used for both inside and outside measurements.

The vernier caliper consists of a **graduated** beam, or frame, with a fixed measuring jaw and a vernier slide assembly. A movable jaw, vernier plate, clamping screws, and adjusting screw make up the slide assembly. The slide assembly moves as a unit along the beam.

The vernier scale on a vernier caliper reads accurately to 0.001 inch. The large numbers on the beam indicate inches; the small numbers indicate hundredths (0.01) of an inch. The scale lines between the hundredths are the tenths. One side of the vernier plate is marked for inside measurements and the other for outside measurements on some calipers. Other brands have equal scales for inside and outside measurement.

Follow these steps to use a vernier caliper:

*Step 1* Clean the nibs, using a lint-free cloth.

*Step 2* Close the caliper, and check to see that the zeros are aligned.

*Step 3* Open the caliper, and place it around or in the workpiece, depending on whether you are taking an outside or inside measurement.

*Step 4* Place the nibs lightly in contact with the workpiece so that there is no play or movement between the nibs and the workpiece.

*Step 5* Read the beam at the first read point to find the inch, tenths, hundredths, and thousandths of an inch values.

*Step 6* Find the line on the vernier scale that lines up exactly with any of the lines along the beam. The number on the line of the vernier scale that lines up with the beam is the second read point and is the thousandths of an inch value. *Figure 11* shows how to read a vernier scale.

*Step 7* Add the inch, hundredth, and thousandth values to determine the total measurement.

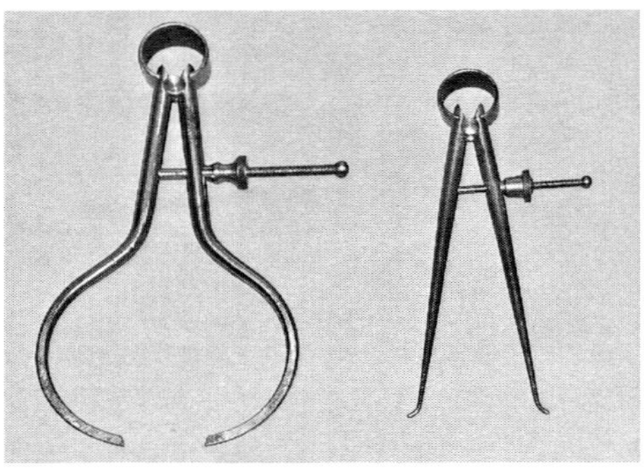


302F08.EPS

*Figure 8* ◆ Inside and outside calipers.

302F09.EPS

*Figure 9* ◆ Measuring space.



Instructor's Notes:

2.6 MILLWRIGHT ◆ LEVEL THREE

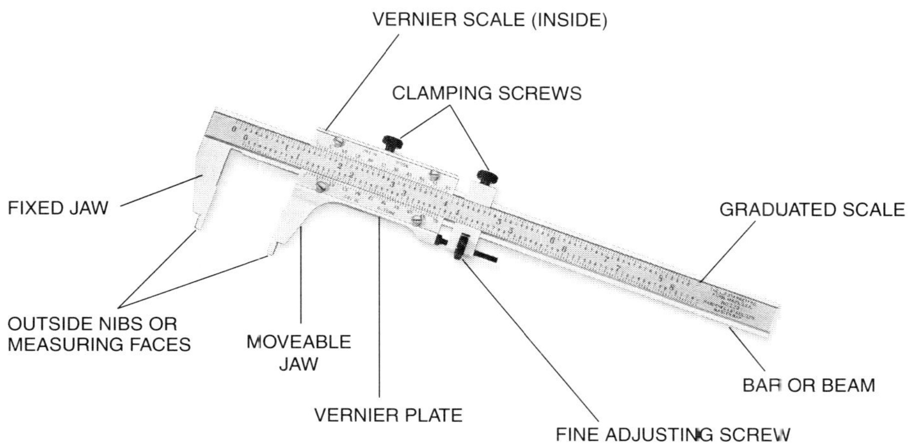


*Figure 10* ◆ Vernier caliper.

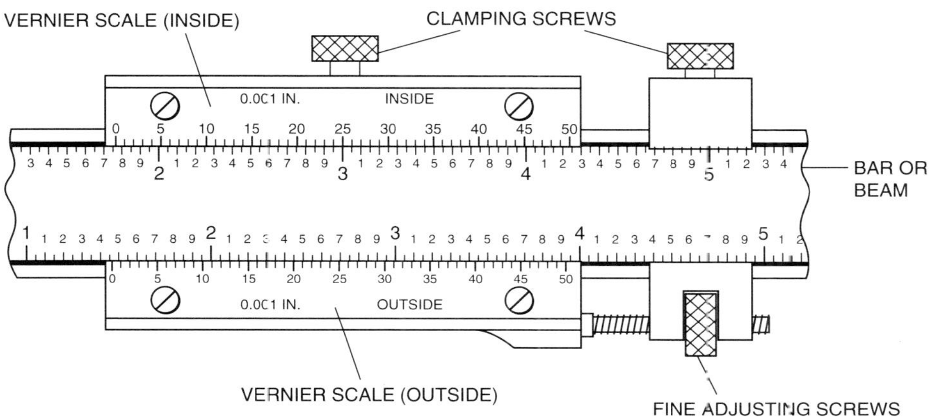


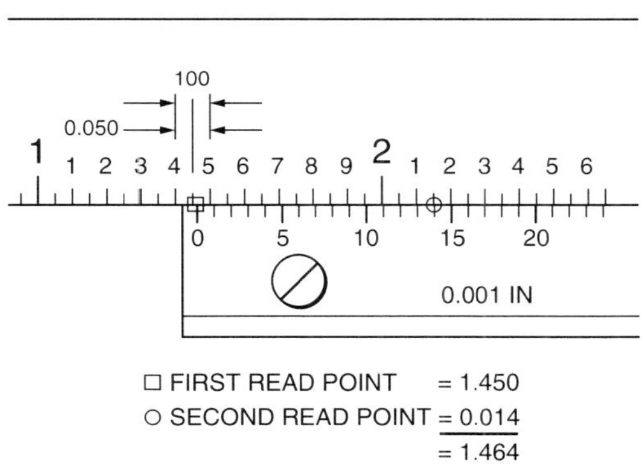


*Figure 11* ◆ Reading a vernier scale.



MODULE 15302-08 ◆ PRECISION MEASURING TOOLS 2.7

**Show Transparency 7 (Figure 12). Discuss dial calipers and review the steps for using them.**

**Demonstrate how to use calipers.**

**Have trainees practice using calipers. Note the proficiency of each trainee. This laboratory corresponds to Performance Task 2.**

**Have trainees review Sections 4.0.0–4.4.0.**

**Ensure that you have everything required to teach this session.**

**See the Teaching Tip for Section 4.0.0 at the end of this module.**

**Discuss how a micrometer is used.**

**Describe the parts of a micrometer.**

### 3.3.0 Dial Calipers

Like vernier calipers, dial calipers (*Figure 12*) are used to make both inside and outside measurements. The inch and hundredths values are read from the scale and dial; the hundredths and thousandths values are read from the dial.

Like other precision instruments, the dial caliper must be zeroed before it is used. Follow these steps to use a dial caliper.

*Step 1* Clean the measuring faces or nibs, using a lint-free cloth.

*Step 2* Close the caliper lightly.

*Step 3* Read the dial to see if the needle is aligned with zero. If the needle is not on the zero mark, clean the nibs again; loosen the locknut on the dial face, and turn the dial face, so that it reads zero; then tighten the locknut.

*Step 4* Open the caliper, and place it around or in the workpiece, depending on whether you are taking an outside or inside measurement.

*Step 5* Bring the nibs lightly in contact with the workpiece so that there is no play or movement between the nibs and the workpiece.

*Step 6* Read the beam to obtain the inch and hundredths values. This is the first read point. Each brand of calipers may place the read point differently, so find zero with the calipers closed.

*Step 7* Read the dial face to find the thousandths value. This is the second read point. The hundredths values are numbers 10 through 90. The thousandths values are the scale marks between the numbers.

*Step 8* Add the inch, hundredth, and thousandth values to determine the total measurement.

## 4.0.0 ◆ MICROMETERS

A micrometer is a measuring tool used to take precise measurements of nonmoving parts. The basic parts of a micrometer are as follows:

- *Frame* – The part that grips the micrometer.
- *Anvil* – A stationary part that is placed against one edge of the surface to be measured. The anvil is one of the measuring faces.
- *Spindle* – The part that is adjusted to touch the opposite edge of the surface to be measured. The spindle is one of the measuring faces.
- *Sleeve* – The part with numbers that indicate the largest measurement of a surface. The sleeve measurement is added to the thimble measurement to obtain the true dimension.
- *Thimble* – The part with numbers that indicate the fine measurement of a surface.

*Figure 13* shows several types of outside micrometers, including two digital models. Other varieties of outside micrometers, such as the blade type and deep throat micrometers serve special purposes. The functional system is the same.

Most types of micrometers are available in models with digital readout screens, usually called digital mikes. The readouts are usually electronic screens with the level of complexity of readout varying widely.

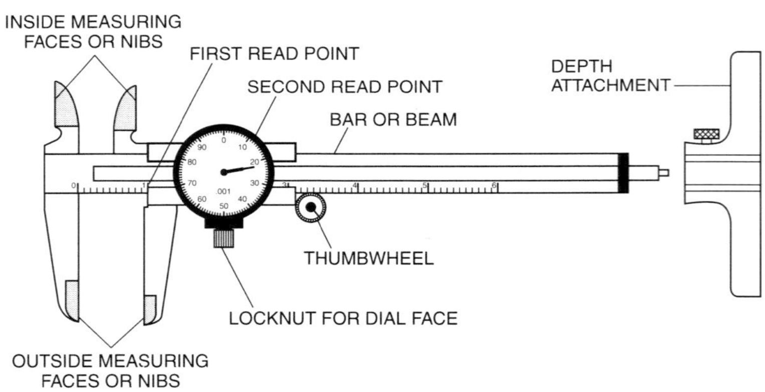


*Figure 12* ◆ Dial caliper.



Instructor's Notes:

2.8 MILLWRIGHT ◆ LEVEL THREE

Each manufacturer makes a variety of tips (*Figure 14*) and options to be used with micrometers. Each job may require that these be used; therefore, it is important to be familiar with the manufacturer's catalog.

Millwrights use the following types of micrometers:

- Outside micrometers
- Inside micrometers
- Depth micrometers
- Height gauges

### 4.1.0 Outside Micrometers

Outside micrometers are the most commonly used micrometers. They can be used to set inside calipers and to measure outside diameters and the width or diameter of stock. Outside micrometers come in a variety of sizes.

Micrometers with interchangeable anvils allow measurements at different sizes with one frame. Manufacturer's instructions will explain the procedure to change anvils. It is necessary to calibrate the micrometer and rezero it any time you change the anvil. A standard is necessary if the zero will occur where the anvil is not in contact with the spindle. It is also necessary to check the zero if the tip is changed.

Follow these steps to use a zero to one outside micrometer. For larger sizes, calibrate with the standards.

*Step 1* Close the micrometer lightly to ensure that it reads zero (0.0000) when it is closed. If the micrometer does not read zero when it is closed, ensure that the measuring faces are clean, and bring the faces together. Insert the spanner wrench provided with the micrometer in the small slot of the sleeve according to the instructions in the manufacturer's manual. Turn the sleeve until the zero line coincides with the zero on the thimble, using the spanner wrench.

*Step 2* Ensure that the surfaces of the workpiece and the micrometer measuring faces are clean.

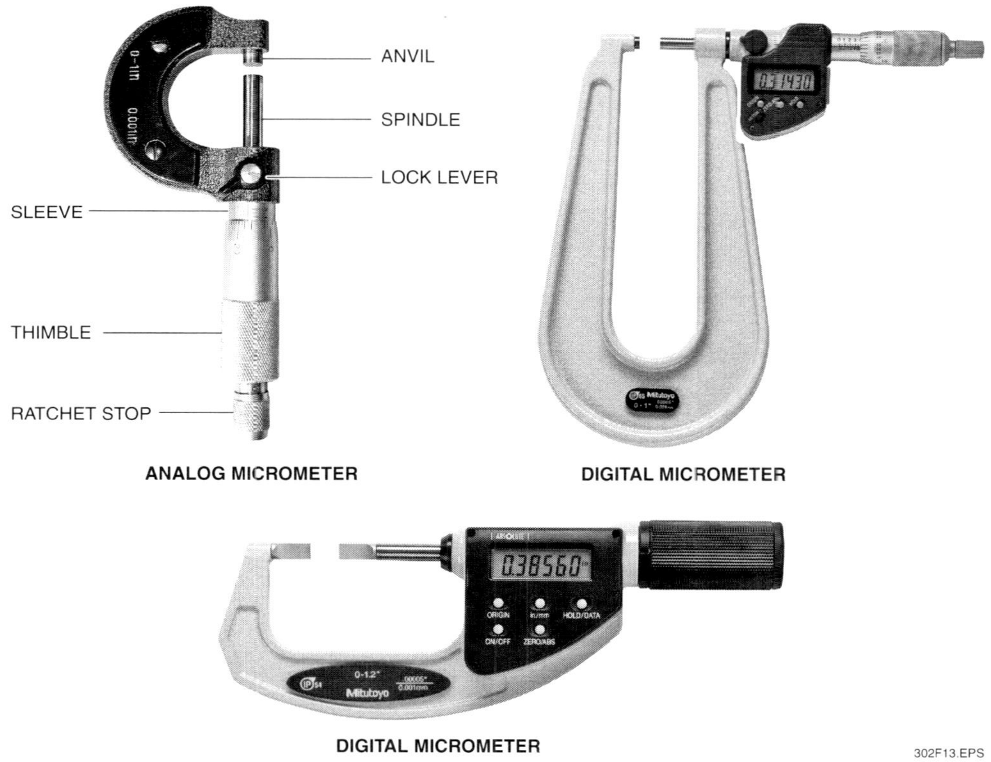


*Figure 13* ◆ Outside micrometers.

**List types of micrometers commonly used by millwrights.**

**Provide micrometers for the trainees to examine.**

**Show Transparency 8 (Figure 14). Discuss how micrometer tips are used.**

**Discuss outside micrometers and explain the steps for calibrating them.**



MODULE 15302-08 ◆ PRECISION MEASURING TOOLS 2.9

*Step 3* Hold the micrometer by its frame, and position the measuring faces near or around the area to be measured.

*Step 4* Turn the thimble until the measuring faces touch the workpiece without slipping or clamping too tightly. Snug the micrometer against the workpiece until the ratchet disengages, using the ratchet stop. Do not overtighten, because this will bend the frame of the micrometer.

*Step 5* Hold the frame firmly, and turn the locknut to lock the thimble.

> **CAUTION**
> Do not overtighten the locknut.

*Step 6* Remove the micrometer from the workpiece carefully, and find a strong light source.

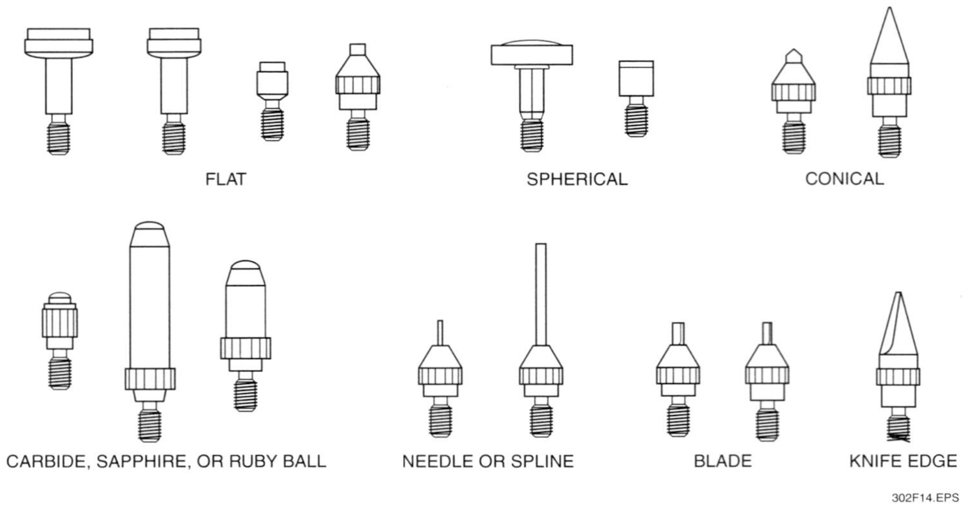


*Figure 14* ◆ Micrometer tips.

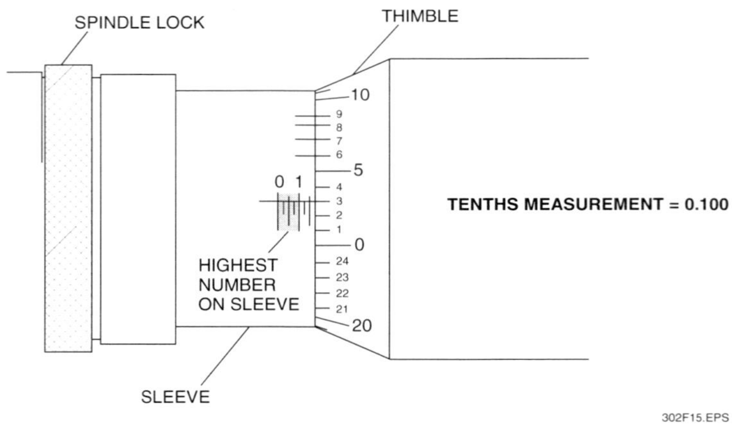


*Figure 15* ◆ Determining tenths measurement.



Instructor's Notes:

2.10 MILLWRIGHT ◆ LEVEL THREE

*Step 7* Read the largest **increment** on the sleeve. This number is the tenths of an inch value. *Figure 15* shows how to determine the tenths measurement.

*Step 8* Count the number of scale lines visible up to the thimble in increments of 0.025, 0.050, or 0.075 hundredths of an inch. *Figure 16* shows how to determine the hundredths measurement.

*Step 9* Read the number on the thimble. The thimble is graduated from 0 to 0.024. This is the thousandths of an inch value (x.xxx). *Figure 17* shows how to determine the thousandths measurement.

> **NOTE**
> The line that exactly matches one of the scale lines on the thimble is read as a value from 1 to 9 on the sleeve. If the micrometer does not have a vernier scale, continue to Step 11.

*Step 10* Read the ten-thousandths value (x.xxxx) on the vernier scale (*Figure 18*).

*Step 11* Add the tenths, hundredths, thousandths, and ten-thousandths values to determine the total measurement (*Figure 19*).

## 4.2.0 Inside Micrometers

Inside micrometers (*Figure 20*) are straight micrometers with the anvil and spindle on opposite ends. They are used to measure the inside diameter of cylinders or rings, to measure between parallel surfaces, and to set outside calipers.

Follow these steps to use an inside micrometer:

*Step 1* Clean the measuring faces and work surfaces to be measured.

*Step 2* Insert the micrometer into the area of or between the surfaces to be measured.

*Step 3* Turn the thimble until both measuring faces touch the work surfaces.

**Show Transparencies 9-13 (Figures 15-19). Explain how the measurement scales on an outside micrometer are read.**

**Describe an inside micrometer and explain how it is used.**

**Review the steps for using an inside micrometer.**

**Show Transparency 14 (Figure 20).**

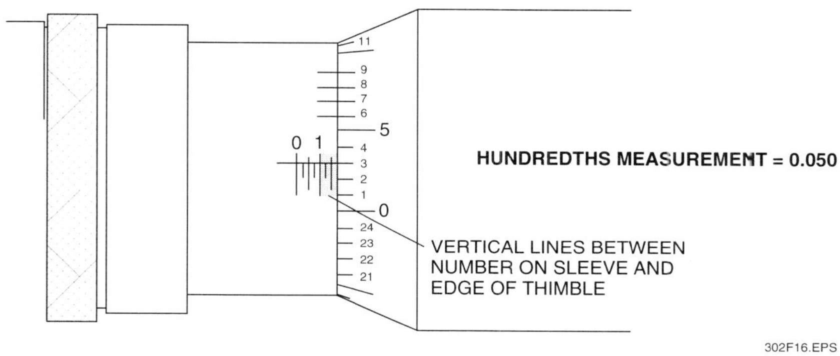


*Figure 16* ◆ Determining hundredths measurement.

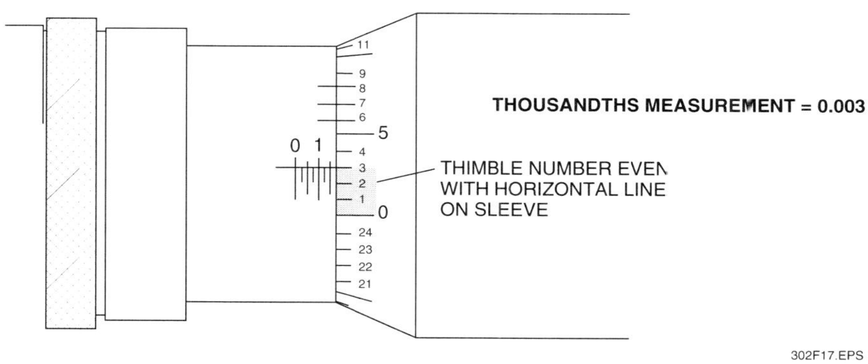


*Figure 17* ◆ Determining thousandths measurement.



MODULE 15302-08 ◆ PRECISION MEASURING TOOLS 2.11

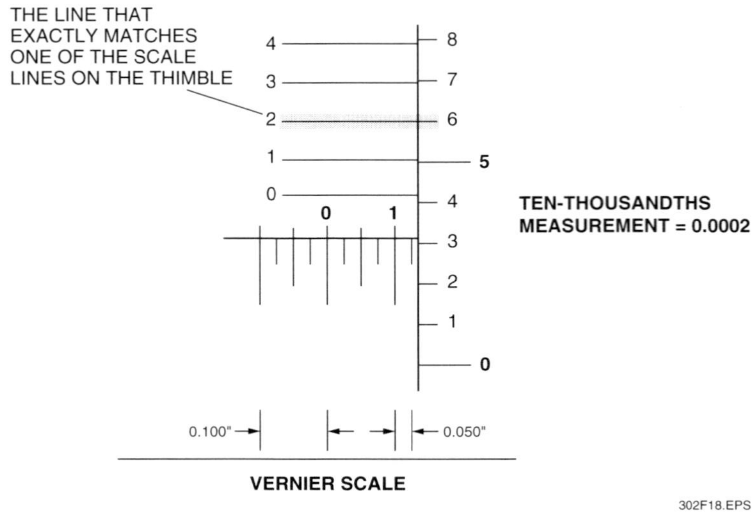


*Figure 18* ◆ Determining ten-thousandths measurement.

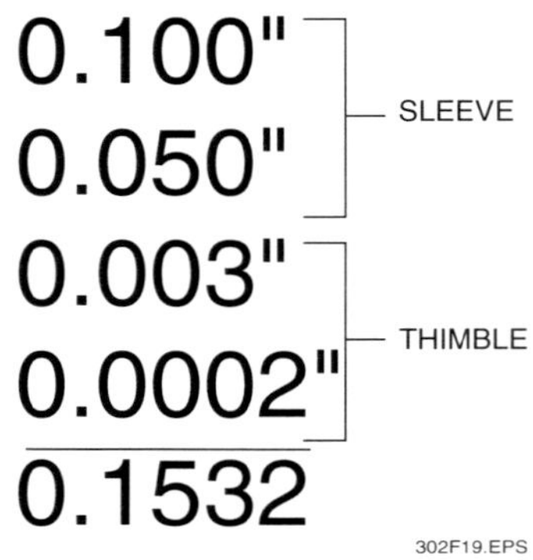


*Figure 19* ◆ Adding measurements.

***Step 4*** Turn the lockscrew to lock the thimble.

**CAUTION**
Do not overtighten the lockscrew.

***Step 5*** Remove the micrometer from the workpiece, and find a strong light source.

***Step 6*** Read the largest number visible on the sleeve. This is the hundredths of an inch value.

***Step 7*** Count the scale lines visible up to the thimble. The scale lines are graduated in 0.025-inch increments. These are the thousandths of an inch values.

***Step 8*** Read the number on the thimble. The thimble is graduated from 0 to 0.024. This is the thousandths of an inch value (x.xxx).

***Step 9*** Read the ten-thousandths value (x.xxxx) on the vernier scale.



Instructor's Notes:

2.12 MILLWRIGHT ◆ LEVEL THREE

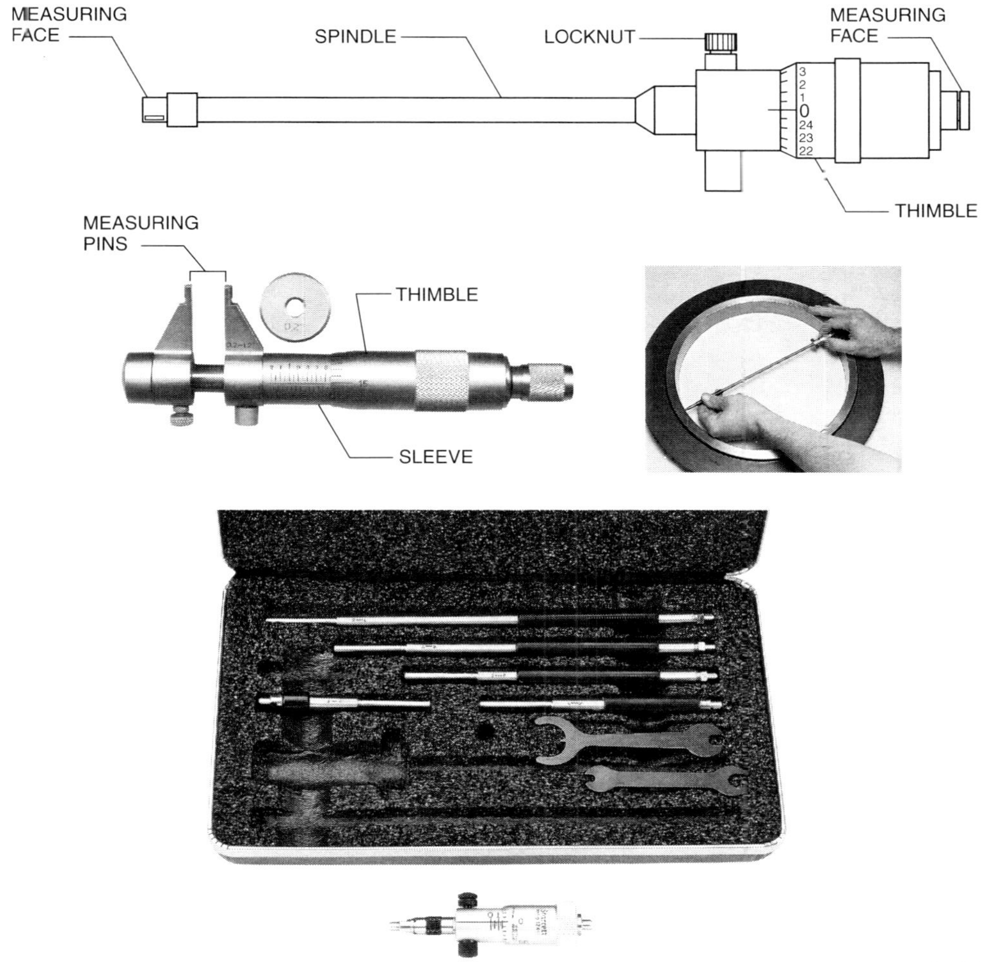


*Figure 20* ◆ Inside micrometers.

> **NOTE**
>
> The line on the sleeve that lines up exactly with one of the scale lines on the thimble is the ten-thousandths of an inch value. Add the inch, hundredth, thousandth, and ten-thousandth values to determine the total measurement. If the micrometer does not have a vernier scale, continue to Step 10.
>
> You will always want to check the reading on an inside micrometer with that of a calibrated outside micrometer.

***Step 10*** Add the inch, hundredth, thousandth, and ten-thousandth values to determine the total measurement.



MODULE 15302-08 ◆ PRECISION MEASURING TOOLS 2.13

**Explain that depth micrometers are also called micrometer depth gauges.**

**Review the steps for using a depth micrometer.**

**Show trainees how to use micrometers.**

**Have trainees practice using micrometers. Note the proficiency of each trainee. This laboratory corresponds to Performance Task 3.**

**Show Transparency 15 (Figure 22). Review the steps for using a height gauge to scribe a line.**

### 4.3.0 Depth Micrometers

Depth micrometers (*Figure 21*) are also called micrometer depth gauges. They are used to measure the depth of holes, slots, or projections. Their measuring range can be increased by using longer measuring rods. Remember that the scale is reversed from the outside micrometer.

Follow these steps to a use a depth micrometer:

*Step 1* Clean the area to be measured, the measuring face, and the base of the depth micrometer.

*Step 2* Turn the thimble slightly past zero to ensure that the depth micrometer will zero.

*Step 3* Place the depth micrometer on a surface plate; hold the base firmly, and turn the thimble. This will zero the micrometer.

> **NOTE**
> The depth micrometer can also be zeroed by placing a gauge block on the surface plate and measuring it with the depth micrometer to check its accuracy against the gauge block.

*Step 4* Place the base over the hole or feature to be measured.

*Step 5* Hold the base firmly, and turn the thimble until the measuring face lightly touches the bottom of the feature.

*Step 6* Turn the thimble to move the spindle in and out slightly to feel where the contact point actually is.

*Step 7* Turn the spindle lock to lock the spindle, and read the micrometer.

### 4.4.0 Height Gauges and Surface Plates

Height gauges are used for measuring and marking vertical distances from a surface plate and can measure distances from 6 to 40 inches. Height gauges have a vernier, dial, or digital scale and have a jaw attachment with a box clamp to hold scribes, probes, dial indicators, or other tools.

Height gauges must be used on top of a surface plate to provide accuracy. A surface plate is made from a large block of granite or steel and is precision-ground and polished to a flatness within 8 ten-thousandths to 5 millionths of an inch, depending on the size and grade.

When not in use, height gauges should be stored and the surface plate covered to prevent it from being damaged. *Figure 22* shows a height gauge and surface plate.

Follow these steps to use a height gauge to scribe a line:

*Step 1* Clean the surface plate and height gauge base, using a lint-free cloth.

*Step 2* Choose the correct tool, and install it on the jaw of the height gauge.

*Step 3* Turn the lock screw to set the height and lock the head in place.

*Step 4* Read the vernier scale to determine the height of the workpiece.

*Step 5* Hold the workpiece firmly, and slide the scribe across the workpiece. If the workpiece is hardened steel, use toolmaker's ink to scribe the line.

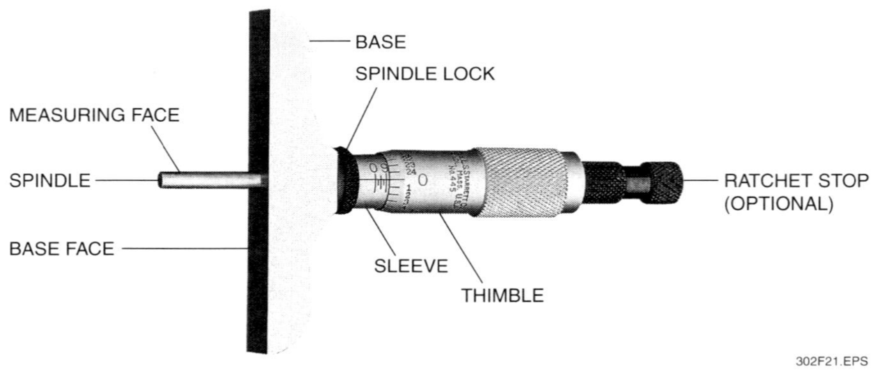


*Figure 21* ◆ Depth micrometer.



*Instructor's Notes:*

2.14 MILLWRIGHT ◆ LEVEL THREE

Follow these steps to use a height gauge for measuring:

*Step 1* Clean the surface plate and height gauge base, using a lint-free cloth.

*Step 2* Attach a dial indicator or use the jaw attachment itself. If using a dial indicator, the gauge must be zeroed with the indicator attached.

*Step 3* Position the workpiece firmly, and bring the height gauge close to the measurement point.

*Step 4* Tighten the upper lock screw lightly.

*Step 5* Turn the fine adjustment dial to move the head so that it just touches the point of measurement with the probe or jaw.

*Step 6* Tighten the lower lock screw on the head lightly.

*Step 7* Read the beam and vernier scale or the dial indicator to obtain the total measurement.

**Review the steps for using a height gauge for measuring.**

**Have trainees review Sections 5.0.0–6.0.0.**

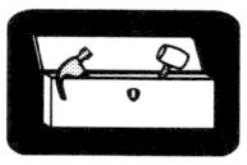

**Ensure that you have everything required to teach this session.**

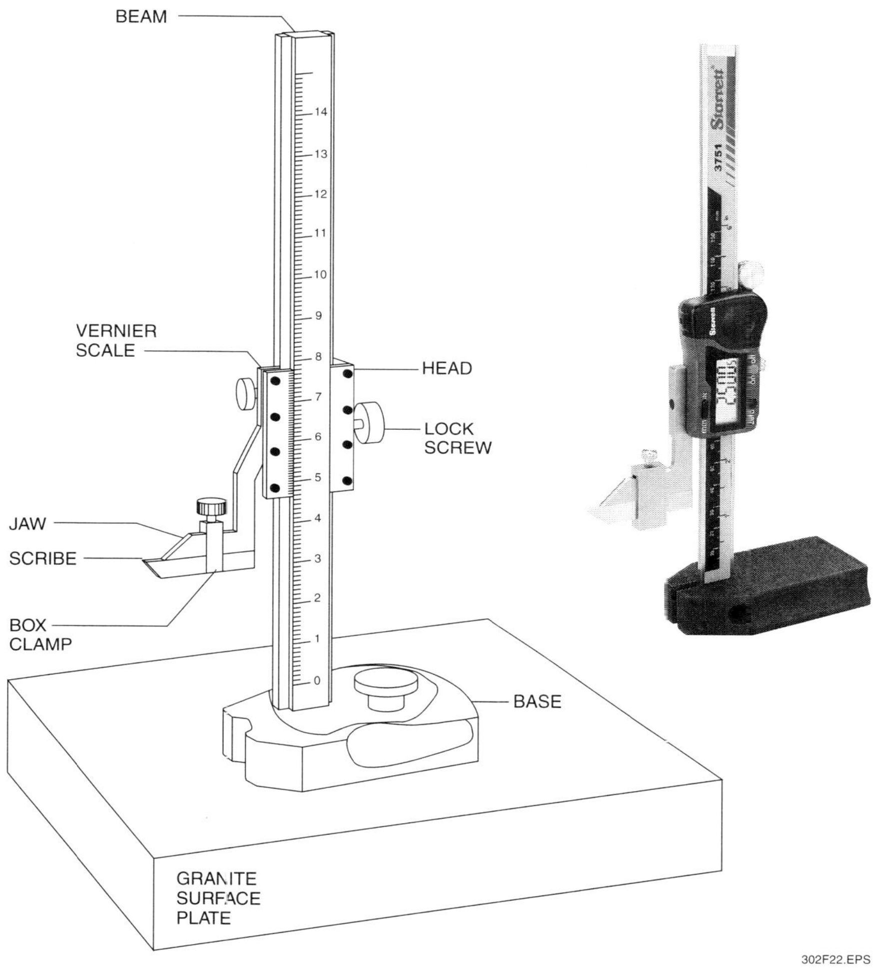


*Figure 22* ◆ Height gauge and surface plate.



MODULE 15302-08 ◆ PRECISION MEASURING TOOLS 2.15

**Discuss dial indicators. Review the steps for using a dial indicator.**

**Show Transparencies 16 and 17 (Figures 23 and 24).**

**Have trainees practice using dial indicators. Note the proficiency of each trainee. This laboratory corresponds to Performance Task 4.**

**Discuss universal bevel protractors and explain how they are used.**

**Show Transparency 18 (Figure 25).**

**Have trainees practice using universal bevel protractors. Note the proficiency of each trainee. This laboratory corresponds to Performance Task 5.**

**Have trainees review Section 7.0.0.**

**Ensure that you have everything required to teach this session.**

**Discuss gauge blocks and how they are used.**

## 5.0.0 ◆ DIAL INDICATORS

A dial indicator is a direct-reading instrument used to measure machined parts for accuracy or surfaces of machinery to determine runout or accuracy of alignment. Dial indicators are also used when installing bearings and seals, setting up lathes and milling machines, and checking the concentricity of a diameter. A dial indicator consists of a graduated dial with an indicator hand and a contact point attached to a spring-loaded plunger. Any movement of the plunger causes the pointer on the dial to move. Dial indicators range in size from 1 to 4½ inches in face diameter. The dials are usually divided into one of the following increments:

- Hundredths of a millimeter (0.01 millimeter)
- Two-thousandths of a millimeter (0.002 millimeter)
- Thousandths of an inch (0.001 inch)
- Ten-thousandths of an inch (0.0001 inch)

The two most common types of dial indicators are the balanced type and the continuous-reading type. The numbers on the balanced type start at zero and increase in both directions. The numbers on the continuous-reading type start at zero and continue around the dial clockwise. The dials can be reset to zero by rotating the entire dial face. *Figure 23* shows two types of dial indicators.

Follow these steps to use a dial indicator.

*Step 1* Clean the surfaces that the indicator will touch.

*Step 2* Determine where to attach the base, and clamp the indicator base onto the workpiece or machine, using a C-clamp or a magnetic base.

> **NOTE**
> For quick setup, dial indicators can be placed in a magnetic base holder. A magnet in the base holds to any flat or round steel or iron surface, eliminating the time necessary to clamp the dial indicator to a machine. The magnetic force is turned on and off by pressing a push button.

*Step 3* Mount the indicator to the indicator base, using the attachments on the indicator holding rod, and tighten all locknuts securely (*Figure 24*).

*Step 4* Move the workpiece or the indicator, depending on what is being measured.

*Step 5* Loosen the clamp on the base post, and push the indicator into the workpiece by at least half the total travel of the indicator to preload the indicator.

*Step 6* Tighten the locknuts securely.

*Step 7* Loosen the dial face lock screw, and rotate the face to read absolute zero to zero the indicator.

*Step 8* Tighten the dial face lock screw. Take care not to bottom the indicator plunger since it will damage the indicator. If the indicator is too close, reposition it with less preload.

*Step 9* Check all other adjustments to ensure that they are tight and secure.

*Step 10* Rotate the shaft or move the indicator slowly to obtain a total runout. This records the most extreme position of measurement in both directions of travel.

*Step 11* Position the indicator base and take new measurements to get at least three points of measurement for an average.

## 6.0.0 ◆ UNIVERSAL BEVEL PROTRACTORS

The universal bevel protractor (*Figure 25*) consists of a graduated disc with a fixed blade and an adjustable blade. It can be used to lay out or measure any angle by reading the angle of the stock and blades shown on the protractor scale. The protractor may be marked from 0 to 180 degrees or from 0 to 90 degrees on each side. Some universal bevel protractors have a vernier attachment for more precise readings.

## 7.0.0 ◆ GAUGE BLOCKS

Gauge blocks (*Figure 26*) are rectangular, square, or angled gauges made of hardened steel, tungsten carbide, or a special ceramic. They are used to calibrate precision measuring tools. Gauge blocks are commonly referred to as standards because they are used to calibrate most other precision measuring tools. Gauge blocks can be used individually or together by stacking them in a process called wringing.

Gauge blocks can be used in combinations to produce different lengths but should be used in combinations of as few blocks as possible. When gauge blocks are stacked together properly, they are said to be wrung together. This means that their surfaces are flat and smooth and stick together as though magnetized. Gauge blocks must be wrung before they can be used together.

Follow these steps to wring gauge blocks:



Instructor's Notes:

2.16 MILLWRIGHT ◆ LEVEL THREE

*Step 1* Clean the gauge blocks, using a commercial gauge block cleaner and a lint-free cloth or chamois.

*Step 2* Place two gauge blocks together, overlapping them about ⅛ inch.

*Step 3* Slide the blocks together while pressing lightly. A slight resistance can be felt when the blocks are pressed together. *Figure 27* shows sliding the gauge blocks together.

**Show Transparency 19 (Figure 26).**

**Review the steps to properly wring gauge blocks.**

**Show Transparencies 20 and 21 (Figures 27 and 28).**

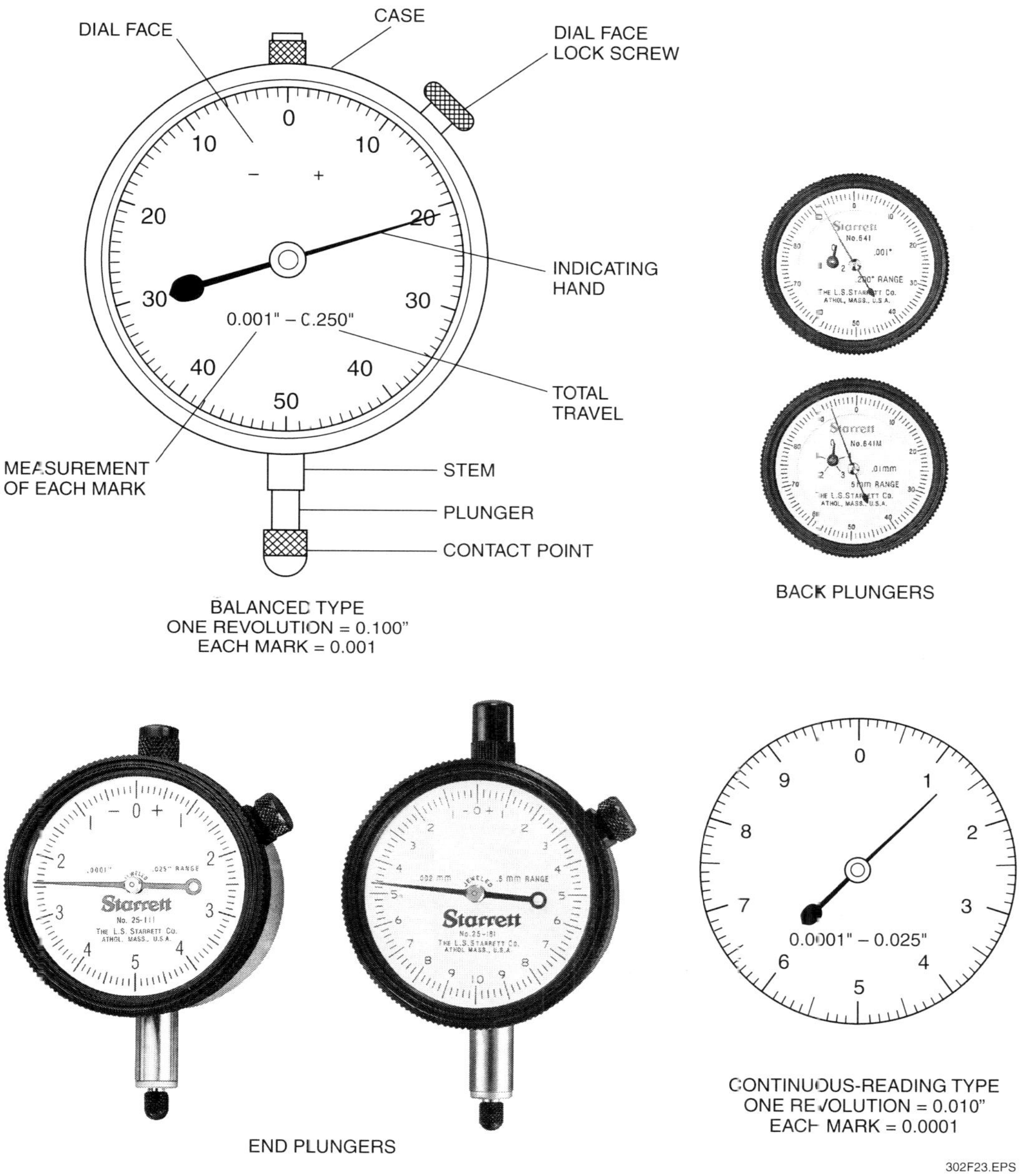


*Figure 23* ◆ Dial indicators.



MODULE 15302-08 ◆ PRECISION MEASURING TOOLS 2.17

**Discuss the four grades of gauge blocks and the accuracy level of each grade.**

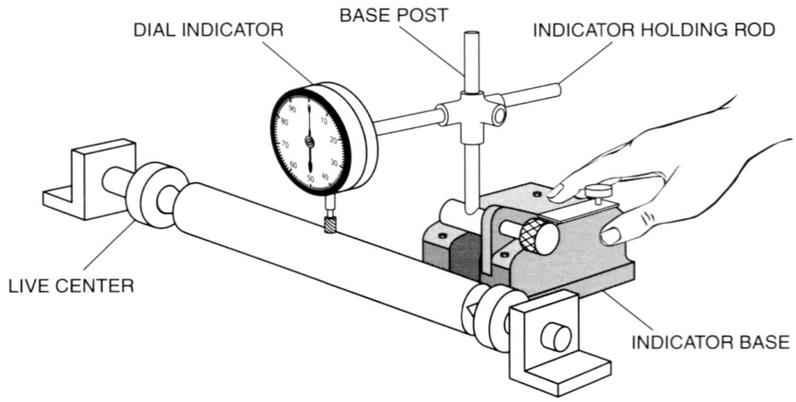


*Figure 24* ◆ Indicator mounted on base.

*Step 4* Position the blocks so that they are aligned. *Figure 28* shows aligning the gauge blocks.

*Step 5* Release one block while holding the wrung blocks over the other hand to ensure that the blocks are properly wrung. To separate the blocks, slide them apart. Clean the blocks and return them to their case.

**CAUTION**

To prevent damage, do not allow the blocks to drop to the floor or other surface.

Gauge blocks are made to dimensions accurate to a few millionths of an inch and are classified by their level of accuracy. The four grades of gauge blocks are the following:

- *Grade AA* – These gauge blocks are laboratory masters used for checking other gauge blocks. They are used as infrequently as possible under temperature-controlled laboratory conditions. Grade AA blocks are accurate to ± 0.000002 inch.
- *Grade A+* – These gauge blocks are used in the laboratory or laboratory-like conditions for precision work where tolerances are 50 millionths of an inch or less.
- *Grade A* – These gauge blocks are used for gauging other measuring instruments. They are generally not used as inspection tools. Grade A blocks are used for tolerances of 50 millionths of an inch or more.

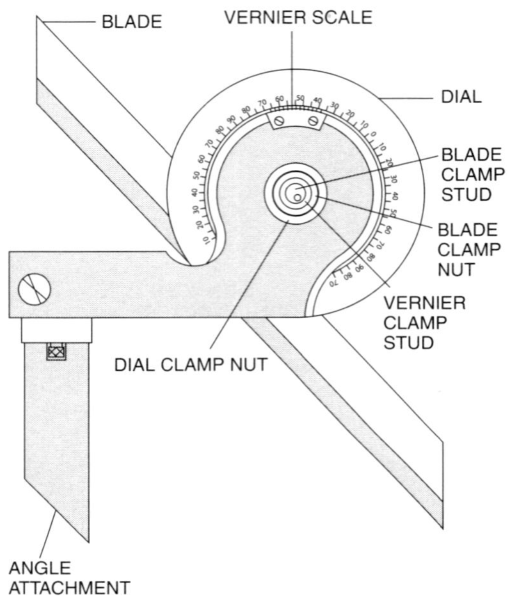


*Figure 25* ◆ Universal bevel protractor.



Instructor's Notes:

2.18 MILLWRIGHT ◆ LEVEL THREE

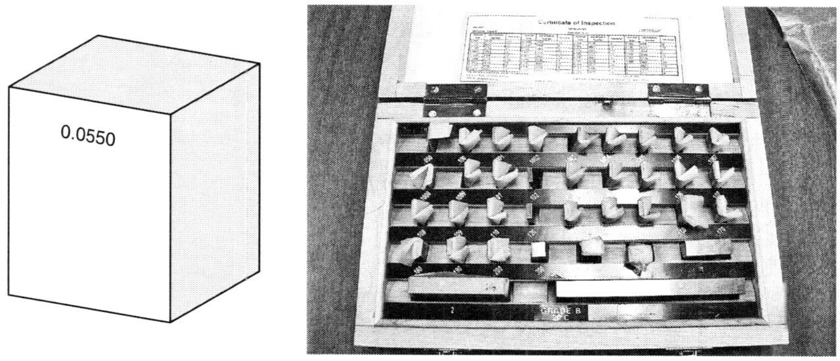


302F26.EPS

*Figure 26* ◆ Gauge block.

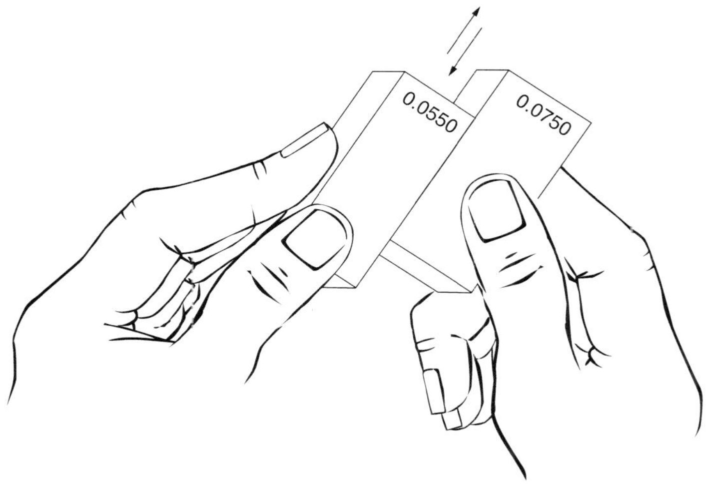


302F27.EPS

*Figure 27* ◆ Sliding gauge blocks.

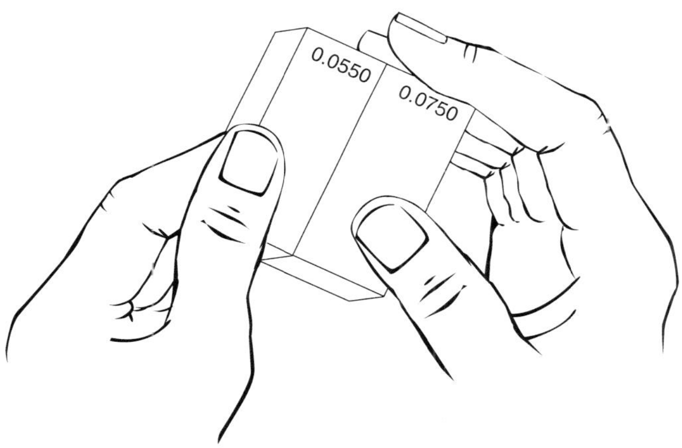


302F28.EPS

*Figure 28* ◆ Aligning gauge blocks.



MODULE 15302-08 ◆ PRECISION MEASURING TOOLS 2.19

**Show Transparency 22 (Table 1). Explain that the most commonly used type of gauge blocks are Jo blocks.**

**Review the guidelines for the proper care of gauge blocks.**

**Show trainees how to use gauge blocks to calibrate precision measuring tools.**

**Have trainees practice using gauge blocks to calibrate precision measuring tools. Note the proficiency of each trainee. This laboratory corresponds to Performance Task 6.**

**Have trainees review Sections 8.0.0–8.3.0.**

**Ensure that you have everything required to teach this session.**

**Introduce speed measurement tools and explain how they are used.**

**Discuss stroboscopes and how they work.**

**Show Transparency 23 (Figure 29).**

- *Grade B* – These gauge blocks are working blocks used in shop conditions for direct measurement when other tools, such as verniers or micrometers, are not adequate.

The most commonly used type of gauge blocks are the Johansson, or Jo, blocks. These are designed in an arithmetical progression size series. Each set contains 111 gauge blocks in four series. They can be combined in any length from 2 to 202 millimeters, in steps of 0.001 millimeters. *Table 1* lists the sizes of Jo blocks in an 81-block set.

Follow these guidelines to properly care for gauge blocks:

- Do not expose gauge blocks to changes in temperature.
- Keep gauge blocks clean. To clean the blocks, use a lint-free cloth or chamois and the cleaning fluid recommended by the manufacturer.
- Never touch the lapped surfaces of gauge blocks or rest them on anything but their sides.
- Use only the oil recommended by the manufacturer for wringing gauge blocks. Never use the oils from your skin.
- Use gauge blocks only when another measuring tool will not perform the job. Gauge blocks wear down as much as a millionth of an inch for every 200 wringings.
- Never carry gauge blocks in your hands. Always carry gauge blocks in a carrying case to prevent nicks or scratches.
- Check each gauge block after it is cleaned for nicks, burrs, scratches, and warpage, using a magnifying glass or an optical flat.

## 8.0.0 ◆ SPEED MEASUREMENT TOOLS

Speed measurement tools are used to measure equipment speed, to balance rotating machinery, and to determine the source of vibration. Common speed measurement tools are stroboscopes, stroboscopic tachometers, and mechanical tachometers.

### 8.1.0 Stroboscopes

Stroboscopes (*Figure 29*) shine a short flash of intense light when triggered by a **transducer**. The transducer is attached to a piece of equipment. A target on the rotating part of the equipment being analyzed is lighted each time the motion causes the transducer to trigger the stroboscope. The source of a vibration can be determined because the vibrating part does not appear to move. If the rotating part appears to move when lighted, the vibration may be unsteady or may be coming from another source nearby.

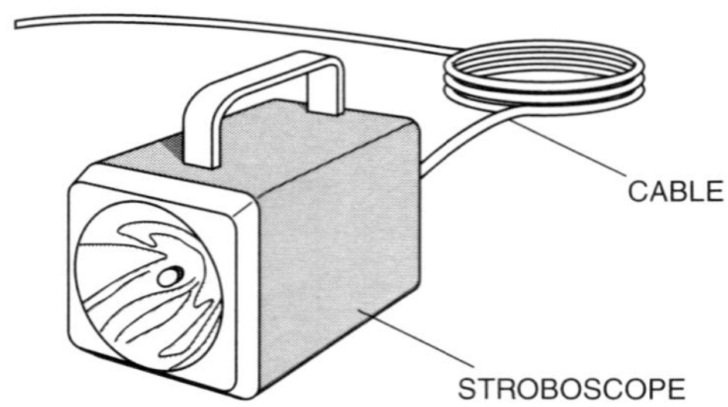


302F29.EPS

*Figure 29* ◆ Stroboscope.

**Table 1** Sizes of Jo Blocks

| | | | | | | | | | |
|---|---|---|---|---|---|---|---|---|---|
| **First Series:** one ten-thousandth series (9 blocks) | | | | | | | | | |
| 0.1001" | 0.1002" | 0.1003" | 0.1004" | 0.1005" | 0.1006" | 0.1007" | 0.1008" | 0.1009" | |
| **Second Series**: one-thousandth series (49 blocks) | | | | | | | | | |
| 0.101" | 0.102" | 0.103" | 0.104" | 0.105" | 0.106" | 0.107" | 0.108" | 0.109" | 0.110" |
| 0.111" | 0.112" | 0.113" | 0.114" | 0.115" | 0.116" | 0.117" | 0.118" | 0.119" | 0.120" |
| 0.121" | 0.122" | 0.123" | 0.124" | 0.125" | 0.126" | 0.127" | 0.128" | 0.129" | 0.130" |
| 0.131" | 0.132" | 0.133" | 0.134" | 0.135" | 0.136" | 0.137" | 0.138" | 0.139" | 0.140" |
| 0.141" | 0.142" | 0.143" | 0.144" | 0.145" | 0.146" | 0.147" | 0.148" | 0.149" | |
| **Third Series:** fifty-thousandth series (19 blocks) | | | | | | | | | |
| 0.050" | 0.100" | 0.150" | 0.200" | 0.250" | 0.300" | 0.350" | 0.400" | 0.450" | 0.500" |
| 0.550" | 0.600" | 0.650" | 0.700" | 0.750" | 0.800" | 0.850" | 0.900" | 0.950" | |
| **Fourth Series:** inch series (4 blocks) | | | | | | | | | |
| 1.000" | 2.000" | 3.000" | 4.000" | | | | | | |

302T01.EPS



Instructor's Notes:

2.20 MILLWRIGHT ◆ LEVEL THREE

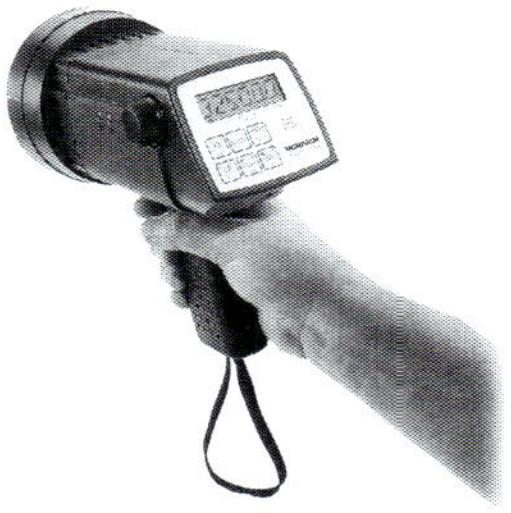

302F30.EPS

*Figure 30* ◆ Stroboscopic tachometer.

## 8.2.0 Stroboscopic Tachometers

Stroboscopic tachometers (*Figure 30*) have a stroboscopic lamp and a scale that reads in flashes per minute or in revolutions per minute. The speed of a rotating device is measured by pointing the stroboscopic lamp at the device and then adjusting the flashing rate until the device does not appear to move. The speed is read on the scale of the instrument. It is also possible that the rate of the flash may be an exact multiple of the rate of rotation. It is for that reason that you bring the frequency up slowly to the point where the apparent motion stops.

## 8.3.0 Mechanical Tachometers

The mechanical tachometer (*Figure 31*) is a gear-driven instrument that measures the rpm or the angular speed of a rotating shaft. The end of the tachometer shaft is placed against a rotating part of the equipment to measure rpm of less than 10,000. Tachometers usually have cone and concave rubber tips, a circumference wheel, and an extension shaft. Other variations include the resonant reed and infrared tachometers. Infrared tachometers are not required to make contact with the rotating shaft.

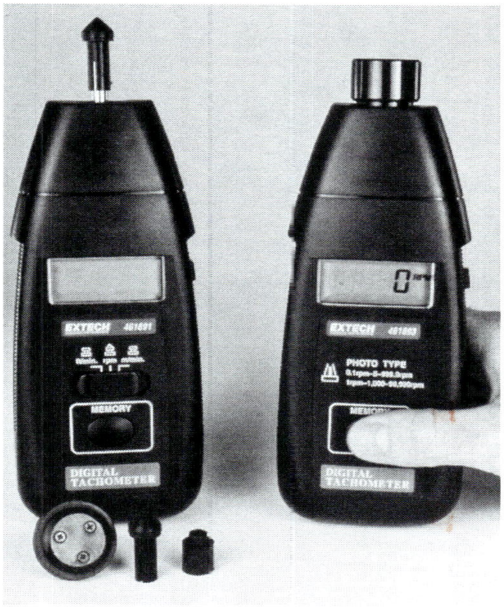


302F31.EPS

*Figure 31* ◆ Mechanical tachometer.

## 9.0.0 ◆ PYROMETERS

A pyrometer is an instrument used to accurately measure temperatures in the higher ranges. Pyrometers can measure temperatures from around 0°C and up to 3,000°C. There are three basic types of pyrometers: optical, thermocouple (which is based on the change in electrical current), and infrared. Of the three, the infrared (*Figure 32*) has the most advantages. It can read from 20 inches to 50 feet away and is as accurate as the contact probes of the thermocouple type. It can read moving objects, take readings in dangerous areas, or be rigged to record or send temperature data remotely.

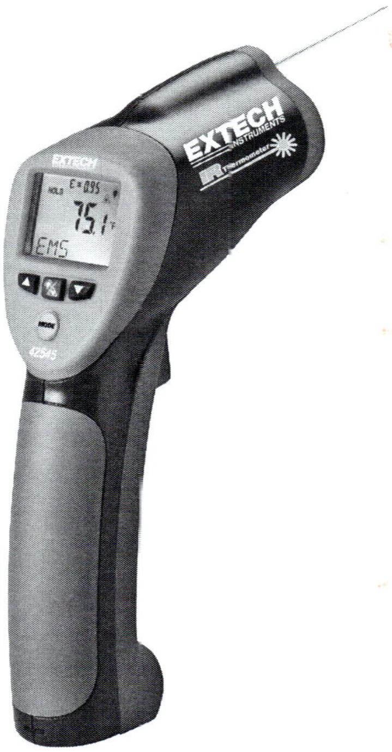


302F32.EPS

*Figure 32* ◆ Infrared pyrometer.



Describe stroboscopic tachometers and mechanical tachometers.

Show trainees how to use speed measurement tools.

Have trainees practice using speed measurement tools. Note the proficiency of each trainee. This laboratory corresponds to Performance Task 7.

Have trainees review Section 9.0.0.

Ensure that you have everything required to teach this session.

Explain the use of pyrometers.

Show trainees how to use pyrometers.

Have trainees practice using pyrometers. Note the proficiency of each trainee. This laboratory corresponds to Performance Task 8.

MODULE 15302-08 ◆ PRECISION MEASURING TOOLS 2.21

Have trainees complete the Review Questions, and go over the answers prior to administering the Module Examination.

## Review Questions

1. The master level is a spirit level that is accurate to _____ inch.
   a. 0.05
   b. 0.01
   c. 0.005
   d. 0.0005

2. The level used for equipment that is more than 25 feet long is the _____ level.
   a. master
   b. machinist's
   c. electronic
   d. optical

3. The large numbers on a vernier caliper beam indicate _____.
   a. inches
   b. tenths
   c. hundredths
   d. thousandths

4. Like other precision instruments, the dial caliper must be _____ before it is used.
   a. greased
   b. zeroed
   c. grounded
   d. washed

5. The first step in zeroing a dial caliper is to open the measuring faces all the way.
   a. True
   b. False

6. The stationary measuring face of a micrometer is the _____.
   a. anvil
   b. spindle
   c. thimble
   d. ratchet stop

7. The first step in checking the accuracy of an outside micrometer is to _____.
   a. open it all the way
   b. use the spanner to turn the sleeve
   c. close it all the way
   d. tighten the ratchet

8. When measuring a piece with an outside micrometer, snug the micrometer against the workpiece until the _____.
   a. frame bends
   b. ratchet disengages
   c. sleeve won't turn anymore
   d. anvil cuts into the workpiece

9. The largest increment on the sleeve of an inch micrometer is _____
   a. tenths of an inch
   b. inches
   c. hundredths of an inch
   d. thousandths of an inch

10. The thimble of a zero-to-one outside micrometer is graduated from _____.
    a. 1 to 10
    b. 0 to 0.024
    c. 0 to 0.0024
    d. 1 to 24

11. You should check the reading on an inside micrometer with a(n) _____.
    a. Jo block
    b. outside micrometer
    c. caliper
    d. height gauge

12. A surface plate is used under a _____.
    a. depth micrometer
    b. height gauge
    c. inside micrometer
    d. vernier caliper

13. A dial indicator is used to determine _____.
    a. outside diameter
    b. inside diameter
    c. accuracy of alignment
    d. length

14. The two most common types of dial indicators are the _____ type and the continuous-reading type.
    a. precision
    b. balanced
    c. intermittent
    d. vernier

2.22 MILLWRIGHT ◆ LEVEL THREE

Instructor's Notes:

## Review Questions

15. Gauge blocks used for checking other gauge blocks are Grade _____.
    a. B
    b. A+
    c. AA
    d. C

16. Stacking two or more gauge blocks properly is called _____.
    a. sliding
    b. wringing
    c. twisting
    d. sticking

17. Gauge blocks are commonly referred to as _____.
    a. standards
    b. rulers
    c. size blocks
    d. check blocks

18. Stroboscopes shine a short flash of intense light when triggered by a _____.
    a. transistor
    b. translator
    c. thermistor
    d. transducer

19. The speed of a rotating device is measured by pointing a stroboscopic lamp at the device and adjusting the flashing rate until the device appears to _____.
    a. move backwards
    b. move slowly
    c. stop moving
    d. move sideways

20. Mechanical tachometers are used to measure rpms of _____.
    a. less than 1,000
    b. less than 10,000
    c. over 50,000
    d. over 100,000



MODULE 15302-08 ◆ PRECISION MEASURING TOOLS 2.23

Summarize the major concepts presented in the module.

Administer the Module Examination. Record the results on Craft Training Report Form 200, and submit the results to the Training Program Sponsor.

Administer the Performance Test, and fill out Performance Profile Sheets for each trainee. If desired, trainee proficiency noted during laboratory sessions may be used to complete the Performance Test. Record the results on Craft Training Report Form 200, and submit the results to the Training Program Sponsor.

## Summary

The consistency of accurate measurements helps ensure quality and is vital when performing tasks. Being practiced in using precision measuring tools ensures that each job is performed precisely. A keen sense of touch is as important for this performance as the proper use and calibration of the tools. With practice and proper handling of these tools, the sense of touch is developed as a skill within itself and will guarantee that each task is performed correctly the first time.

## Notes



Instructor's Notes:

2.24 MILLWRIGHT ◆ LEVEL THREE

## Trade Terms Introduced in This Module

*Graduated:* Marked with degrees of measurement.

*Increment:* One of a set of regular, consecutive additions.

*Plumb:* Vertically aligned.

*Transducer:* A device that converts an input signal to a different type of output signal.



MODULE 15302-08 ◆ PRECISION MEASURING TOOLS 2.25

# Resources & Acknowledgments

## Figure Credits

L.S. Starrett Company, 302F01 (photo), 302F10, 302F20C-D, 302F21, 302F22 (photo), 302F23 (photos)

Brunson Instrument Co., 302F03 (top)

Trimble Navigation Limited, 302F03 (bottom)

Photo courtesy of Ludeca, Inc., 302F06 www.ludeca.com

Brooks Automation, Inc., 302F07 www.brooks.com/microtool, (719) 471-9888

Topaz Publications, Inc., 302F08, 302F09, 302F13 (analog micrometer)

Mitutoyo, 302F13 (digital micrometers)

Photo courtesy of Fred V. Fowler Co., 302F20B (photo)

Ed LePage, 302F26 (photo)

Photo courtesy of Monarch Instrument, 302F30

Extech Instruments, 302F31, 302F32



Instructor's Notes:

2.26 MILLWRIGHT ◆ LEVEL THREE

# MODULE 15302-08 — TEACHING TIPS

The following are suggested activities or instructional methods to help you teach the material in this module.

## General

When you call on someone to answer a question, the rest of the class relaxes or even tunes out because they expect that the question and answer will take place only between you and the trainee you called on. Instead, use this technique to involve more trainees in answering questions and to keep them on their toes.

1. Ask trainees to define a term or explain a concept.
2. After one trainee has answered, ask a trainee seated nearby if the answer is right. Then ask whether a trainee in the back of the room agrees.
3. Ask trainees to explain why they think an answer is right or wrong.
4. Use the session to clear up incorrect ideas and encourage trainees to learn from their mistakes.

## Section 3.0.0 *Calipers*

This exercise familiarizes trainees with how to use calipers. This exercise corresponds to Performance Task 2.

1. Set up several workstations, each with a different type of caliper and sample measurements to be transferred.
2. Have trainees practice using the calipers and transferring the measurements.
3. Answer any questions they may have.

## Section 4.0.0 *Micrometers*

This exercise familiarizes trainees with how to use micrometers. This exercise corresponds to Performance Task 3.

1. Set up several workstations, each with a different type of micrometer and sample nonmoving parts to be measured.
2. Have trainees practice using the micrometers and recording their measurements.
3. Answer any questions they may have.

## MODULE 15302-08 — ANSWERS TO REVIEW QUESTIONS

| Answer | | Section |
|---|---|---|
| 1. | d | 2.1.0 |
| 2. | d | 2.3.0 |
| 3. | a | 3.2.0 |
| 4. | b | 3.3.0 |
| 5. | b | 3.3.0 |
| 6. | a | 4.0.0 |
| 7. | c | 4.1.0 |
| 8. | b | 4.1.0 |
| 9. | a | 4.1.0 |
| 10. | b | 4.1.0 |
| 11. | b | 4.2.0 |
| 12. | b | 4.4.0 |
| 13. | c | 5.0.0 |
| 14. | b | 5.0.0 |
| 15. | c | 7.0.0 |
| 16. | b | 7.0.0 |
| 17. | a | 7.0.0 |
| 18. | d | 8.1.0 |
| 19. | c | 8.2.0 |
| 20. | b | 8.3.0 |

## CONTREN® LEARNING SERIES — USER UPDATE

The NCCER makes every effort to keep these textbooks up-to-date and free of technical errors. We appreciate your help in this process. If you have an idea for improving this textbook, or if you find an error, a typographical mistake, or an inaccuracy in NCCER's Contren® textbooks, please write us, using this form or a photocopy. Be sure to include the exact module number, page number, a detailed description, and the correction, if applicable. Your input will be brought to the attention of the Technical Review Committee. Thank you for your assistance.

*Instructors* – If you found that additional materials were necessary in order to teach this module effectively, please let us know so that we may include them in the Equipment/Materials list in the Annotated Instructor's Guide.

**Write:** Product Development and Revision
National Center for Construction Education and Research
3600 NW 43rd St., Bldg. G, Gainesville, FL 32606

**Fax:** 352-334-0932

**E-mail:** curriculum@nccer.org

Craft ________ Module Name ________

Copyright Date ________ Module Number ________ Page Number(s) ________

Description ________

(Optional) Correction ________

(Optional) Your Name and Address ________

Module 15303-08

# *Installing Packing*

## NCCER STANDARDIZED CRAFT TRAINING PROGRAM

The National Center for Construction Education and Research (NCCER) provides a standardized national program of accredited craft training. Key features of the program include instructor certification, competency-based training, and performance testing. The program provides trainees, instructors, and companies with a standard form of recognition through a National Craft Training Registry. The program is described in full in the *Guidelines for Accreditation*, published by NCCER. For more information on standardized craft training, contact the NCCER by writing us at 3600 NW 43rd St., Bldg. G, Gainesville, FL 32606; calling 352-334-0911; or emailing info@nccer.org. More information may be found at our website, www.nccer.org.

## HOW TO USE THIS ANNOTATED INSTRUCTOR'S GUIDE

Each page presents two sections of information. The larger section displays each page exactly as it appears in the Trainee Module. The narrow column ties suggested trainee and instructor actions to each page and provides icons (detailed below) to call your attention to material, safety, audiovisual, or testing requirements. The bottom of each page includes space for your notes.

The **Audiovisual** icon indicates an appropriate time to show a transparency or other audiovisual aid.

The **Classroom** icon prompts you to define a term, stress a point, ask trainees to explain a concept, or give examples.

The **Demonstration** icon directs you to show trainees how to perform tasks.

The **Examination** icon tells you to administer the written module examination.

The **Homework** icon is placed where you may wish to assign reading for the next class, assign a project, or advise trainees to prepare for an examination.

The **Laboratory** icon is used when trainees are to practice performing tasks.

The **Materials** icon is a reminder for you to gather materials needed for classes, labs, and testing.

The **Performance Testing** icon tells you to administer a performance test or a portion thereof.

The **Safety** icon is used to emphasize safety issues. It is often keyed to *Caution* and *Warning!* statements in the Trainee Module.

The **Teaching Tip** icon indicates additional guidance is available, such as how to conduct an exercise, get the most educational value from a field trip, or encourage class participation. Teaching Tips may expand on a feature (*Think About It*, *Did You Know?*) or provide *Quick Quizzes* or similar exercises. You will be referred to the Teaching Tips section at the back of the module if there is additional material.

The **Combination** icon indicates that the laboratory listed corresponds with a performance task. If desired, you can note the proficiency of the trainees during the laboratory, and use it to satisfy performance testing requirements.

## PREPARATION

Before teaching this module, you should review the Objectives, Performance Tasks, Materials and Equipment List, and Module Outline. Be sure to allow ample time to prepare your own training or lesson plan and gather all required materials and equipment.

## MODULE OVERVIEW

This module describes the types of packing and packing materials found in a typical stuffing box, and explains how to remove and install packing.

## PREREQUISITES

Prior to training with this module, it is recommended that the trainee shall have successfully completed the following: *Core Curriculum; Millwright Level One; Millwright Level Two;* and *Millwright Level Three,* Modules 15301-08 and 15302-08.

## OBJECTIVES

Upon completion of this module, the trainee will be able to do the following:

1. Identify and explain the types of packing.
2. Identify and explain packing materials.
3. Remove packing.
4. Install packing.

## PERFORMANCE TASKS

Under the supervision of the instructor, the trainee should be able to do the following:

1. Identify the types of packing.
2. Identify packing materials.
3. Remove packing.
4. Install compression packing.
5. Install lip-type packing.

## MATERIALS AND EQUIPMENT LIST

Overhead projector and screen
Transparencies
Blank acetate sheets
Transparency pens
Whiteboard/chalkboard
Markers/chalk
Pencils and scratch paper
Appropriate personal protective equipment
Basic trainee tools and packing tools
Samples of packing configurations, including:
- Square-braid
- Braid-over-braid
- Interlocking braid
- Twisted
- Multi-cord, wrapped, and laminated
- Metal
- Graphite ribbon

Samples of lip-type packings, including:
- V-packing
- U-ring packing
- Cup packing
- Flange packing

Samples of packing materials, including:
- Graphite yarn
- Teflon®
- Carbon yarn
- Vegetable fiber
- Duck and rubber
- Plastic
- Metal
- Fiberglass
- Aramid yarn

Quick Quiz*
Module Examinations**
Performance Profile Sheets**

* Located in the back of this module
**Located in the Test Booklet

## SAFETY CONSIDERATIONS

Ensure that the trainees are equipped with appropriate personal protective equipment and know how to use it properly. This module requires that the trainees use hand tools. Ensure that trainees are briefed on shop safety procedures.

## ADDITIONAL RESOURCES

This module is intended to present thorough resources for task training. The following reference works are suggested for both instructors and motivated trainees interested in further study. These are optional materials for continued education rather than for task training.

EPM, Inc., www.epm.com/styles.htm

Draco Mechanical Supply, Inc., www.dracomech.com/comppack/pumpinstall.htm

## TEACHING TIME FOR THIS MODULE

An outline for use in developing your lesson plan is presented below. Note that each Roman numeral in the outline equates to one session of instruction. Each session has a suggested time period of 2½ hours. This includes 10 minutes at the beginning of each session for administrative tasks and one 10-minute break during the session. Approximately 10 hours are suggested to cover *Installing Packing*. You will need to adjust the time required for hands-on activity and testing based on your class size and resources. Because laboratories often correspond to Performance Tasks, the proficiency of the trainees may be noted during these exercises for Performance Testing purposes.

| Topic | Planned Time |
|---|---|
| **Session I. Introduction; Packing Configurations** | |
| A. Introduction | ____________ |
| B. Packing Configurations | ____________ |
| C. Laboratory<br>Trainees practice identifying types of packing. This laboratory corresponds to Performance Task 1. | ____________ |
| **Session II. Packing Materials** | |
| A. Packing Materials | ____________ |
| B. Laboratory<br>Trainees practice identifying packing materials. This laboratory corresponds to Performance Task 2. | ____________ |
| **Session III. Removing and Installing Packing** | |
| A. Removing Packing | ____________ |
| B. Laboratory<br>Trainees practice removing packing. This laboratory corresponds to Performance Task 3. | ____________ |
| C. Installing Packing | ____________ |
| D. Laboratory<br>Trainees practice installing compression packing. This laboratory corresponds to Performance Task 4. | ____________ |
| E. Laboratory<br>Trainees practice installing lip-type packing. This laboratory corresponds to Performance Task 5. | ____________ |

**Session IV. Review and Testing**

A. Module Review ______

B. Module Examination ______

1. Trainees must score 70% or higher to receive recognition from NCCER.
2. Record the testing results on Craft Training Report Form 200, and submit the results to the Training Program Sponsor.

C. Performance Testing ______

1. Trainees must perform each task to the satisfaction of the instructor to receive recognition from NCCER. If applicable, proficiency noted during laboratory exercises can be used to satisfy the Performance Testing requirements.
2. Record the testing results on Craft Training Report Form 200, and submit the results to the Training Program Sponsor.

Millwright Level Three

15303-08

# Installing Packing

Assign reading of Module 15303-08.

# 15303-08
# *Installing Packing*

*Topics to be presented in this module include:*



## Overview

Packing is used to control leakage around stems in valves and pumps. In this module, you will learn how to remove and to install packing. You will learn the types of packing available and the strengths and weaknesses of each of them.

Instructor's Notes:

## Objectives

When you have completed this module, you will be able to do the following:

1. Identify and explain the types of packing.
2. Identify and explain packing materials.
3. Remove packing.
4. Install packing.

## Trade Terms

| | |
|---|---|
| Alkalies | Filament |
| Aqueous | Gland |
| Braiding | Migrating |
| Brine | Oxidizer |
| Caustic | Reciprocating |
| Corrosive | Skive |
| Crimped | Stuffing box |
| Ester | |

## Required Trainee Materials

1. Pencil and paper
2. Appropriate personal protective equipment

## Prerequisites

Before you begin this module, it is recommended that you successfully complete *Core Curriculum; Millwright Level One; Millwright Level Two;* and *Millwright Level Three,* Modules 15301-08 and 15302-08.

This course map shows all of the modules in the third level of the *Millwright* curriculum. The suggested training order begins at the bottom and proceeds up. Skill levels increase as you advance on the course map. The local Training Program Sponsor may adjust the training order.

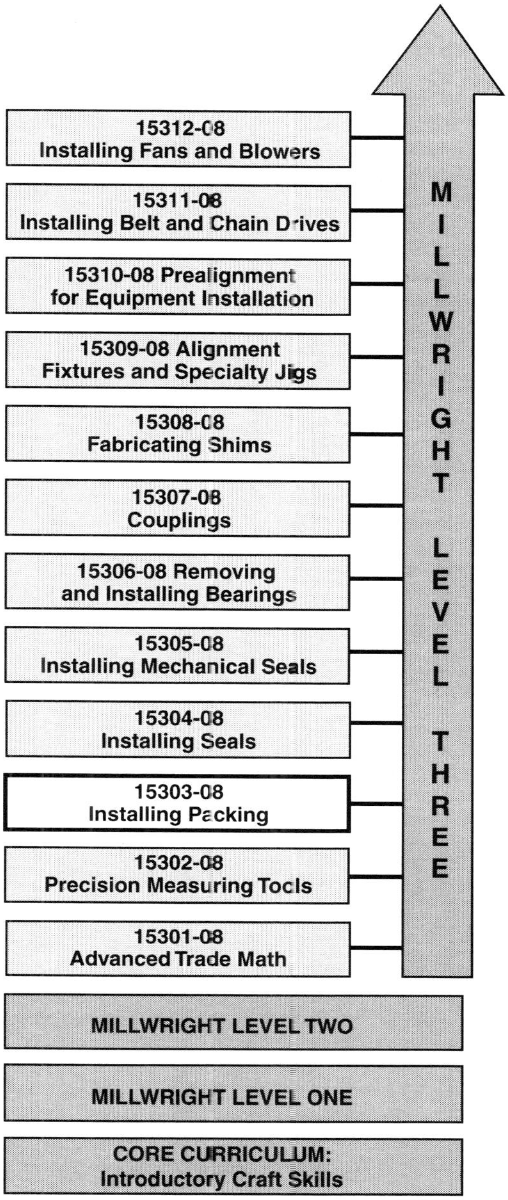


303CMAP.EPS



**Ensure that you have everything required to teach the course. Check the Materials and Equipment list at the front of this module.**

**See the general Teaching Tip at the end of this module.**

**Explain that terms shown in bold are defined in the Glossary at the back of this module.**

**Show Transparency 1, Objectives, and Transparency 2, Performance Tasks. Review the goals of the module, and explain what will be expected of the trainees.**

**Review the modules covered in Level Three and explain how this module fits in.**

**See the Teaching Tips for Sections 1.0.0–5.0.0 and 2.0.0 at the end of this module.**

MODULE 15303-08 ◆ INSTALLING PACKING 3.1

**Introduce packing and explain how it is used.**

**Discuss common types of packing.**

**Explain why millwrights need to be able to identify and work with packing.**

**Provide samples of different packing configurations for trainees to examine.**

**Show Transparency 3 (Figure 1). Describe square-braid packing.**

**Show Transparency 4 (Figure 2). Describe braid-over-braid packing.**

## 1.0.0 ◆ INTRODUCTION

Packings are rings of **braided** or molded material that are inserted into a pump or valve **stuffing box** to minimize or eliminate gas or fluid leakage. The packing is installed between the rotating or **reciprocating** shaft and the body of the pump or valve. It creates a seal when the **gland** is tightened against the outermost packing ring in the packing set. When the gland is tightened, pressure is transmitted to the packing and the rings expand radially against the side of the stuffing box. This creates the seal between the shaft and the housing.

The type of packing installed in a system or piece of equipment is determined by the application in which it is used, the material it is made from, and its construction. Packing comes in various types, sizes, and materials designed for specific applications. Packing can be purchased in standard sized die-cut rings or as continuous square-sided or round rope of various thicknesses to match the clearance between the shaft and stuffing box body. Some of these packings require regular replacement to ensure proper performance of the system or equipment in which they are used. Millwrights must be able to identify and select the correct type of packing for the application. They must also be able to remove and replace different types of packing correctly. Incorrect selection or installation of the packing can result in premature failure of the system or its equipment. The following sections identify and explain different types of packing and procedures used to remove, inspect, and replace packing.

## 2.0.0 ◆ PACKING CONFIGURATIONS

Packing is used in almost every type of equipment that has shafts or stems that pass through casings. Some of the many applications are pumps, valves, air compressors, soot blowers, and vacuum pumps. There are many types of packing to suit the large number of applications. Some of the most common types of packing are the following:

- Square-braid
- Braid-over-braid
- Interlocking braid
- Twisted
- Multi-cord, wrapped, and laminated
- Metal
- Graphite ribbon
- Lip-type

### 2.1.0 Square-Braid Packing

Square-braid packing (*Figure 1*) consists of fibers braided by passing one strand of fiber over and under another strand running in the same direction. No strand passes completely through the packing. The result is a square packing. Because no strand is intertwined with all of the other strands, the packing is very flexible. Because of its flexibility, square-braid packing is commonly used on reciprocating shafts.

### 2.2.0 Braid-Over-Braid Packing

Packing made up of a series of small braids braided over each other is called braid-over-braid (*Figure 2*). The braid-over-braid construction creates dense packing that works well in high-pressure applications. Again, the packing is very flexible.

### 2.3.0 Interlocking Braid Packing

Packing constructed by passing each strand of fiber through the body of the packing at an approximate 45-degree angle is called interlock-

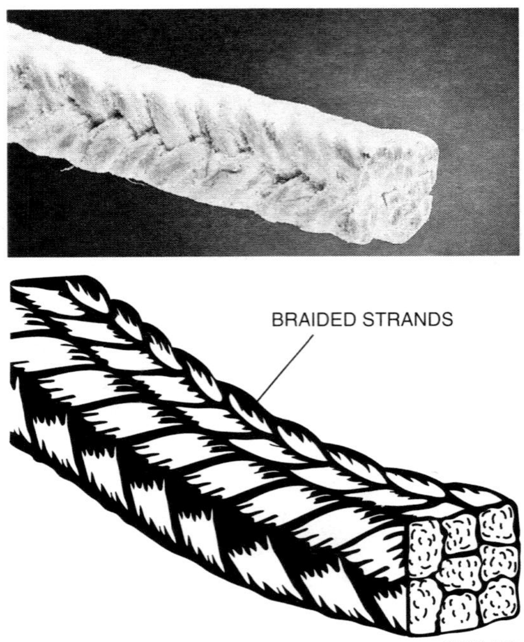


*Figure 1* ◆ Square-braid packing.



Instructor's Notes:

3.2 MILLWRIGHT ◆ LEVEL THREE

ing braid (*Figure 3*). The result is packing that is braided internally and externally in an interlocking pattern. The interlocking construction creates dense packing with very few voids. This type of packing is very strong, and resists being deformed or squeezed out of the packing gland of rotating and reciprocating equipment.

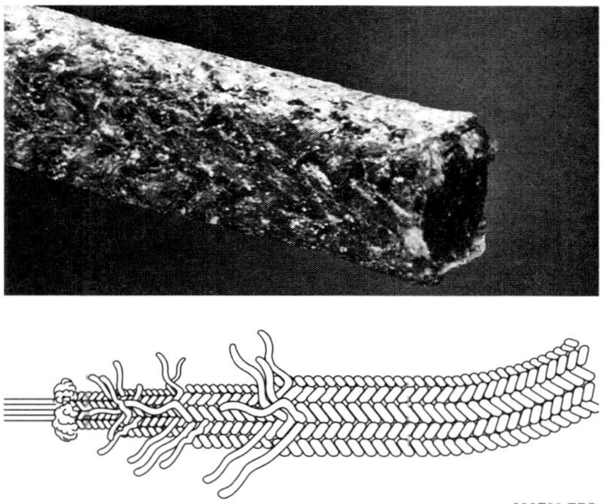


*Figure 2* ◆ Braid-over-braid packing.

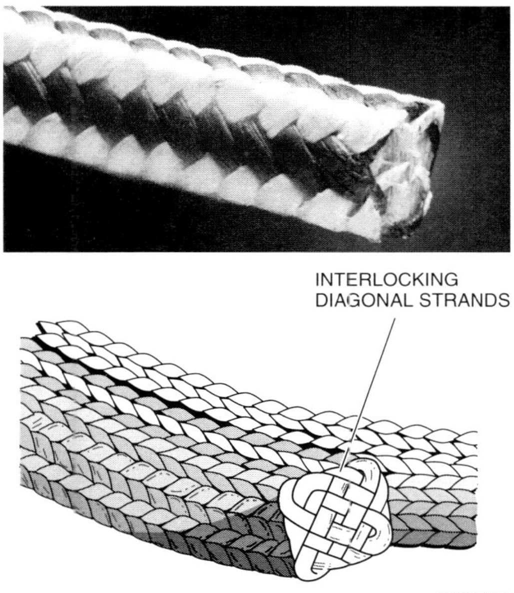


*Figure 3* ◆ Interlocking braid packing.

## 2.4.0 Twisted Packing

Packing made by twisting strands together to form a round rope is called twisted packing (*Figure 4*). This type of packing is flexible, durable, and temperature resistant. It is typically used for low-pressure valve stems and rotary shafts in process systems where the system is subject to high temperatures. Twisted packing can be untwisted and some strands removed, and the remainder retwisted to make the packing smaller.

## 2.5.0 Multicore Braid Packing

Packing constructed around four equally spaced cords is called multicore braid (*Figure 5*). The construction creates versatile packing that is excellent for pumps and valves with worn glands or imbalanced shafts. This type of packing is recommended for use with rotating and reciprocating shafts, mixers, and agitators. It is used in most general service applications and processes with water, hydraulic fluids, and oils.

Laminated packing is available both as die cut rings and as stock packing. The laminations may be as simple as layers of rubber and asbestos or as high-tech as graphite foil layered with Teflon®. Laminated packing is especially resistant to release of vapors and liquids, because of the laminated structure. A variant on lamination is folded packing, held together with some sort of resin.

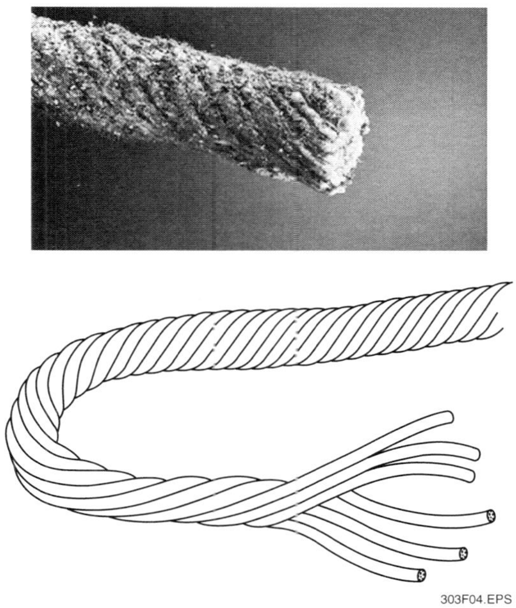


*Figure 4* ◆ Twisted packing.

**Show Transparency 5 (Figure 3). Describe interlocking braid packing.**

**Show Transparency 6 (Figure 4). Describe twisted packing.**

**Discuss multicore braid packing and its applications.**



MODULE 15303-08 ◆ INSTALLING PACKING 3.3

**Discuss metal packings.**

**Describe graphite ribbon packing.**

**Review the types of lip-type packing and how each is used.**

**Show Transparency 7 (Figure 7). Describe V-packing.**

**Show Transparency 8 (Figure 8). Describe U-ring packing.**

**Show Transparency 9 (Figure 9). Describe cup packing.**

A third type of layered packing is wrapped packing, using fibers of one material to enclose and physically strengthen another fiber, to achieve a combination of different qualities. An example might be a braided carbon core wrapped in graphite mesh.

## 2.6.0 Metal Packings

Packings made of, or reinforced with, various metal foils of lead, steel, or special alloys are called metal packings. These packings are used in a number of applications, however, only soft metals are used on rotating shafts because the harder metals will eventually wear down the shaft. Metal packings are usually **crimped** or spiral-wound (*Figure 6*).

## 2.7.0 Graphite Ribbon Packing

Graphite ribbon packing usually comes in rolls that can be used to form rings of almost any required thickness. To form packing rings from ribbon packing, the ribbon is wound around the shaft until it will just fit into the stuffing box. The packing is then pushed into the box using a gland follower and a split bushing. The gland follower bolts are tightened to compress the packing into a ring. This process is repeated until enough rings are formed to fill the stuffing box.

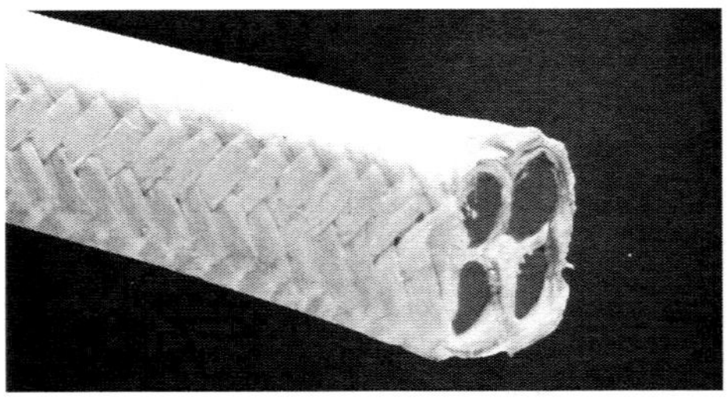

303F05.EPS

*Figure 5* ◆ Multicore braid packing.

## 2.8.0 Lip-Type Packings

Lip-type packings work by the pressure of the fluid being sealed pushing against lips and pressing them lightly against the sealing surface. These packings are often also called automatic or hydraulic packings. They are used to seal against reciprocating motion. Lip-type packings are made with slightly flared lips to provide sealing under zero pressure. They are molded or formed in the shape of a V, U, cup, or flange.

### *2.8.1 V-Packing*

The most commonly used lip-type packing is V-packing. It can be used for low- or high-pressure reciprocating motion and it can be installed on pistons and rods. V-packing is a multiple packing that is installed in stacks or nested sets of three to six V-rings. It is installed with a support ring or adapter at the top and the bottom (*Figure 7*).

### *2.8.2 U-Ring Packing*

U-ring packings (*Figure 8*) are used for rod and piston packing of hydraulic cylinders and plungers where space is limited and where low friction is needed.

### *2.8.3 Cup Packing*

Another lip-type packing is called cup packing (*Figure 9*). Cup packing is commonly used for piston seals inside cylinders.

> **NOTE**
> V-packings and cup packings are directional packings. These must be installed in the right relationship to pressure.

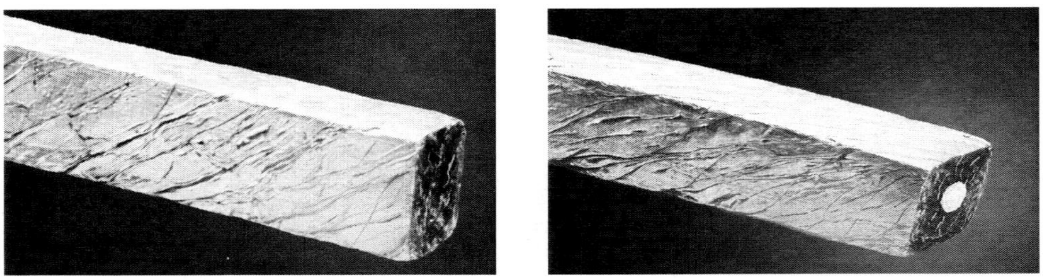

303F06.EPS

*Figure 6* ◆ Crimped and spiral-wound metal packings.



Instructor's Notes:

3.4 MILLWRIGHT ◆ LEVEL THREE

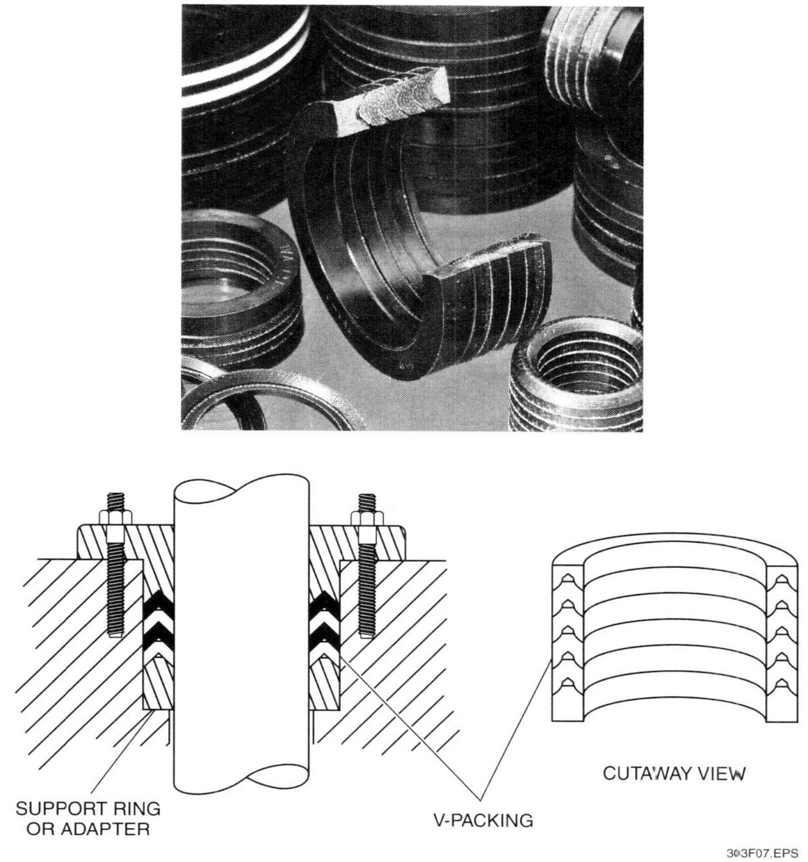


*Figure 7* ◆ V-packing.

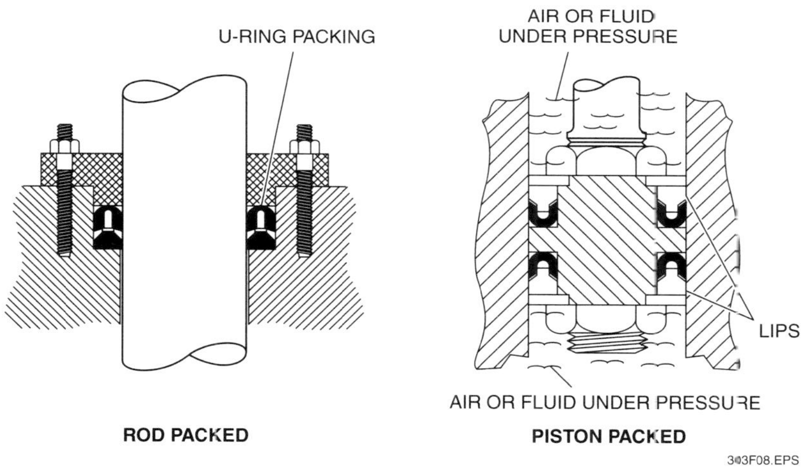


*Figure 8* ◆ U-ring packing.



MODULE 15303-08 ◆ INSTALLING PACKING 3.5

Show Transparency 10 (Figure 10). Describe flange packing.

Discuss die cut packing.

See the Teaching Tip for Section 2.0.0 at the end of this module.

Have trainees practice identifying types of packing. Note the proficiency of each trainee. This laboratory corresponds to Performance Task 1.

Assign reading of Sections 3.0.0–3.9.0.

Ensure that you have everything required to teach this session.

List the most common materials used for making packing.

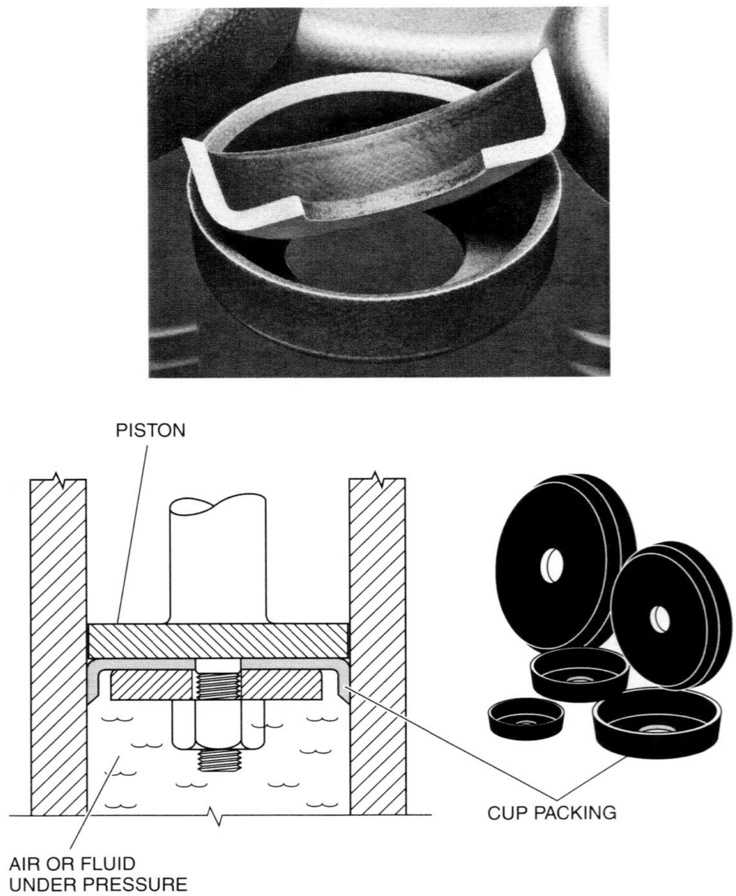


*Figure 9* ◆ Cup packing.

### *2.8.4 Flange Packing*

The least common lip-type packing is flange packing (*Figure 10*). It seals on the inside diameter only and is used only to retain low pressures. The flange must be sealed from the outside, usually by a gland.

### 2.9.0 Die Cut Packing

Die cut packing rings are pre-sized rings used to pack valves and pumps. They frequently come in sets of five, with two high-density top and bottom rings and three center sealing rings. The top and bottom rings provide strength and prevent the center rings from squeezing out. Like other types of packing, they are available in many sizes and materials for use in a wide variety of applications and processes.

## 3.0.0 ◆ PACKING MATERIALS

A number of materials are used to manufacture packing. The type of material used depends on the application and is usually determined by the design engineer. Some of the most common materials used for making packing are the following:

- Graphite yarn
- Teflon®
- Carbon yarn
- Vegetable fiber
- Duck and rubber
- Plastic
- Metal
- Fiberglass
- Aramid yarn



Instructor's Notes:

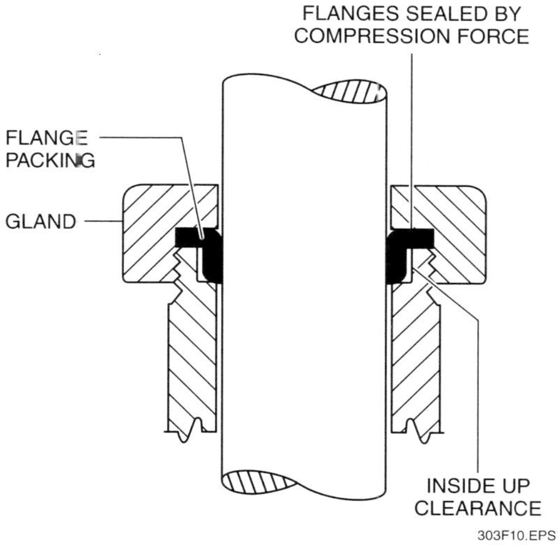


*Figure 10* ◆ Flange packing.

### 3.1.0 Graphite Yarn

Graphite is a good heat conductor and dissipates heat in the stuffing box, permitting higher shaft speeds and less leakage than some other packings. It is commonly used on all rotating and reciprocating shafts, valves, and agitators, and can normally be used in applications where there are strong **caustics**, acids, and chemicals. It is also suitable for high-pressure steam applications and can run on little or no leakage.

### 3.2.0 Teflon®

There are several types of Teflon® materials that are used to manufacture packing. Teflon® is commonly used in the most severe applications of high **corrosives** and **oxidizers**. It is also used extensively in the food-processing industry and anywhere an FDA-approved material is required. Teflon® use is limited to temperature ranges up to 500°F.

### 3.3.0 Carbon Yarn

Carbon yarn packing can be used on rotating and reciprocating shafts and in high-temperature valves. It is used in many types of chemical services except with strong oxidizers. Carbon yarn packing can be treated with graphite particles to improve heat conductivity or with Teflon® to prevent carbon **filaments** from **migrating** into the pumping system. Carbon yarn packing is generally limited to shaft speeds up to 3,000 feet per minute (fpm) and temperatures up to 650°F.

### 3.4.0 Vegetable Fiber

Vegetable fiber packings can be used on rotating and reciprocating shafts, plungers, hydraulic rams, and stern tubes. They are recommended for use with **brine**, cold water, and cold oil. They are suitable for shaft speeds up to 1,885 fpm and temperatures up to 220°F.

### 3.5.0 Duck and Rubber

Duck and rubber packing material is made of cross-laminated cotton duck and heat-resistant rubber. It can be used on pistons, end rings, and reciprocating equipment in systems that contain air **alkalies**, oil phosphate **esters**, solvents, or water. It is limited to temperatures up to 250°F and pressures up to 350 psi.

### 3.6.0 Plastic

Plastic packing material consists of synthetic yarns and binders, and graphite flakes are sometimes added to minimize friction. Plastic packings can be used on rods, valve stems, and centrifugal pump shafts. They are recommended for use with steam, gases, and **aqueous** solutions, except mineral-acid solutions. They are limited to temperatures up to 600°F and speeds up to 3,600 rpm.

### 3.7.0 Metal

Metallic packing material consists of thin ribbons of metal foil that are crimped or spiral-wound into the shape of packing. Metallic packing is usually lubricated with graphite and oil. It is used on centrifugal and reciprocating pump rods and valve stems and is recommended for use with steam oils, gasoline, air, ammonia, and water. Metallic packing should not be used on soft metal shafts and rods, such as brass and bronze, because it will cause them to wear quickly. It is limited to temperatures of up to 450°F and speeds up to 3,600 rpm.

### 3.8.0 Fiberglass

Fiberglass material is used to make packing in many shapes and styles. It is incombustible and is used in different forms to seal furnace doors, manhole covers, generator doors, kettles, and tanks and is recommended for service with hot air, gases, molten metal, and dry steam. It works well in temperatures up to 1,000°F.

### 3.9.0 Aramid Yarn

Aramid yarn is a very strong packing material and is highly chemical resistant. It is commonly

**Discuss the applications of graphite yarn, Teflon®, carbon yarn, and vegetable fiber.**

**Discuss the applications of duck and rubber, plastic, metal, fiberglass, and aramid yarn.**

**See the Teaching Tip for Section 3.0.0 at the end of this module.**

**Have trainees practice identifying packing materials. Note the proficiency of each trainee. This laboratory corresponds to Performance Task 2.**



MODULE 15303-08 ◆ INSTALLING PACKING 3.7

**Have trainees review Sections 4.0.0–5.2.4.**

**Ensure that you have everything required to teach this session.**

**Discuss the procedure for removing and replacing packing.**

**Show Transparency 11 (Figure 11). Review the steps for removing old packing from a stuffing box assembly.**

**Show Transparency 12 (Figure 12). Explain how to remove the gland nuts.**

**Show Transparency 13 (Figure 13). Explain how to remove the gland.**

**Show Transparency 14 (Figure 14). Explain that the packing puller should be screwed firmly into the packing ring.**

braided with other types of yarns, such as PTFE (Teflon®), for use in a wide variety of applications. Aramid packing is recommended for reciprocating and rotary shafts, for valve applications in pulp and paper systems, and for petrochemical processes. It can be used with water, oils, greases, weak acids, and alkaline solutions.

## 4.0.0 ◆ REMOVING PACKING

The working life of packing depends on the type of packing material it is made from and the service for which the packing is used. The working life of packing can range from a few weeks to many years. The most common symptom of packing failure is excessive leakage. When the packing fails, it must be replaced.

Before the packing can be replaced, the old packing must be removed from the stuffing box assembly (*Figure 11*). Removing packing is a relatively simple procedure, but it is important that the procedure be performed properly. The procedure for removing packing varies slightly from one piece of equipment to another, but the steps are basically the same. Refer to the equipment manufacturer's manual for the equipment that you are working on for particular procedures.

The following steps explain how to remove packing from the stuffing box assembly of a pump:

*Step 1* Clean the area around the stuffing box of the pump.

*Step 2* Remove the gland nuts, and place them in a safe place (*Figure 12*).

*Step 3* Remove the gland. If the pump has a two-piece gland, it can be completely removed from the shaft (*Figure 13*). A solid one-piece gland must be slid back as far as possible on the shaft toward the motor to gain access to the packing.

*Step 4* Push a packing puller or packing hook into the stuffing box until it touches the first ring of packing. Two packing pullers are often used at the same time to pull the packing on both sides.

*Step 5* Twist the packing puller to screw it firmly into the packing ring (*Figure 14*).

> **CAUTION**
> Use the packing puller carefully to avoid scratching or scoring the shaft. The best way to prevent damaging the shaft is to keep the puller angled away from the shaft at all times. Even minor damage to the shaft will cause accelerated packing wear and failure.

*Step 6* Pull the first ring of packing carefully out of the stuffing box. If the puller is properly screwed into the ring, the ring usually comes out in one piece. If the

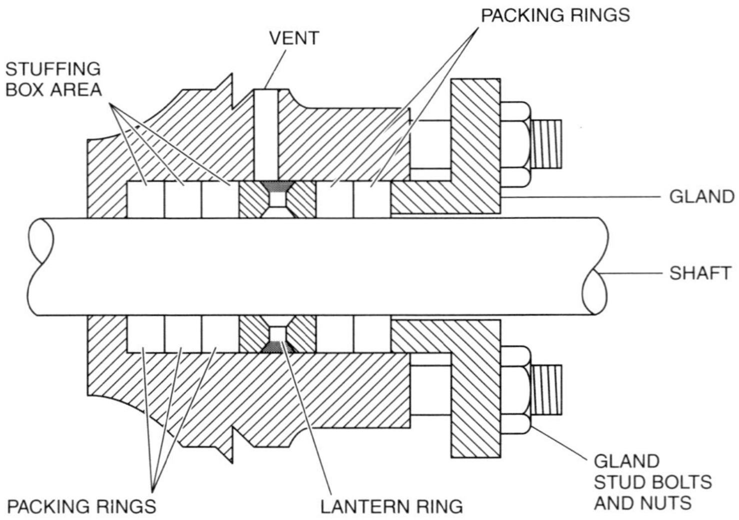


*Figure 11* ◆ Stuffing box assembly.



Instructor's Notes:

3.8 MILLWRIGHT ◆ LEVEL THREE

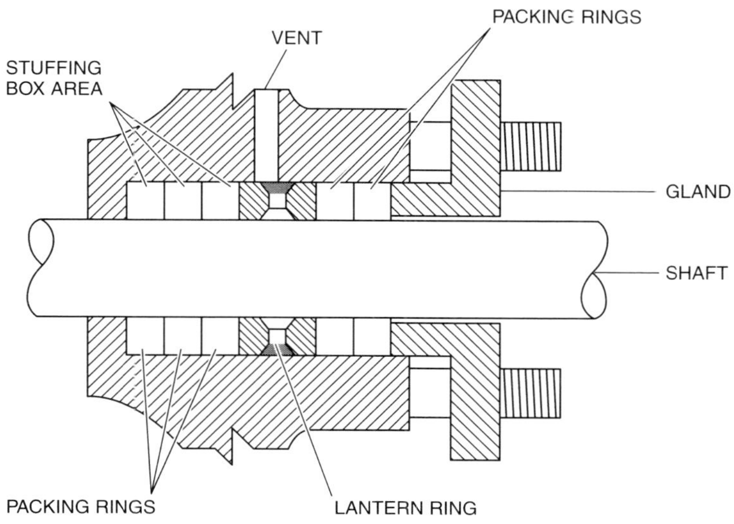


*Figure 12* ◆ Gland nuts removed.

ring breaks during removal, you will need to remove all of the smaller pieces. You may need to use other packing removal tools.

*Step 7* Repeat Steps 4 through 6 to remove the additional rings in front of the lantern ring. The lantern ring is a spacer that is placed in the stuffing box for lubrication. It is placed in line with the lubrication inlet port. Lubricant enters the port to lubricate and cool the shaft and the packing. Note how many rings you remove so that you will know how many to replace before and after the lantern ring.

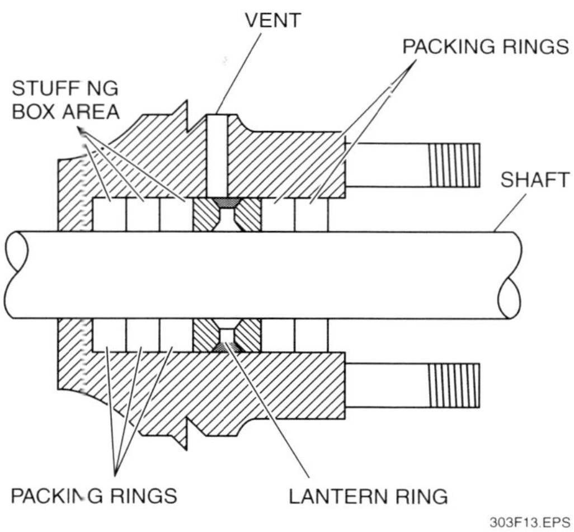


*Figure 13* ◆ Gland removed.

*Step 8* Insert a hooked wire or similar tool into the stuffing box; move it around the lantern ring until the hook goes into a hole; then pull the wire to remove the lantern ring. The lantern ring has holes in at least two places for removal. Lantern rings can usually be reused.

*Step 9* Repeat Steps 4 through 6 to remove the remaining packing rings.

*Step 10* Inspect the stuffing box to ensure that all of the packing has been removed and that there are no small pieces left in the box.

*Step 11* Thoroughly clean the shaft and the stuffing box, using a solvent and a rag.

*Step 12* Inspect the stuffing box and the shaft to ensure that there are no burrs or excessive wear.

**CAUTION**

If there are any burrs, they must be filed off the shaft and stuffing box before new packing is installed. If there is excessive wear on the shaft or in the stuffing box, it must be reported and evaluated for repair before new packing is installed.

**Demonstrate how to remove packing.**

**Have trainees practice removing packing. Note the proficiency of each trainee. This laboratory corresponds to Performance Task 3.**



MODULE 15303-08 ◆ INSTALLING PACKING 3.9

Discuss the importance of proper packing installation.

Review the steps for installing compression packing.

Show Transparency 15 (Figure 15). Explain how to wrap the packing tightly around the shaft.

Show Transparency 16 (Figure 16). Explain how to mark the packing for cutting.

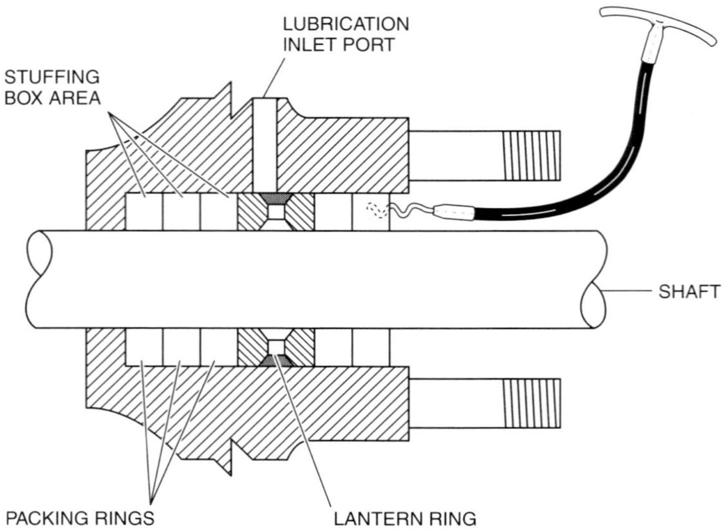


*Figure 14* ◆ Packing puller screwed into packing.

*Step 13* Inspect the packing to determine why it failed. The following are some common reasons why packing fails:

- It is the wrong type or is mismatched to the speed or application.
- It is installed too tightly.
- It is the wrong size.
- It has been exposed to excessive heat.
- It is old. Packing should be renewed annually.

## 5.0.0 ◆ INSTALLING PACKING

Following the proper installation procedures for packing is critical to the life expectancy and efficiency of the packing. Improperly installing packing can result in a loss of up to 75 percent of the working life of the packing and may cause excessive wear to the equipment shaft and stuffing box. The following sections explain how to install compression packing, lip-type packing, and injection packing.

### 5.1.0 Installing Compression Packing

Compression packing must be installed according to the proper procedure and in the proper sequence to achieve a successful packing job. Follow these steps to install compression packing:

*Step 1* Inspect the stuffing box to ensure that all old packing has been removed.

*Step 2* Inspect the stuffing box and the shaft to ensure that they are clean and that there are no burrs, nicks, scratches, or excessive wear. Any burrs, nicks, or debris found in the stuffing box or on the shaft must be cleaned or dressed, using a fine file. If there are deep scratches or excessive wear, an evaluation must be made as to the need for repairs before packing is installed.

*Step 3* Select the proper type and size packing for the job. The size of the packing is determined by the size of the stuffing box. The type of packing is usually determined by an engineer and specified on your work order.

*Step 4* Wrap the packing tightly around the shaft the number of revolutions equal to the number of rings required for the job (*Figure 15*).

*Step 5* Mark the packing for cutting. Mark the packing in a position that will make maximum use of the packing material. The mark should be started near the end of the packing and continued straight along the shaft (*Figure 16*).

*Step 6* Remove the packing from the shaft.

*Step 7* Place the packing on a hard surface for cutting.



Instructor's Notes:

3.10 MILLWRIGHT ◆ LEVEL THREE

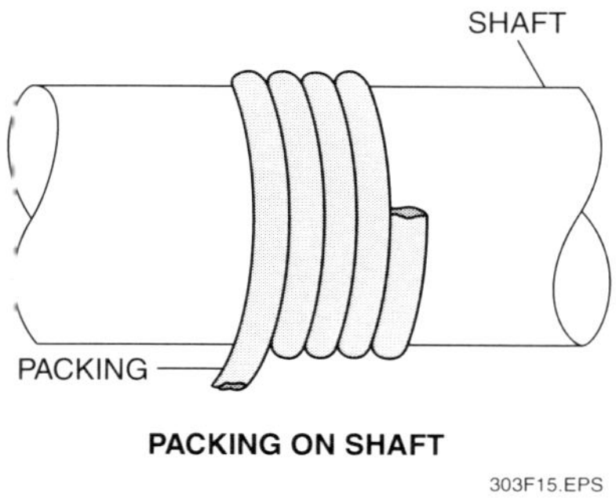


*Figure 15* ◆ Packing wrapped around shaft.

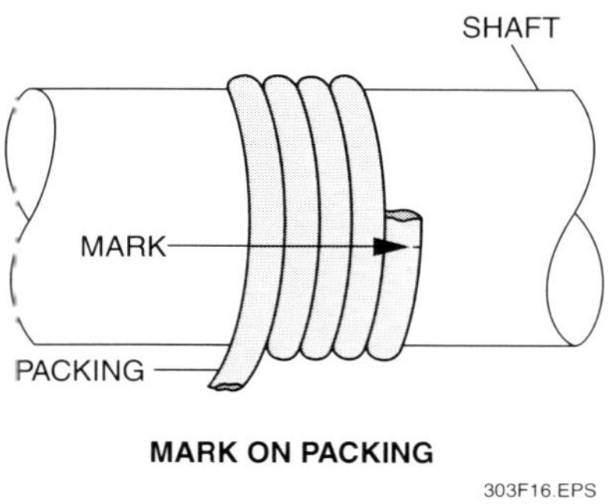


*Figure 16* ◆ Mark on packing.

**NOTE**
Do not cut the packing on the shaft being packed because the cutting tool may scratch the shaft. Cut the packing on a round steel shaft or pipe of the approximate size of the shaft being packed.

*Step 8* Cut on the mark, using a sharp knife, to cut the packing into rings. Some packing manufacturers recommend a straight cut; others recommend a **skive** (angled) cut. Check the manufacturer's specifications and follow their recommendations for cutting the packing.

*Step 9* Lubricate the packing rings on the inner and outer surfaces with a light coat of oil.

*Step 10* Slip one ring onto the shaft, and position it so that the cut is at the top of the shaft (*Figure 17*).

**CAUTION**
Use an S twist to slip the rings onto the shaft. Do not bend the ring open because this might break the packing.

*Step 11* Slide the ring to the back of the stuffing box, keeping the cut at the top of the shaft.

**CAUTION**
There are several tools that are used to install packing. A blunt-ended brass tool is recommended to prevent cutting or damaging the packing or the shaft. It is important that the ring fit all the way in the back of the box so that the subsequent rings will fit properly.

*Step 12* Slip the second ring onto the shaft, and push it into the box, keeping the cut on the bottom of shaft. The cut in each ring should be rotated 180 degrees from the cut in the ring before. Staggering the cuts in the rings provides a better seal than stacking the rings with the cuts in line with one another (*Figure 18*).

*Step 13* Install additional rings until the correct number have been installed in front of the lantern ring. If the correct number of rings is not known, measure the stuffing box to the lubrication inlet port to determine the correct number. The lantern ring must fit in line with the lubrication inlet port.

*Step 14* Install the lantern ring. *Figure 19* shows the correct lantern ring position prior to tightening the gland nuts.

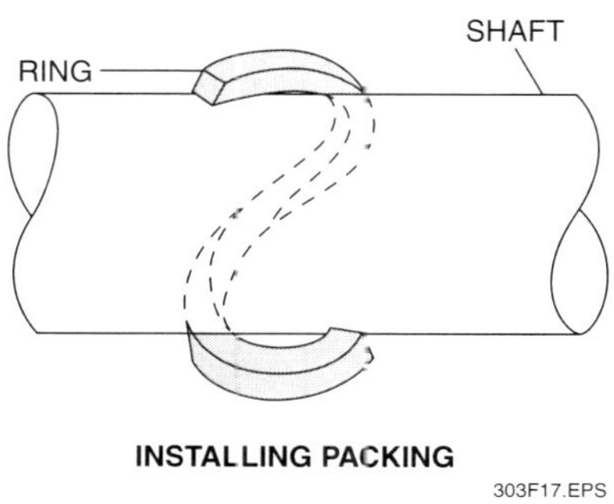


*Figure 17* ◆ Packing ring on shaft.



**Show Transparency 17 (Figure 17). Explain how to slip the ring onto the shaft with the cut at the top.**

**Show Transparency 18 (Figure 18). Describe the proper way to stack the rings.**

**Show Transparency 19 (Figure 19). Discuss the correct lantern ring position prior to tightening the gland nuts.**

MODULE 15303-08 ◆ INSTALLING PACKING 3.11

Demonstrate how to install compression packing.

Have trainees practice installing compression packing. Note the proficiency of each trainee. This laboratory corresponds to Performance Task 4.

Discuss the inspection points for installing lip-type packing.

Describe the qualities of V-packing.

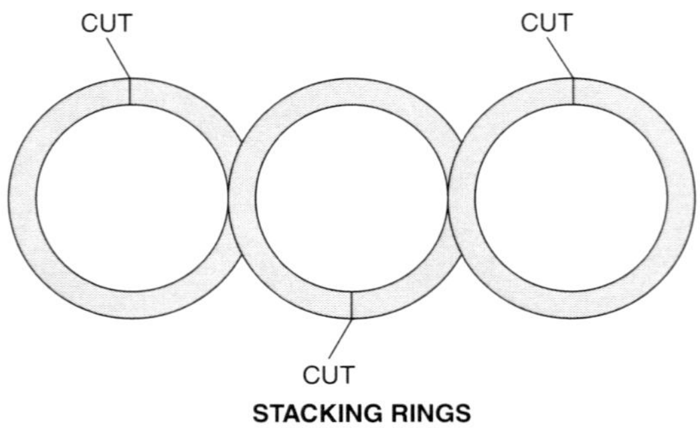


*Figure 18* ◆ Proper way to stack rings.

*Step 15* Install remaining packing.

*Step 16* Install the gland. The gland should extend out of the stuffing box about one-third of the total packing depth. This allows for the packing to compress about two-thirds of its original volume when it is worn.

*Step 17* Install the gland nuts.

*Step 18* Tighten the gland nuts to apply light pressure on the packing. Tighten the gland nuts lightly until the pump is started up; then adjust the nuts until there is slight leakage around the packing. Some leakage is required on pump packing for lubrication and cooling. Check the pump in a few hours to see if any additional adjustments are required. Do not overtighten the gland nuts on the valve. Review the application for requirements.

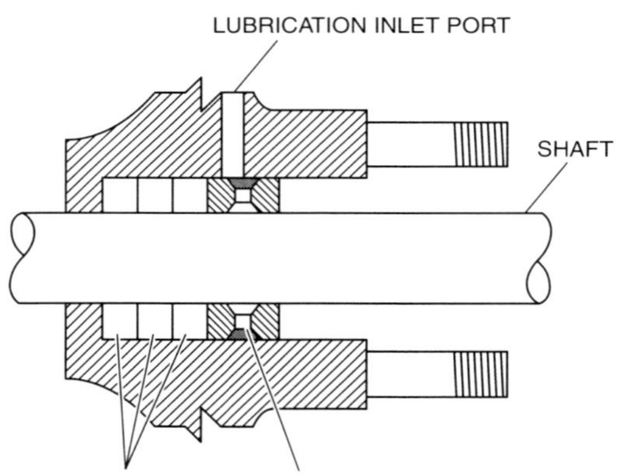


*Figure 19* ◆ Lantern ring position.

## 5.2.0 Installing Lip-Type Packing

Lip-type packing must be installed so it can expand and contract freely. High mechanical pressure is not required for this sealing. Study the operation of the packing and install it correctly. Always inspect the installation when installing packing. Some important inspection points are the following:

- Inspect all old packing to see if there are any indications of developing problems.
- Inspect the packing for any extrusions or distortion that may indicate excessive clearances.
- Inspect the packing surfaces for wear, cuts, and grooves that can cause rapid wear.
- Inspect the packing surface for wear at places of reversing directions or stopping.
- Inspect the new packing to ensure that it is the proper size and correct material.
- Inspect all fillers, adapters, support rings, and pedestal rings for wear.
- Inspect the fluid for contaminants that can cause the packing and sealing surfaces to wear rapidly.

### *5.2.1 Installing V-Packing*

V-packing is the most commonly used lip-type packing (*Figure 20*). It combines flexible lip sealing with the compression style of packing and uses packing rings with male and female support rings. V-packing has an advantage over the other forms of packing. When the packing is worn and begins to leak, it can be tightened to spread open the packing, forcing it to seal. This gives added life to the installation.

The packing rings are supported by male and female support rings that are made of metal. When they are made of any other material, they are called adapter rings. The female support ring



Instructor's Notes:

3.12 MILLWRIGHT ◆ LEVEL THREE

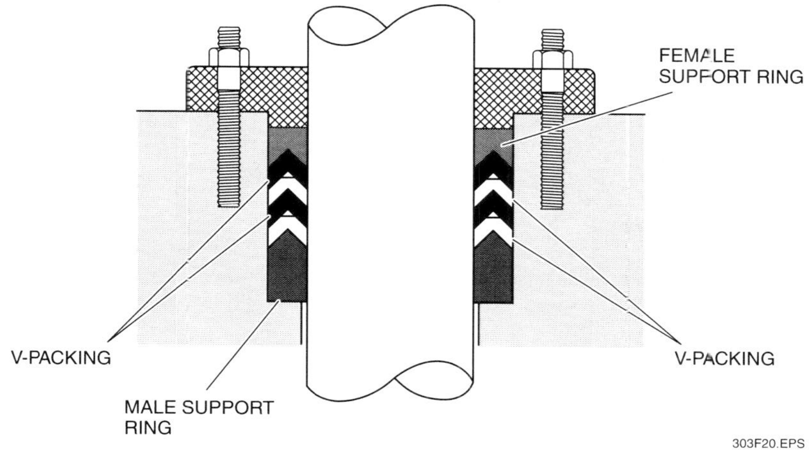


*Figure 20* ◆ V-packing installation.

wears out before the male support ring; therefore, excessive wear on the female support ring may cause extrusion of the packing. The female support ring should be examined for wear each time the packing is replaced. The male support ring almost never wears out, and excessive clearances on the male ring will not affect the life of the installation.

Follow these steps to install V-packing:

*Step 1* Remove the nuts that hold the packing gland.

*Step 2* Remove the female support ring, and check it for wear.

*Step 3* Remove the old V-packing.

*Step 4* Remove the male support ring, and inspect it for any damage.

*Step 5* Install the male support ring.

*Step 6* Install the new V-packing.

*Step 7* Install the female support ring.

*Step 8* Install packing gland.

*Step 9* Tighten the nuts, and check to ensure that there is no leakage.

### 5.2.2 Installing U-Ring Packing

U-ring packing seals both the shaft and housing surfaces against leakage. Instead of a gland follower for support, the U-ring packing has a pedestal ring. The pedestal ring helps keep the packing lips separated for maximum sealing (*Figure 21*).

U-ring packing may be used in several different types of installations. Some common guidelines to follow in all installations of U-rings are the following:

- Ensure that the pedestal ring is not too large. Too large a pedestal ring will force the lips against the surfaces to be sealed and increase wear.
- Ensure that there is clearance between the packing lips and the bottom of the pedestal ring to allow the lips to seal.

U-ring packing usually has a flat back support that allows for more flexibility and better sealing. The packing is held firmly against this support by the pedestal ring. The pedestal ring is usually drilled to allow for equal pressure and even wear of the packing. *Figure 22* shows a typical U-ring packing installation.

If the packing is installed incorrectly (*Figure 23*), pressure becomes trapped between the packings. This area becomes filled with fluid under high pressure. When the packing is in direct contact with the surfaces to be sealed under high pressure, the packing will wear rapidly even under normal operation.

### 5.2.3 Installing Cup Packing

Cup packings are commonly used for piston seals inside cylinders (*Figure 24*). Follow these steps to install cup packing:

*Step 1* Remove the nut and the follower.

*Step 2* Remove and inspect the old packing.

*Step 3* Install the new packing.

*Step 4* Install the follower and nut.



**Review the steps for installing V-packing.**

**Show Transparency 20 (Figure 20).**

**Describe U-ring packing and explain how it is installed.**

**Show Transparency 21 (Figure 21).**

**Show Transparency 22 (Figure 22). Describe a typical U-ring packing installation.**

**Show Transparency 23 (Figure 23). Discuss the problems that can occur if U-ring packing is installed incorrectly.**

**Describe cup packing. Review the steps for installing cup packing.**

**Show Transparency 24 (Figure 24).**

**Describe the precautions that should be taken when installing cup packing.**

**Explain how flange packing is installed.**

**Show Transparency 25 (Figure 25).**

**Demonstrate how to install lip-type packing.**

**Have trainees practice installing lip-type packing. Note the proficiency of each trainee. This laboratory corresponds to Performance Task 5.**

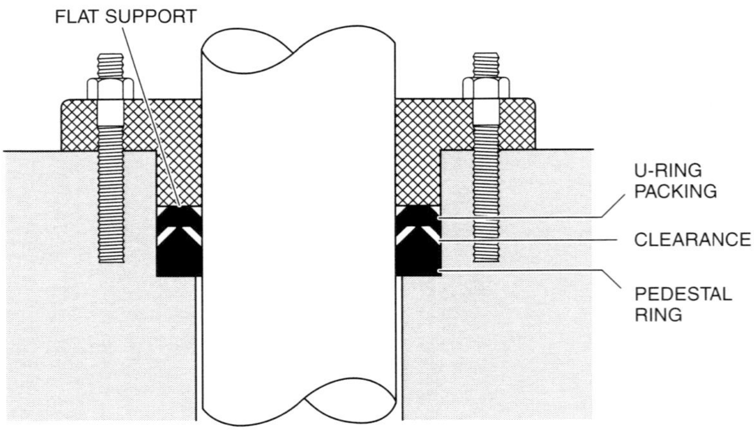


*Figure 21* ◆ U-ring packing installation.

*Step 5* Tighten the nut to hold the packing in place.

In cup packing installations, several precautions should be taken. These include the following:

- Do not overtighten the packing follower because this will reduce the amount of flex the cup lips have.
- Make sure the follower is not too large. If it is, the lips will not be able to flex. This will allow for rapid wear and leakage.
- Inspect the back support for wear. If the back support has excessive clearances, the packing will extrude into the clearances. If cup packing is allowed to extrude, the shoulder will wear rapidly, resulting in leakage and packing failure. If excessive clearances are found, the packing must be replaced as soon as possible.

### *5.2.4 Installing Flange Packing*

Flange packing is the least common lip-type packing. It seals on one side by lip action and seals on the other side by compression. This type of packing is normally used on low-pressure applications. To prevent extrusion, inspect the packing to ensure that the clearance between the compression nut and shaft does not become excessive (*Figure 25*).

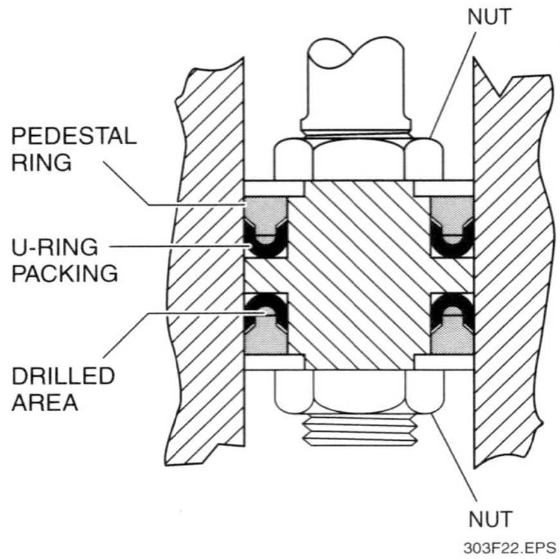


*Figure 22* ◆ U-ring packing installation.

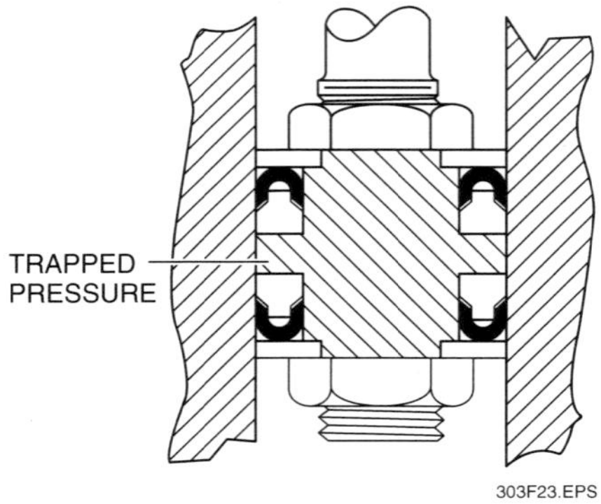


*Figure 23* ◆ Incorrect installation of U-ring packing.



Instructor's Notes:

3.14 MILLWRIGHT ◆ LEVEL THREE

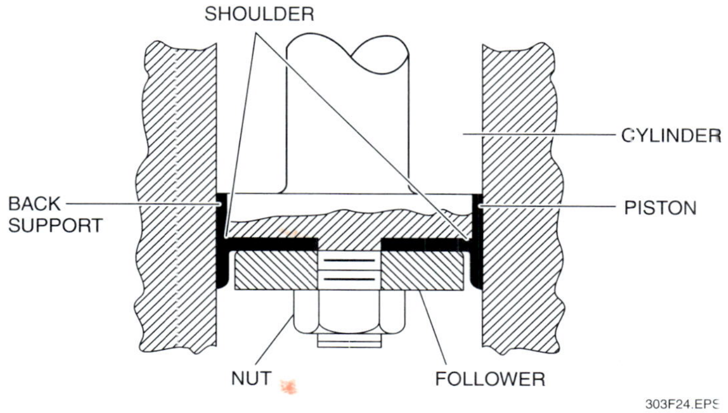


*Figure 24* ◆ Cup packing installation.

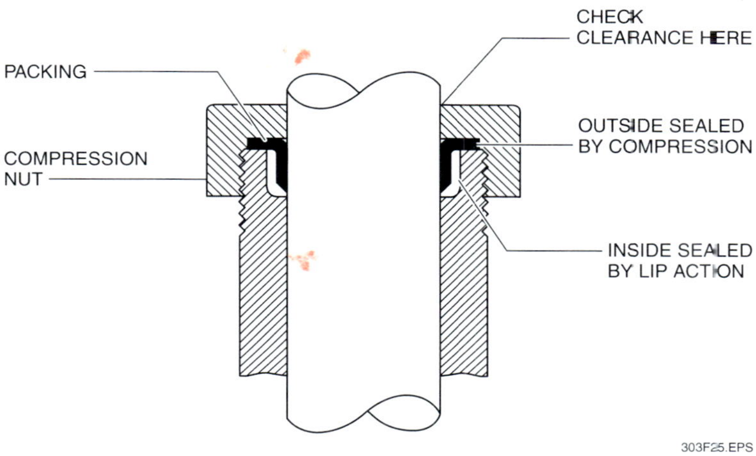


*Figure 25* ◆ Flange packing installation.



MODULE 15303-08 ◆ INSTALLING PACKING 3.15

**Have trainees complete the Review Questions, and go over the answers prior to administering the Module Examination.**

## Review Questions

1. A fiber packing in which one strand of fiber is braided over and under another strand running in the same direction is ______.
   a. square-braid
   b. metal
   c. twisted braid
   d. braid-under-braid

2. A very flexible packing in which no strand is intertwined with all of the other strands is ______.
   a. square-braid
   b. braid-over-braid
   c. pigtail braid
   d. twisted

3. The packing commonly used on reciprocating shafts because of its flexibility is ______.
   a. square-braid
   b. cotton rope
   c. rubber and duck laminated
   d. twisted metal

4. A fiber packing constructed from a series of small braids braided over each other is ______.
   a. metal
   b. braid-over-braid
   c. twisted braid
   d. wrapped

5. A type of packing that resists being deformed or squeezed out of the packing gland by high pressures inside a pump is ______.
   a. injection braid
   b. braid-over-braid
   c. interlocking braid
   d. accordion-fold

6. A kind of packing typically used for low-pressure valve stems and rotary shafts in process systems where there are high temperatures is ______.
   a. metal
   b. wrapped
   c. twisted
   d. laminated

7. A type of packing that is excellent for pumps and valves with worn glands or imbalanced shafts is ______.
   a. graphite ribbon
   b. metal
   c. multicore
   d. interlocking-braid

8. Lip-type packing installed in stacks or nested sets is ______ packing.
   a. metal
   b. V-
   c. folded
   d. ring

9. Lip-type packing used for piston seals inside cylinders is ______ packing.
   a. cup
   b. wrapped
   c. twisted
   d. V-

10. The packing material that is a good heat conductor and dissipates heat in the stuffing box is ______ yarn.
   a. graphite
   b. techron
   c. cotton
   d. epoxy

11. The material normally used in applications where there are strong caustics, acids, and chemicals is ______ yarn.
   a. epoxy
   b. Teflon®
   c. carbon
   d. graphite

12. A type of material limited to temperature ranges up to 500°F is ______.
   a. graphite
   b. carbon yarn
   c. Teflon®
   d. foil wrapped

13. A packing material recommended for use with brine and cold water is ______.
   a. duck and rubber
   b. vegetable fiber
   c. plastic
   d. fiberglass



Instructor's Notes:

3.16 MILLWRIGHT ◆ LEVEL THREE

## Review Questions

14. The material recommended for use with steam, gases, and aqueous solutions, *except* mineral-acid solutions, is ______.
    a. fiberglass
    b. metal
    c. carbon
    d. plastic

15. Incombustible packing recommended for service with hot air, gases, molten metal, and dry steam is ______ packing.
    a. fiberglass
    b. aramid
    c. Teflon®
    d. carbon



MODULE 15303-08 ◆ INSTALLING PACKING 3.17

**Summarize the major concepts presented in the module.**

**Administer the Module Examination. Record the results on Craft Training Report Form 200, and submit the results to the Training Program Sponsor.**

**Administer the Performance Test, and fill out Performance Profile Sheets for each trainee. If desired, trainee proficiency noted during laboratory sessions may be used to complete the Performance Test. Record the results on Craft Training Report Form 200, and submit the results to the Training Program Sponsor.**

## Summary

Packings are used to prevent leakage from the stem area or shaft of a valve or pump. Depending on pressure, temperature, and process materials, many different materials can be used. The material to be used will be specified by the manufacturer or engineer. If repacking a valve or pump, you will frequently duplicate the material previously used. However, the packing should be verified for material, size, and application.

Packing is removed carefully, so as not to harm the stem. Use a pick or packing tool, if you cannot get the packing out directly. Always use correct isolation procedures before working on stuffing boxes or packing glands.

Generally, packing is either installed as die cut rings or cut by hand to fit the stem and installed as if it were a ring, maintaining the staggered cuts of the layers. Be sure to follow manufacturers' instructions. If a valve or pump is leaking, tighten the gland or stuffing box to specifications. If it still leaks, shut down and start over.

## Notes



*Instructor's Notes:*

3.18 MILLWRIGHT ◆ LEVEL THREE

## Trade Terms Introduced in This Module

***Alkalies:*** Various soluble mineral salts found in water.

***Aqueous:*** Watery; containing or dissolved in water.

***Braiding:*** Twisting or interweaving three or more strands of fiber to form one rope-like strand.

***Brine:*** Salty water.

***Caustic:*** Capable of burning, corroding, dissolving, or eating away by chemical action.

***Corrosive:*** Inclined to produce corrosion.

***Crimped:*** Pressed or pinched into small, regular folds.

***Ester:*** A compound formed by eliminating water and bonding an alcohol and an organic acid.

***Filament:*** A fine or thinly spun thread, fiber, or wire.

***Gland:*** A part used to compress packing in a stuffing box.

***Migrating:*** Moving from one place to another.

***Oxidizer:*** A substance that supports the combustion of a fuel or propellant.

***Reciprocating:*** Moving back and forth.

***Skive:*** A cut made at an angle.

***Stuffing box:*** The housing that holds the packing in a pump, valve, or piece of equipment.



MODULE 15303-08 ◆ INSTALLING PACKING 3.19

# Resources & Acknowledgments

## Additional Resources

This module is intended to present thorough resources for task training. The following reference works are suggested for further study. These are optional materials for continued education rather than for task training.

EPM, Inc., www.epm.com/styles.htm

Draco Mechanical Supply, Inc., www.dracomech.com/comppack/pumpinstall.htm

## Figure Credits

UTEX Industries, Inc., 303F01 (photo), 303F02 (photo), 303F03 (photo), 303F04 (photo), 303F05, 303F06, 303F07 (photo), 303F09 (photo)



Instructor's Notes:

3.20 MILLWRIGHT ◆ LEVEL THREE

# MODULE 15303-08 — TEACHING TIPS

The following are suggested activities or instructional methods to help you teach the material in this module.

## General

When you call on someone to answer a question, the rest of the class relaxes or even tunes out because they expect that the question and answer will take place only between you and the trainee you called on. Instead, use this technique to involve more trainees in answering questions and to keep them on their toes.

1. Ask trainees to define a term or explain a concept.
2. After one trainee has answered, ask a trainee seated nearby if the answer is right. Then ask whether a trainee in the back of the room agrees.
3. Ask trainees to explain why they think an answer is right or wrong.
4. Use the session to clear up incorrect ideas and encourage trainees to learn from their mistakes.

## Sections 1.0.0 –5.0.0

### *Trade Terms Quiz*

This Quick Quiz will familiarize trainees with packing trade terms commonly used by millwrights. You will need photocopies of the quiz provided on the following page. Trainees will need pencils. If you allow trainees to use the Trainee Guide, decrease the amount of time you give them to complete the quiz.

1. Make a photocopy of the quiz for each trainee.
2. Give trainees between 5 and 10 minutes to complete the quiz.
3. Go over the answers to the quiz.
4. Ask trainees if they have questions.

### *Answers to Quick Quiz*

1. g
2. b
3. j
4. e
5. d
6. a
7. h
8. i
9. c
10. f

## Quick Quiz

### *Trade Terms*

For each description listed, identify the term that the text best describes. Write the corresponding letter in the blank provided.

_____ 1. A part used to compress packing in a stuffing box is a(n) _____.

_____ 2. Twisting or interweaving three or more strands of fiber to form one rope-like strand is called _____.

_____ 3. The housing that holds the packing in a pump, valve, or piece of equipment is known as the _____.

_____ 4. Metal packings are usually _____ or spiral-wound.

_____ 5. Teflon® is used in the most severe applications of strong _____.

_____ 6. Plastic packing materials are recommended for use with _____ solutions.

_____ 7. A substance that supports the combustion of a fuel or propellant is an _____.

_____ 8. A cut made at an angle is called a(n) _____.

_____ 9. Graphite yarn is commonly used where there are strong _____.

_____ 10. A compound formed by eliminating water and bonding an alcohol and an organic acid is _____.

a. aqueous
b. braiding
c. caustics
d. corrosives
e. crimped
f. ester
g. gland
h. oxidizer
i. skive
j. stuffing box

## Section 2.0.0 *Packing Configurations*

This exercise will familiarize trainees with various packing configurations. Trainees will need pencils and paper. Allow 20 to 30 minutes for this exercise.

1. Set up workstations with various types of packing configurations.
2. Have trainees circulate to each workstation, identifying the type of packing configuration and its applications.
3. Once all trainees have visited each station, have one trainee report on each type of packing configuration.
4. Ask trainees if they have questions.

## Section 3.0.0 *Packing Materials*

This exercise will familiarize trainees with various types of packing materials. Trainees will need pencils and paper. Allow 20 to 30 minutes for this exercise.

1. Set up workstations with various types of packing materials.
2. Have trainees circulate to each workstation, identifying the type of packing material and its applications.
3. Once all trainees have visited each station, have one trainee report on each type of packing material.
4. Ask trainees if they have questions.

## MODULE 15303-08 — ANSWERS TO REVIEW QUESTIONS

| Answer | Section |
|---|---|
| 1. a | 2.1.0 |
| 2. a | 2.1.0 |
| 3. a | 2.1.0 |
| 4. b | 2.2.0 |
| 5. c | 2.3.0 |
| 6. c | 2.4.0 |
| 7. c | 2.5.0 |
| 8. b | 2.8.1 |
| 9. a | 2.8.3 |
| 10. a | 3.1.0 |
| 11. d | 3.1.0 |
| 12. c | 3.2.0 |
| 13. b | 3.4.0 |
| 14. d | 3.6.0 |
| 15. a | 3.8.0 |

## CONTREN® LEARNING SERIES — USER UPDATE

The NCCER makes every effort to keep these textbooks up-to-date and free of technical errors. We appreciate your help in this process. If you have an idea for improving this textbook, or if you find an error, a typographical mistake, or an inaccuracy in NCCER's Contren® textbooks, please write us, using this form or a photocopy. Be sure to include the exact module number, page number, a detailed description, and the correction, if applicable. Your input will be brought to the attention of the Technical Review Committee. Thank you for your assistance.

*Instructors* – If you found that additional materials were necessary in order to teach this module effectively, please let us know so that we may include them in the Equipment/Materials list in the Annotated Instructor's Guide.


**Write:** Product Development and Revision
National Center for Construction Education and Research
3600 NW 43rd St., Bldg. G, Gainesville, FL 32606

**Fax:** 352-334-0932

**E-mail:** curriculum@nccer.org


Craft _______________ Module Name _______________

Copyright Date _______________ Module Number _______________ Page Number(s) _______________

Description

_______________

_______________

_______________

(Optional) Correction

_______________

_______________

(Optional) Your Name and Address

_______________

_______________

Module 15304-08

# *Installing Seals*

## NCCER STANDARDIZED CRAFT TRAINING PROGRAM

The National Center for Construction Education and Research (NCCER) provides a standardized national program of accredited craft training. Key features of the program include instructor certification, competency-based training, and performance testing. The program provides trainees, instructors, and companies with a standard form of recognition through a National Craft Training Registry. The program is described in full in the *Guidelines for Accreditation*, published by NCCER. For more information on standardized craft training, contact the NCCER by writing us at 3600 NW 43rd St., Bldg. G, Gainesville, FL 32606; calling 352-334-0911; or emailing info@nccer.org. More information may be found at our website, www.nccer.org.

## HOW TO USE THIS ANNOTATED INSTRUCTOR'S GUIDE

Each page presents two sections of information. The larger section displays each page exactly as it appears in the Trainee Module. The narrow column ties suggested trainee and instructor actions to each page and provides icons (detailed below) to call your attention to material, safety, audiovisual, or testing requirements. The bottom of each page includes space for your notes.

The **Audiovisual** icon indicates an appropriate time to show a transparency or other audiovisual aid.

The **Classroom** icon prompts you to define a term, stress a point, ask trainees to explain a concept, or give examples.

The **Demonstration** icon directs you to show trainees how to perform tasks.

The **Examination** icon tells you to administer the written module examination.

The **Homework** icon is placed where you may wish to assign reading for the next class, assign a project, or advise trainees to prepare for an examination.

The **Laboratory** icon is used when trainees are to practice performing tasks.

The **Materials** icon is a reminder for you to gather materials needed for classes, labs, and testing.

The **Performance Testing** icon tells you to administer a performance test or a portion thereof.

The **Safety** icon is used to emphasize safety issues. It is often keyed to *Caution* and *Warning!* statements in the Trainee Module.

The **Teaching Tip** icon indicates additional guidance is available, such as how to conduct an exercise, get the most educational value from a field trip, or encourage class participation. Teaching Tips may expand on a feature (*Think About It*, *Did You Know?*) or provide *Quick Quizzes* or similar exercises. You will be referred to the Teaching Tips section at the back of the module if there is additional material.

The **Combination** icon indicates that the laboratory listed corresponds with a performance task. If desired, you can note the proficiency of the trainees during the laboratory, and use it to satisfy performance testing requirements.

## PREPARATION

Before teaching this module, you should review the Objectives, Performance Tasks, Materials and Equipment List, and Module Outline. Be sure to allow ample time to prepare your own training or lesson plan and gather all required materials and equipment.

## MODULE OVERVIEW

This module explains the applications, removal, and installation procedures for various types of seals.

## PREREQUISITES

Prior to training with this module, it is recommended that the trainee shall have successfully completed the following: *Core Curriculum; Millwright Level One; Millwright Level Two;* and *Millwright Level Three,* Modules 15301-08 through 15303-08.

## OBJECTIVES

Upon completion of this module, the trainee will be able to do the following:

1. Identify and explain types of seals.
2. Identify and explain seal materials.
3. Remove and install seals.

## PERFORMANCE TASKS

Under the supervision of the instructor, the trainee should be able to do the following:

1. Identify seals and seal materials.
2. Remove and install seals.

## MATERIALS AND EQUIPMENT LIST

Overhead projector and screen
Transparencies
Blank acetate sheets
Transparency pens
Whiteboard/chalkboard
Markers/chalk
Pencils and scratch paper
Appropriate personal protective equipment
Various sized O-rings
Lip seals
Oil seals
Labyrinth seals
Cup seals
Various types of seal materials, including:
- Buna-N
- Silicone
- Neoprene
- Plastic and elastomer compounds
- Metal O-rings
- Leather

Module Examinations*
Performance Profile Sheets*

* Located in the Test Booklet

## SAFETY CONSIDERATIONS

Ensure that the trainees are equipped with appropriate personal protective equipment and know how to use it properly. This module requires that the trainees use hand tools. Ensure that trainees are briefed on shop safety procedures.

## ADDITIONAL RESOURCES

This module is intended to present thorough resources for task training. The following reference work is suggested for both instructors and motivated trainees interested in further study. This is optional material for continued education rather than for task training.

Flowserve Corporation,
www.flowserve.com/eim/Services/Services/PumpsServiceandRepairInformation

## TEACHING TIME FOR THIS MODULE

An outline for use in developing your lesson plan is presented below. Note that each Roman numeral in the outline equates to one session of instruction. Each session has a suggested time period of 2½ hours. This includes 10 minutes at the beginning of each session for administrative tasks and one 10-minute break during the session. Approximately 5 hours are suggested to cover *Installing Seals.* You will need to adjust the time required for hands-on activity and testing based on your class size and resources. Because laboratories often correspond to Performance Tasks, the proficiency of the trainees may be noted during these exercises for Performance Testing purposes.

| Topic | Planned Time |
|---|---|
| **Session I. Types of Seals; Seal Materials; Removing and Installing Seals** | |
| A. Introduction | ____________ |
| B. Types of Seals | ____________ |
| C. Seal Materials | ____________ |
| D. Laboratory<br>Trainees practice identifying seals and seal materials. This laboratory corresponds to Performance Task 1. | ____________ |
| E. Removing and Installing Seals | ____________ |
| F. Laboratory<br>Trainees practice removing and installing seals. This laboratory corresponds to Performance Task 2. | ____________ |
| **Session II. Review and Testing** | |
| A. Module Review | ____________ |
| B. Module Examination<br>1. Trainees must score 70% or higher to receive recognition from NCCER.<br>2. Record the testing results on Craft Training Report Form 200, and submit the results to the Training Program Sponsor. | ____________ |
| C. Performance Testing<br>1. Trainees must perform each task to the satisfaction of the instructor to receive recognition from NCCER. If applicable, proficiency noted during laboratory exercises can be used to satisfy the Performance Testing requirements.<br>2. Record the testing results on Craft Training Report Form 200, and submit the results to the Training Program Sponsor. | ____________ |

15304-08

# Installing Seals

Assign reading of Module 15304-08.

# 15304-08
# *Installing Seals*

*Topics to be presented in this module include:*



## Overview

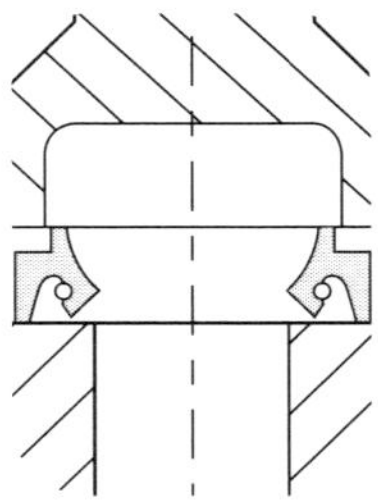

Seals are used to control leakage, especially of lubricants and process fluids. This module explains what different kinds of seals are used for and how to install some types. The applications and limitations of types and materials are described, as well as the way different seals work.

*Instructor's Notes:*

## Objectives

When you have completed this module, you will be able to do the following:

1. Identify and explain types of seals.
2. Identify and explain seal materials.
3. Remove and install seals.

## Trade Terms

Bore
Elastomer

## Required Trainee Materials

1. Pencil and paper
2. Appropriate personal protective equipment

## Prerequisites

Before you begin this module, it is recommended that you successfully complete *Core Curriculum; Millwright Level One; Millwright Level Two;* and *Millwright Level Three*, Modules 15301-08 through 15303-08.

This course map shows all of the modules in the third level of the *Millwright* curriculum. The suggested training order begins at the bottom and proceeds up. Skill levels increase as you advance on the course map. The local Training Program Sponsor may adjust the training order.

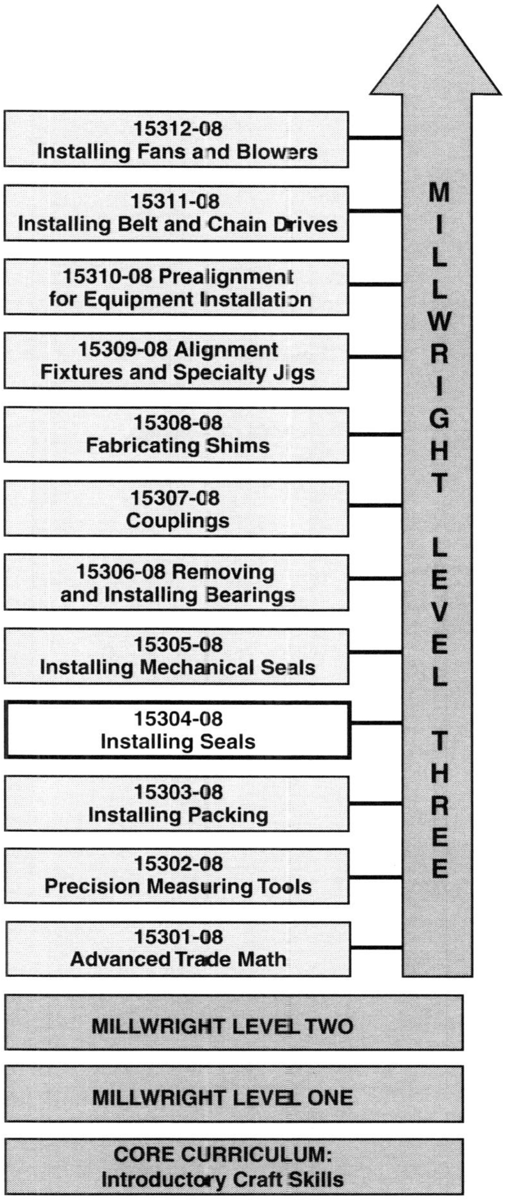


304CMAP.EPS

**Ensure that you have everything required to teach the course. Check the Materials and Equipment list at the front of this module.**

**See the general Teaching Tip at the end of this module.**

**Explain that terms shown in bold are defined in the Glossary at the back of this module.**

**Show Transparency 1, Objectives, and Transparency 2, Performance Tasks. Review the goals of the module, and explain what will be expected of the trainees.**

**Review the modules covered in Level Three and explain how this module fits in.**



MODULE 15304-08 ◆ INSTALLING SEALS 4.1

Describe seals and explain their function.

Discuss why millwrights need to be able to identify types of seals and remove and replace them.

Discuss different types of seals.

Explain which seals are appropriate for static and dynamic applications.

Describe positive-contact and noncontact dynamic seals.

Discuss O-rings.

Show Transparencies 3-5 (Figures 1-3).

Show Transparency 6 (Table 1).

Show Transparency 7 (Figure 4).

## 1.0.0 ◆ INTRODUCTION

Seals are devices or materials used to prevent or control leakage between two surfaces. They are used between moving and stationary parts or between two stationary parts. Seals come in various types and sizes and are made of many materials for different applications. Some seals require regular replacement to ensure proper performance. Properly installing seals is critical to the performance of the seal and the equipment. Incorrect installation of seals can result in premature failure of the seals or damage to the equipment. Millwrights must be able to identify the types of seals and seal materials as well as remove and replace seals.

## 2.0.0 ◆ TYPES OF SEALS

When selecting a seal, determine whether the application is static or dynamic. Static applications are those in which there is no movement between the two joining parts or between the seal and the mating part. O-rings are used for static applications. Dynamic seals are used where there is movement between two mating parts or between one of the parts and the seal. The two types of dynamic applications are positive-contact and noncontact seals. Seals used for dynamic applications include O-rings, lip seals, oil seals, and labyrinth seals.

Positive-contact, or rubbing, seals are used where the seal area is continuously flooded. If properly selected and installed, positive-contact seals can prevent all leakage of most fluids. However, because they are sensitive to fluid temperature, pressure, and velocity, improper use can cause early failure. Positive-contact seals can be used on both rotating and reciprocating shafts.

Noncontact, or controlled-clearance, seals are those that prevent rubbing between rotating and stationary parts. They operate by controlling fluid movement through narrow passages in the seal. Noncontact seals are frictionless and are not sensitive to fluid temperature and velocity. They are most effective for limiting leakage rather than stopping it. Noncontact seals have limited uses when leakage rates are critical. These seals are frequently custom-designed for a particular application and can be expensive.

### 2.1.0 O-Rings

An O-ring is a seal commonly made as a ring. O-rings are sized according to their inside diameter (ID) and cross-section (W) dimensions. The inside diameter means what it says, the distance from wall to wall of the O-ring. The cross-section means the thickness of the ring itself. O-rings can be used as either static or dynamic seals. In a dynamic application, an O-ring is usually placed in a groove or joint that is 135 to 150 percent wider than the cross section of the O-ring. In a static application, an O-ring is installed in a groove approximately 25 percent wider than the cross section of the O-ring. When the O-ring comes in contact with the areas to be sealed, it is slightly distorted in a motion called mechanical squeeze (*Figure 1*).

A dynamic application requires a running fit; a static application needs a complete seal and the O-ring should almost fill the O-ring groove. *Figure 2* shows both applications.

The pressure caused by mechanical squeeze holds the O-ring in contact with the surfaces to be sealed. This pressure also causes the O-ring to roll and slide to the side of the groove away from the pressure, lubricating the O-ring surface and extending its life. *Figure 3* shows how an O-ring seals a joint.

Mechanical squeeze is necessary to maintain a seal. Too much squeeze wears out the O-ring quickly; insufficient squeeze allows leaks to start. The correct squeeze is usually equal to 10 percent of the O-ring diameter. General-purpose O-rings are made with a cross section 10 percent larger than the nominal size. This allows for the initial mechanical squeeze. *Table 1* lists the nominal and actual diameters of standard O-rings. With the harder material types of O-ring, it is possible to gently run a micrometer on it to determine the cross section of the ring.

*Figure 4* shows a housing vent with O-rings.

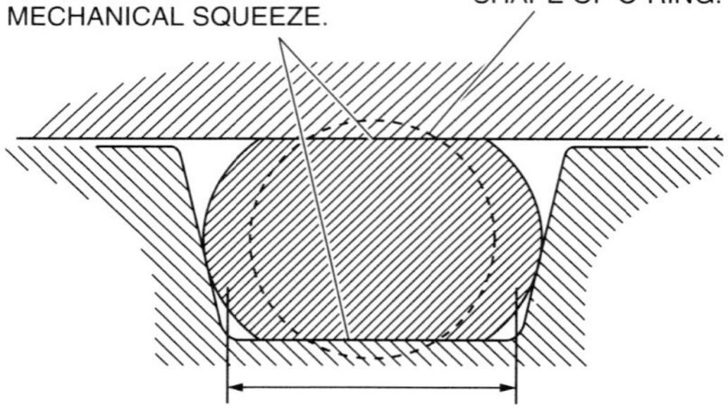


*Figure 1* ◆ Mechanical squeeze.



Instructor's Notes:

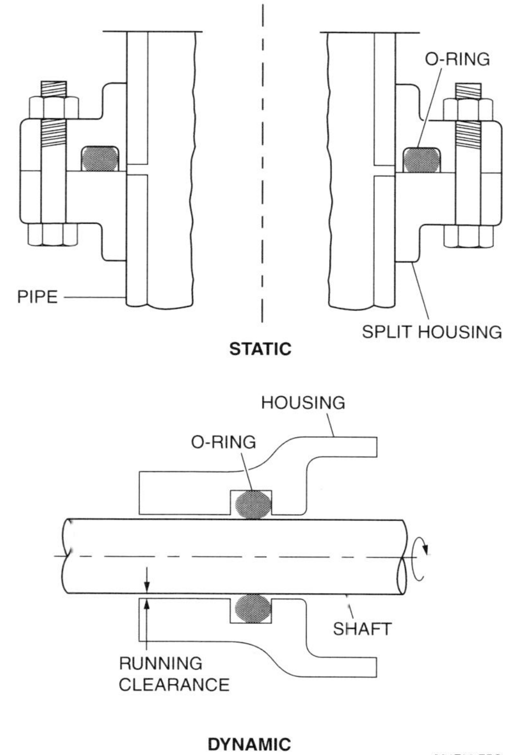


*Figure 2* ◆ Static and dynamic applications.

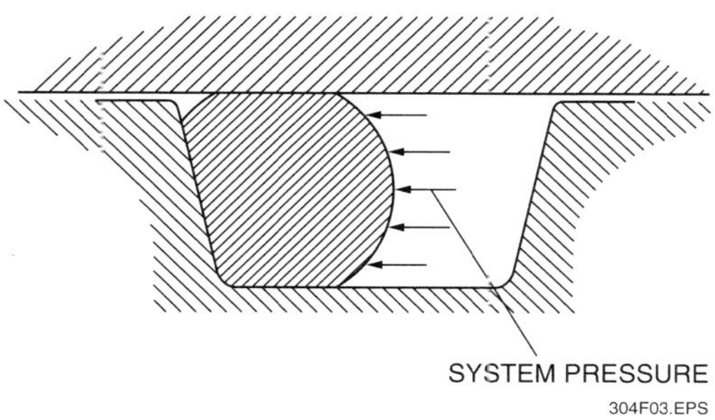


*Figure 3* ◆ O-ring sealing joint.

**Table 1** O-Ring Sizes

| Nominal Cross Section Diameter (inches) | Actual Cross Section Diameter (inches) |
|---|---|
| 1/32 | 0.040 |
| 3/64 | 0.050 |
| 1/16 | 0.070 |
| 3/32 | 0.103 |
| 1/8 | 1.139 |
| 3/16 | 0.210 |
| 1/4 | 0.275 |

304T01.EPS

**Discuss lip seals and explain how they are installed.**

**Discuss oil seals and their application.**

**Show Transparency 8 (Figure 5).**

## 2.2.0 Lip Seals

Lip seals are low-pressure, positive-contact seals used with rotating shafts. Pressure on the lip or a vacuum behind the lip pushes it against the shaft for a tighter seal. Lip seals usually consist of a housing that is used for aligning the seal when it is installed and for supporting the lip, a lip that fits around the shaft, and a spring that holds the lip in contact with the shaft.

Lip seals are usually installed to contain lubricant in a housing when the outside conditions are relatively clean. In dirty conditions, the seal is installed with the lip facing out to prevent foreign matter from getting into the housing. With the lip facing out, the housing should be vented to prevent the lubricant from leaking. A vent and drain can be applied to any inside seal by drilling and tapping two openings into the gland ring outboard of the stationary insert.

In some applications, the area being sealed changes from pressure to vacuum conditions. Double-lip seals are available for these applications to prevent air or dirt from getting in and lubricant from getting out.

## 2.3.0 Oil Seals

The oil seal, or radial lip seal, is a positive-contact seal used on rotating or reciprocating shafts. This seal is used to keep fluids in or to keep dirt and other foreign matter out. Oil seals are usually used to keep fluids in and are installed with the lip facing in. When an oil seal is used to keep contaminants out, it is installed with the lip facing away from the housing. Double-lip seals that perform both functions are also available. The three basic components are the case, the seal, and the metal retainer. *Figure 5* shows a typical oil seal.

Oil seals are manufactured in a wide variety of types for use in various applications. These types can be divided into the following three broad categories:



MODULE 15304-08 ◆ INSTALLING SEALS 4.3

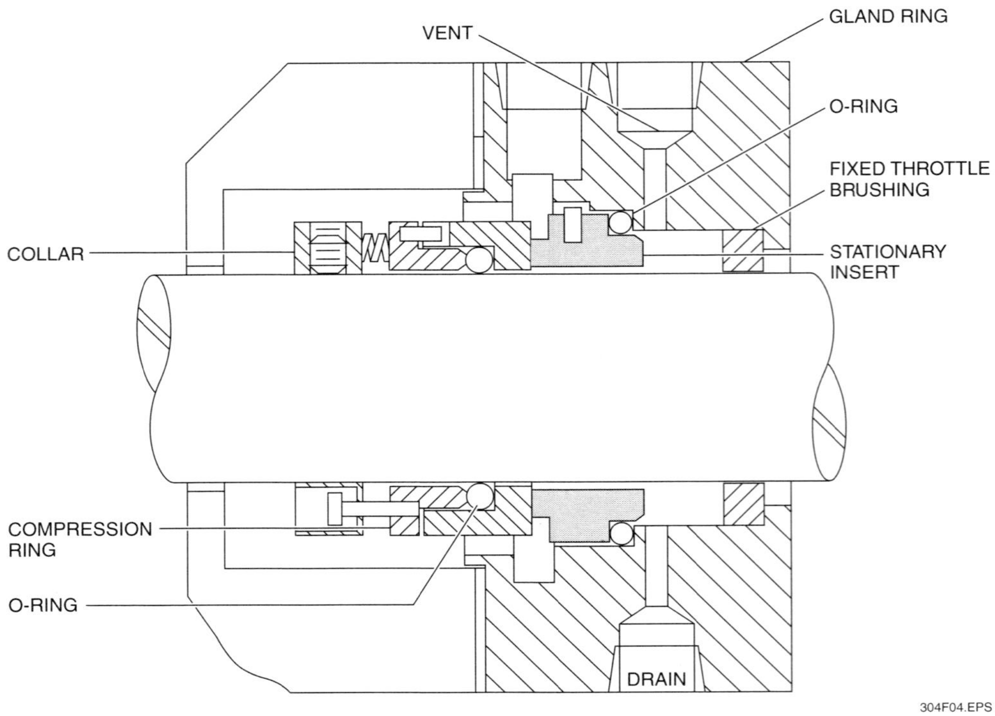


*Figure 4* ◆ Housing vent.

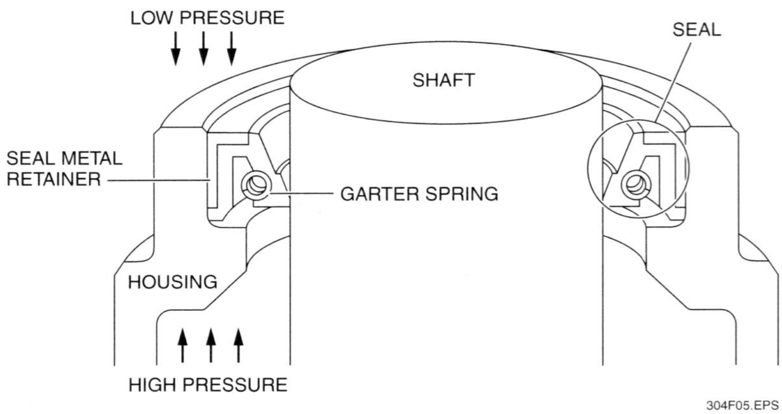


*Figure 5* ◆ Oil seal.



Instructor's Notes:

4.4 MILLWRIGHT ◆ LEVEL THREE

- *Cased seals* – These contain a sealing element in a manufactured metal case.
- *Bonded seals* – These contain a sealing element bonded to a washer or a formed metal case.
- *Dual-element seals* – These contain a double-sealing element. The double element is used for applications where liquids are present on both sides of the seal or for severe service.

## 2.4.0 Labyrinth Seals

Labyrinth seals (*Figure 6*) are noncontact dynamic seals that are used on rotating shafts. Labyrinth seals do not actually touch the surface to be sealed. Instead, they contain a narrow path through which fluid must flow before it can leak. Because small amounts of liquid can leak from a labyrinth seal, they work best with thick or semisolid materials, such as grease. However, they can be used with liquids and gases if some leakage is not a problem.

The noncontact design of the labyrinth seal also provides the following benefits:

- Prevents damage to the shaft
- Helps prevent heat buildup
- Does not require lubrication

## 2.5.0 Cup Seals

Cup seals (*Figure 7*) are positive-contact dynamic seals generally used on cylinder pistons. Sealing occurs when pressure forces the cup lip against the cylinder barrel. Cup seals are pressure-driven in both directions. They are supported by a backing plate and can handle operating pressures of 1,000 to 1,500 psi. To work correctly, cup seals must be clamped tightly in place. In a cylinder piston application, the piston actually holds the cup seals.

## 3.0.0 ◆ SEAL MATERIALS

The design and composition of seals must be compatible with the fluid used in a system, including its use and the operating conditions. The following points should be considered when choosing a seal material:

- *Fluid compatibility* – Few seals are compatible with all fluids. Some fluids and fluid additives can cause certain types of seals to disintegrate. Contact the fluid supplier for a list of compatible seal materials.

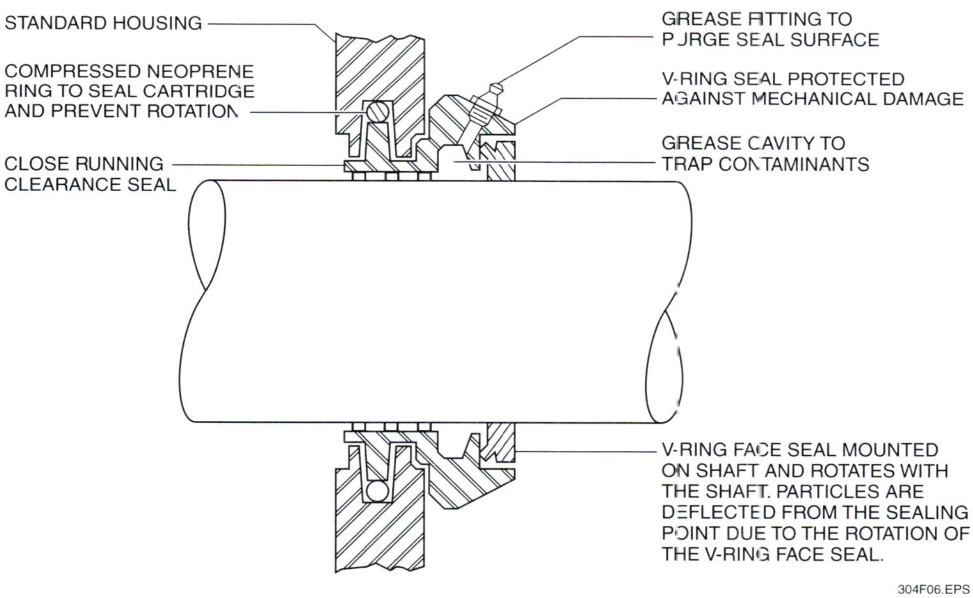


*Figure 6* ◆ Labyrinth seal.



**Discuss the three categories of oil seals.**

**Discuss labyrinth seals and their benefits.**

**Show Transparency 9 (Figure 6).**

**Discuss cup seals and explain how they work.**

**Show Transparency 10 (Figure 7).**

**See the Teaching Tip for Section 2.0.0 at the end of this module.**

**Introduce seal materials.**

**Provide a variety of seal materials for trainees to examine.**

**Discuss the considerations for choosing a seal material.**

**Explain why Buna-N, or nitrile, is the most widely used sealing material.**

**Discuss the uses of silicone, neoprene, and plastic and elastomer compounds.**

**Discuss the uses of metal O-rings and leather.**

**Note that O-rings can be fabricated to size.**

**Have trainees practice identifying seals and seal materials. Note the proficiency of each trainee. This laboratory corresponds to Performance Task 1.**

- *Temperature* – Seals operate within specific temperature ranges. At extremely low temperatures, a seal can become brittle. At high temperatures, a seal can harden, soften, or swell.
- *Pressure* – Fluid pressure puts a strain on seals and can cause leaks. Always use a seal rated to withstand the pressure in a given application.
- *Lubrication* – No seal should be installed or operated without lubrication. Some seals must be soaked in fluid before they are installed; others only need to be coated with fluid.

Each seal manufacturer has applications charts to ensure compatibility with each application; therefore, consult the manufacturer's specifications when choosing a seal for any application. Do not arbitrarily substitute seal materials.

### 3.1.0 Buna-N

Buna-N, or nitrile, is the most widely used sealing material. It is an elastomer that wears well and is inexpensive. Buna-N is compatible with petroleum oil and works at temperatures ranging from –40°F to 230°F. At high temperatures, it retains its shape in most petroleum oils where other materials tend to swell. It does swell in some synthetic fluids.

### 3.2.0 Silicone

Silicone is an elastomer used for rotating shaft seals and static seals in systems with wide temperature ranges. Silicone retains its shape and sealing ability at ranges from –60°F to 500°F. It is compatible with most fluids. At high temperatures, silicone absorbs oil and swells, but this is not a disadvantage for static applications. It is not used for reciprocating applications because it tears too easily.

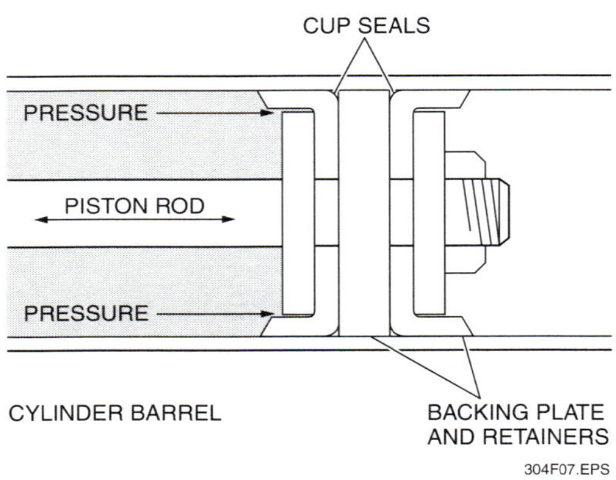


*Figure 7* ◆ Cup seals.

### 3.3.0 Neoprene

Neoprene is an elastomer used only for low-temperature systems using petroleum fluids. It is only suitable at temperatures below 150°F because it tends to stick together above that temperature.

### 3.4.0 Plastic and Elastomer Compounds

Several sealing materials are formed by combining fluorine gas with an elastomer or plastic, including Kel-F, Viton A, and Teflon®. Another synthetic material with similar properties is nylon. Nylon is often used in combination with the elastomers to reinforce them. All these materials can be used at temperatures up to 500°F and are compatible with most fluids.

### 3.5.0 Metal O-Ring

In very high temperature and high-pressure applications, hollow metal O-rings are used. These are made of stainless or spring steel. They are not the best choice for use as dynamic rings because of the expense and wear, and should not be used for rings in contact with very acidic materials.

### 3.6.0 Leather

Leather is an inexpensive and tough seal material used for some cup and lip seals. Some leather seals are saturated with an elastomer to improve the sealing ability. The disadvantages of leather are a tendency to squeal when dry and a limited temperature range. Most leather seals cannot function above 165°F, however, they function well to temperatures of –65°F. Leather is less frequently used now because more easily worked synthetics are available. It is rarely, if ever, used as a new installation, but is occasionally still found in older installations.

### 3.7.0 Fabricating O-Rings

O-rings can be fabricated to size, using a kit that cuts and seals ring material. In cases where rings have to be replaced frequently, and where the more exotic materials are not required, this can be useful.



Instructor's Notes:

4.6 MILLWRIGHT ◆ LEVEL THREE

## 4.0.0 ◆ REMOVING AND INSTALLING SEALS

The working life of a seal depends on the type of seal and the application for which it is used. The most common indication of seal failure is excessive leakage. When a seal fails, it must be removed and replaced. Removing seals is a relatively simple procedure, but it is important that the procedure be performed properly. The procedure for removing seals varies slightly from one piece of equipment to another, but the steps are basically the same. Refer to the equipment manufacturer's manual for specific procedures.

Following proper installation procedures for a seal is critical to the life expectancy and efficiency of the seal. Improperly installing a seal can result in a loss of up to 75 percent of its working life and may cause excessive wear to the equipment. The following sections explain how to remove a seal and how to install O-rings and lip and oil seals.

### 4.1.0 Removing Seals

Before a new seal can be installed, the old seal must be removed. Follow these steps to remove a seal.

*Step 1* Clean the shaft and exterior of the seal, using a clean, soft cloth.

*Step 2* Remove any burrs or sharp edges from the shaft.

*Step 3* Remove the old seal. To remove the old seal, tap it gently using a flat hammer, or cut into the seal using a rounded, blunt chisel. Do not cut all the way through the seal. Cutting through the seal can damage the seal bore.

### 4.2.0 Installing O-Rings

Follow these steps to install an O-ring.

*Step 1* Select the proper size O-ring for the application.

*Step 2* Ensure that the O-ring is not damaged.

*Step 3* Inspect and polish, deburr, and clean the shaft and housing in the seal area. Repair any damaged areas of the housing shaft.

*Step 4* Cover any threads with a cone made of brass shim stock, aluminum, or stiff plastic sheeting.

*Step 5* Lubricate the O-ring with the lubricant specified in the O-ring catalog.

*Step 6* Place the O-ring in the groove. *Figure 8* shows O-rings installed in static and dynamic applications.

### 4.3.0 Installing Lip and Oil Seals

Follow these steps to install lip and oil seals.

*Step 1* Select the proper size seal and an installation tool, such as a press ram or driving tool.

**NOTE**

The outside diameter of the installation tool should not be more than 0.010 inch smaller than the bore diameter. If an installation tool is not available, a tube or short length of pipe that is slightly smaller in diameter than the bore can be used. If a tube is not available, a block of wood placed squarely on the back of the seal can also be used.

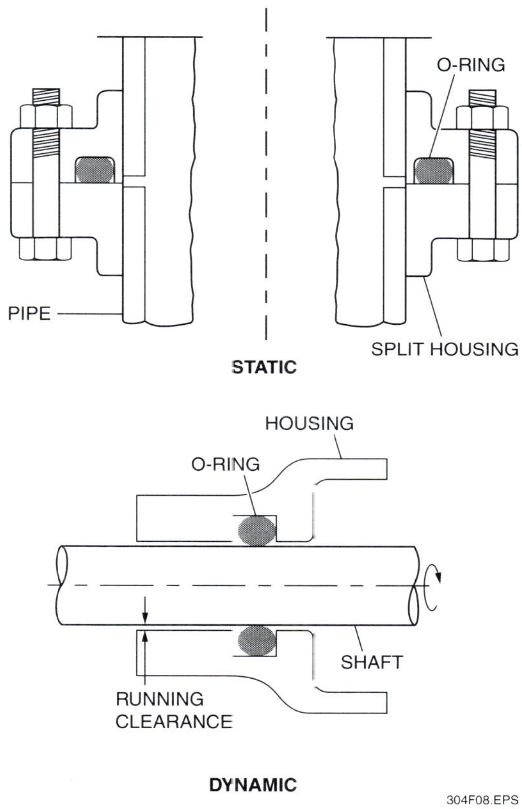


*Figure 8* ◆ Installed O-rings.

**Explain why removing and installing seals properly is important.**

**Review the steps for removing seals.**

**Review the steps for installing O-rings.**

**Show Transparency 11 (Figure 8). Discuss O-rings installed in static and dynamic applications.**

**Review the steps for installing lip and oil seals.**



MODULE 15304-08 ◆ INSTALLING SEALS 4.7

**Show Transparency 12 (Figure 9). Discuss how to mount the seal.**

**Show Transparency 13 (Figure 10). Explain how to use the installation tool to seat the seal.**

**Demonstrate how to remove and install seals.**

**Have trainees practice removing and installing seals. Note the proficiency of each trainee. This laboratory corresponds to Performance Task 2.**

*Step 2* Ensure that the seal has not been damaged, and that its spring is in place.

*Step 3* Inspect and polish, deburr, and clean the shaft and housing in the seal area. Repair any damaged areas of the housing shaft.

*Step 4* Place a mounting thimble on the shaft to protect the seal. If a mounting thimble is not available, cover the edge or shoulder with a protective covering, such as brass shim stock, aluminum, or stiff plastic sheeting. Wrap the protective covering around the shaft, over the end that could damage the seal. *Figure 9* shows how the seal is mounted.

*Step 5* Lubricate the shaft and seal lip, using the lubricant specified in the seal catalog.

*Step 6* Slide the seal over the shaft and into place.

*Step 7* Place an installation tool against the seal (*Figure 10*).

*Step 8* Tap the installation tool, using a mallet to seat the seal.

**CAUTION**

To prevent damaging the seal, never hit it directly with a mallet.

*Step 9* Ensure that the seal is installed squarely with the shaft center line. Press the seal flush with the front of the bore or bottom it against a shoulder.

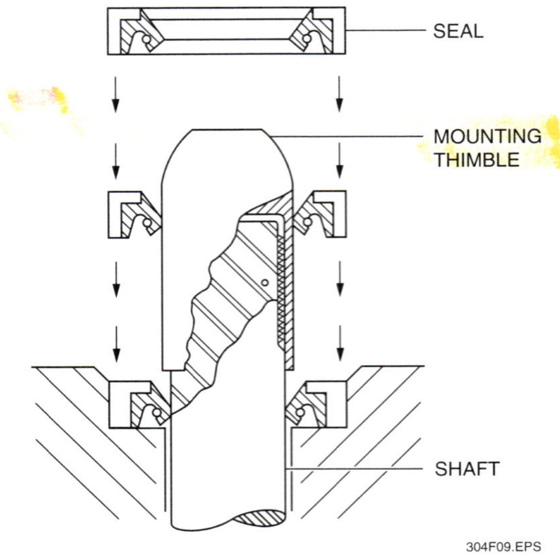


*Figure 9* ◆ Preparing to mount the seal.

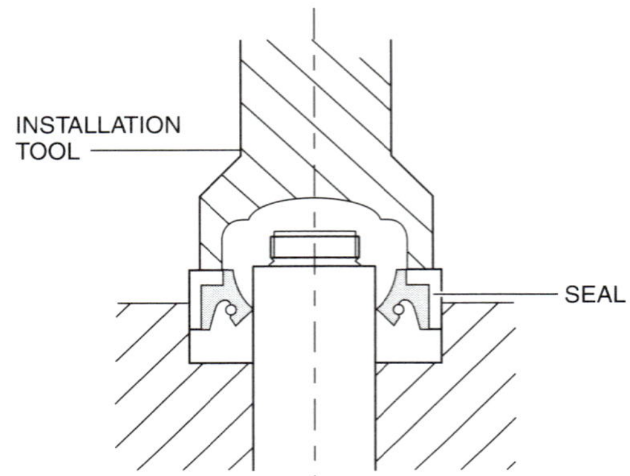


*Figure 10* ◆ Using the installation tool.



Instructor's Notes:

4.8 MILLWRIGHT ◆ LEVEL THREE

## Review Questions

1. Applications in which there is movement between two mating parts are called _____ applications.
   a. positive-contact
   b. controlled-clearance
   c. dynamic
   d. static

2. Seals that can be used in both static and dynamic applications are _____.
   a. cup seals
   b. O-rings
   c. labyrinth seals
   d. lip seals

3. General purpose O-rings are made with a cross section _____ percent larger than nominal size.
   a. 10
   b. 15
   c. 20
   d. 25

4. The three parts to an oil seal are the _____.
   a. case, seal, and metal retainer
   b. housing, seal, and lapped face
   c. case, housing, and O-ring
   d. spring, seal, and clamping hub

5. Positive-contact seals used on rotating or reciprocating shafts are called _____.
   a. oil seals
   b. O-rings
   c. labyrinth seals
   d. cup seals

6. Non-contact seals used on rotating shafts are called _____.
   a. oil seals
   b. O-rings
   c. labyrinth seals
   d. cup seals

7. Seals that work best with thick or semi-solid materials are called _____.
   a. oil seals
   b. O-rings
   c. labyrinth seals
   d. cup seals

8. The factors that should be considered when choosing seal materials include _____.
   a. direction, temperature, hardness, and elevation
   b. fluid compatibility, temperature, pressure, and lubrication
   c. clamping, hardness, pressure, and color
   d. hardness, temperature, pressure, and color

9. The elastomer used only for low-temperature systems using petroleum fluids is _____.
   a. Buna-N
   b. silicone
   c. neoprene
   d. Viton-A®

10. The most common indication of seal failure is excessive _____.
    a. noise
    b. heat
    c. leakage
    d. smell



**Have trainees complete the Review Questions, and go over the answers prior to administering the Module Examination.**

MODULE 15304-08 ◆ INSTALLING SEALS 4.9

Summarize the major concepts presented in the module.

Administer the Module Examination. Record the results on Craft Training Report Form 200, and submit the results to the Training Program Sponsor.

Administer the Performance Test, and fill out Performance Profile Sheets for each trainee. If desired, trainee proficiency noted during laboratory sessions may be used to complete the Performance Test. Record the results on Craft Training Report Form 200, and submit the results to the Training Program Sponsor.

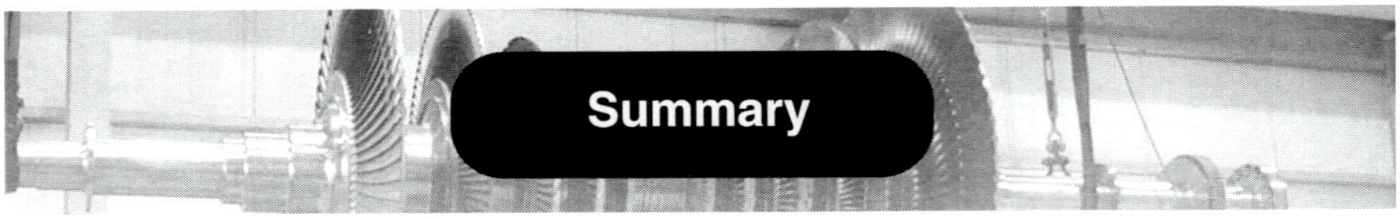

## Summary

Seals are used to control leakage. Different types of seals are specifically designed for particular applications. O-rings are used for many applications, as they are very adaptable to different purposes. Seals may be dynamic, meaning that they are used with surfaces that move with respect to each other, or they may be static, which means that they do not move. If seals are either improperly installed or worn, the result will be leakage, loss of critical fluids or lubricants, and damage to machines.

## Notes



Instructor's Notes:

4.10 MILLWRIGHT ◆ LEVEL THREE

## Trade Terms Introduced in This Module

***Bore:*** The interior diameter of a shaft.

***Elastomer:*** A synthetic material with the elastic qualities of natural rubber.



MODULE 15304-08 ◆ INSTALLING SEALS 4.11

# Resources & Acknowledgments

## Additional Resources

This module is intended to present thorough resources for task training. The following reference work is suggested for further study. This is optional material for continued education rather than for task training.

Flowserve Corporation, www.flowserve.com/eim/Services/Services/PumpsServiceandRepairInformation



Instructor's Notes:

4.12 MILLWRIGHT ◆ LEVEL THREE

# MODULE 15304-08 — TEACHING TIPS

The following are suggested activities or instructional methods to help you teach the material in this module.

## General

When you call on someone to answer a question, the rest of the class relaxes or even tunes out because they expect that the question and answer will take place only between you and the trainee you called on. Instead, use this technique to involve more trainees in answering questions and to keep them on their toes.

1. Ask trainees to define a term or explain a concept.
2. After one trainee has answered, ask a trainee seated nearby if the answer is right. Then ask whether a trainee in the back of the room agrees.
3. Ask trainees to explain why they think an answer is right or wrong.
4. Use the session to clear up incorrect ideas and encourage trainees to learn from their mistakes.

## Section 2.0.0 *Types of Seals*

This exercise will help familiarize trainees with types of seals. Trainees will need pencils and paper. You will need to arrange workstations with a different type of seal at each station. Allow 20 to 30 minutes for this exercise. This exercise corresponds to Performance Task 1.

1. Discuss various types of seals with trainees.
2. Have trainees rotate between the various workstations and identify the type of seal at each station.
3. After all trainees have visited each station, have one trainee identify the type of seal at their station. Answer any questions trainees may have.

# MODULE 15304-08 — ANSWERS TO REVIEW QUESTIONS

| Answer | | Section |
|---|---|---|
| 1. | c | 2.0.0 |
| 2. | b | 2.1.0 |
| 3. | a | 2.1.0 |
| 4. | a | 2.3.0 |
| 5. | a | 2.3.0 |
| 6. | c | 2.4.0 |
| 7. | c | 2.4.0 |
| 8. | b | 3.0.0 |
| 9. | c | 3.3.0 |
| 10. | c | 4.0.0 |

## CONTREN® LEARNING SERIES — USER UPDATE

The NCCER makes every effort to keep these textbooks up-to-date and free of technical errors. We appreciate your help in this process. If you have an idea for improving this textbook, or if you find an error, a typographical mistake, or an inaccuracy in NCCER's Contren® textbooks, please write us, using this form or a photocopy. Be sure to include the exact module number, page number, a detailed description, and the correction, if applicable. Your input will be brought to the attention of the Technical Review Committee. Thank you for your assistance.

*Instructors* – If you found that additional materials were necessary in order to teach this module effectively, please let us know so that we may include them in the Equipment/Materials list in the Annotated Instructor's Guide.

**Write:** Product Development and Revision
National Center for Construction Education and Research
3600 NW 43rd St., Bldg. G, Gainesville, FL 32606

**Fax:** 352-334-0932

**E-mail:** curriculum@nccer.org

Craft ____________________ Module Name ____________________

Copyright Date __________ Module Number __________ Page Number(s) __________

Description

______________________________________________

______________________________________________

______________________________________________

______________________________________________

(Optional) Correction

______________________________________________

______________________________________________

______________________________________________

(Optional) Your Name and Address

______________________________________________

______________________________________________

______________________________________________

Module 15305-08

# *Installing Mechanical Seals*

## NCCER STANDARDIZED CRAFT TRAINING PROGRAM

The National Center for Construction Education and Research (NCCER) provides a standardized national program of accredited craft training. Key features of the program include instructor certification, competency-based training, and performance testing. The program provides trainees, instructors, and companies with a standard form of recognition through a National Craft Training Registry. The program is described in full in the *Guidelines for Accreditation*, published by NCCER. For more information on standardized craft training, contact the NCCER by writing us at 3600 NW 43rd St., Bldg. G, Gainesville, FL 32606; calling 352-334-0911; or emailing info@nccer.org. More information may be found at our website, www.nccer.org.

## HOW TO USE THIS ANNOTATED INSTRUCTOR'S GUIDE

Each page presents two sections of information. The larger section displays each page exactly as it appears in the Trainee Module. The narrow column ties suggested trainee and instructor actions to each page and provides icons (detailed below) to call your attention to material, safety, audiovisual, or testing requirements. The bottom of each page includes space for your notes.

The **Audiovisual** icon indicates an appropriate time to show a transparency or other audiovisual aid.

The **Classroom** icon prompts you to define a term, stress a point, ask trainees to explain a concept, or give examples.

The **Demonstration** icon directs you to show trainees how to perform tasks.

The **Examination** icon tells you to administer the written module examination.

The **Homework** icon is placed where you may wish to assign reading for the next class, assign a project, or advise trainees to prepare for an examination.

The **Laboratory** icon is used when trainees are to practice performing tasks.

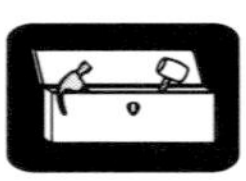

The **Materials** icon is a reminder for you to gather materials needed for classes, labs, and testing.

The **Performance Testing** icon tells you to administer a performance test or a portion thereof.

The **Safety** icon is used to emphasize safety issues. It is often keyed to *Caution* and *Warning!* statements in the Trainee Module.

The **Teaching Tip** icon indicates additional guidance is available, such as how to conduct an exercise, get the most educational value from a field trip, or encourage class participation. Teaching Tips may expand on a feature (*Think About It, Did You Know?*) or provide *Quick Quizzes* or similar exercises. You will be referred to the Teaching Tips section at the back of the module if there is additional material.

The **Combination** icon indicates that the laboratory listed corresponds with a performance task. If desired, you can note the proficiency of the trainees during the laboratory, and use it to satisfy performance testing requirements.

## PREPARATION

Before teaching this module, you should review the Objectives, Performance Tasks, Materials and Equipment List, and Module Outline. Be sure to allow ample time to prepare your own training or lesson plan and gather all required materials and equipment.

## MODULE OVERVIEW

This module explains the function and advantages of mechanical seals. It identifies parts and types of seals, lists mechanical seal classifications, and includes procedures for removing, inspecting, and installing mechanical seals.

## PREREQUISITES

Prior to training with this module, it is recommended that the trainee shall have successfully completed the following: *Core Curriculum; Millwright Level One; Millwright Level Two;* and *Millwright Level Three,* Modules 15301-08 through 15304-08.

## OBJECTIVES

Upon completion of this module, the trainee will be able to do the following:

1. Identify and explain types of mechanical seals.
2. Explain mechanical seal classification.
3. Remove, inspect, and install mechanical seals.

## PERFORMANCE TASKS

Under the supervision of the instructor, the trainee should be able to do the following:

1. Identify types of mechanical seals.
2. Remove, inspect, and install mechanical seals.

## MATERIALS AND EQUIPMENT LIST

Overhead projector and screen
Transparencies
Blank acetate sheets
Transparency pens
Whiteboard/chalkboard
Markers/chalk
Pencils and scratch paper
Appropriate personal protective equipment
Basic trainee tools
Centrifugal pump with mechanical seal

Samples of various mechanical seals, including:

| | |
|---|---|
| Inside | Outside |
| Tandem | Cartridge |
| Balanced | Multiple-spring |
| Welded metal bellows | Elastomer bellows |
| Rotating | Stationary |

Quick Quiz*
Module Examinations**
Performance Profile Sheets**

* Located in the back of this module
**Located in the Test Booklet

## SAFETY CONSIDERATIONS

Ensure that the trainees are equipped with appropriate personal protective equipment and know how to use it properly. This module requires that the trainees use hand tools. Ensure that trainees are briefed on shop safety procedures.

## ADDITIONAL RESOURCES

This module is intended to present thorough resources for task training. The following reference works are suggested for both instructors and motivated trainees interested in further study. These are optional materials for continued education rather than for task training.

Flowserve Corporation, www.flowserve.com/eim/EducationalServices

Mechanical Seals.Net, www.mechanicalseals.net/Mechanical_Seal_Repair.htm

McNally Institute, www.mcnallyinstitute.com/10-html/10-2.html

## TEACHING TIME FOR THIS MODULE

An outline for use in developing your lesson plan is presented below. Note that each Roman numeral in the outline equates to one session of instruction. Each session has a suggested time period of 2½ hours. This includes 10 minutes at the beginning of each session for administrative tasks and one 10-minute break during the session. Approximately 20 hours are suggested to cover *Installing Mechanical Seals.* You will need to adjust the time required for hands-on activity and testing based on your class size and resources. Because laboratories often correspond to Performance Tasks, the proficiency of the trainees may be noted during these exercises for Performance Testing purposes.

| Topic | Planned Time |
|---|---|
| **Session I. Introduction; Basic Design** | |
| A. Introduction | ____________ |
| B. Basic Design | ____________ |
| **Sessions II and III. Mechanical Seal Classifications** | |
| A. Classifying Mechanical Seals by Arrangement | ____________ |
| B. Classifying Mechanical Seals by Design | ____________ |
| C. Laboratory<br>Trainees practice identifying types of mechanical seals. This laboratory corresponds to Performance Task 1. | ____________ |
| **Sessions IV through VI. Replacing Mechanical Seals** | |
| A. Removing Mechanical Seals | ____________ |
| B. Inspecting Mechanical Seals | ____________ |
| C. Installing Mechanical Seals | ____________ |
| D. Laboratory<br>Trainees practice removing, inspecting, and installing mechanical seals. This laboratory corresponds to Performance Task 2. | ____________ |
| **Session VII. Module Review** | |
| A. Module Review | ____________ |
| **Session VIII. Testing** | |
| A. Module Examination<br>1. Trainees must score 70% or higher to receive recognition from NCCER.<br>2. Record the testing results on Craft Training Report Form 200, and submit the results to the Training Program Sponsor. | ____________ |
| B. Performance Testing<br>1. Trainees must perform each task to the satisfaction of the instructor to receive recognition from NCCER. If applicable, proficiency noted during laboratory exercises can be used to satisfy the Performance Testing requirements.<br>2. Record the testing results on Craft Training Report Form 200, and submit the results to the Training Program Sponsor. | ____________ |

15305-08

# Installing Mechanical Seals

Assign reading of Module 15305-08.

# 15305-08
# *Installing Mechanical Seals*

*Topics to be presented in this module include:*



## Overview

In this module, you will learn about the types of mechanical seals. You will learn how the seals work, and how the seals must be installed to work effectively. These are the dynamic structures that make O-rings and simple seals work in machinery.

*Instructor's Notes:*

## Objectives

When you have completed this module, you will be able to do the following:

1. Identify and explain types of mechanical seals.
2. Explain mechanical seal classification.
3. Remove, inspect, and install mechanical seals.

## Trade Terms

| | |
|---|---|
| Carcinogenic | Rotating face |
| Cartridge mount | Seal face loading |
| Deflection | Stationary face |
| Flashing | Stuffing box |
| Fretting | Thermal spike |
| Leaching | Toxic |
| Nonpilcting | Trace element |
| Perimeter | Volatile |

## Required Trainee Materials

1. Pencil and paper
2. Appropriate personal protective equipment

## Prerequisites

Before you begin this module, it is recommended that you successfully complete *Core Curriculum; Millwright Level One; Millwright Level Two;* and *Millwright Level Three*, Modules 15301-08 through 15304-08.

This course map shows all of the modules in the third level of the *Millwright* curriculum. The suggested training order begins at the bottom and proceeds up. Skill levels increase as you advance on the course map. The local Training Program Sponsor may adjust the training order.

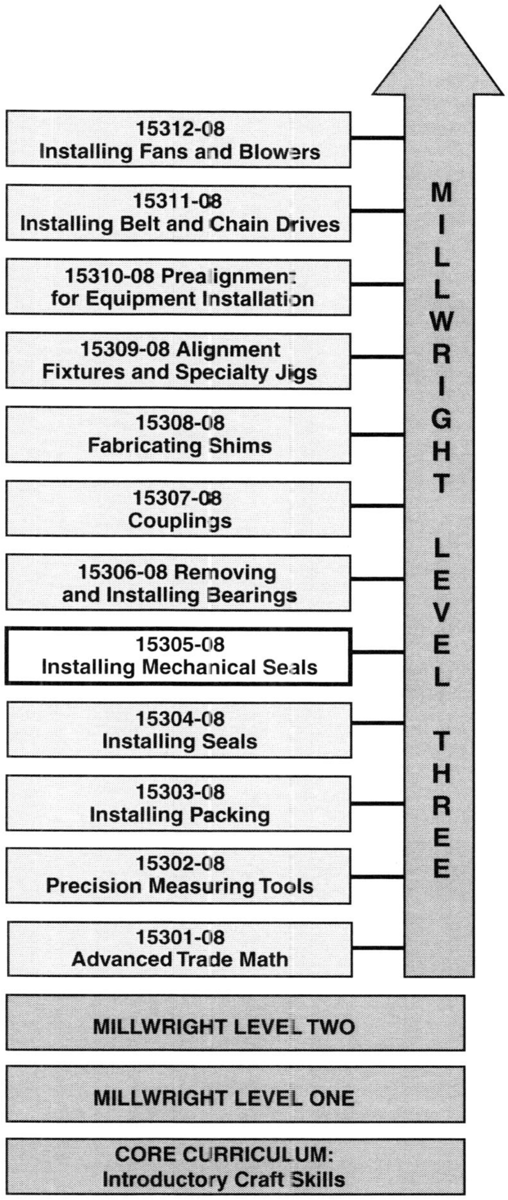


305CMAP.EPS

**Ensure that you have everything required to teach the course. Check the Materials and Equipment list at the front of this module.**

**See the general Teaching Tip at the end of this module.**

**Explain that terms shown in bold are defined in the Glossary at the back of this module.**

**Show Transparency 1, Objectives, and Transparency 2, Performance Tasks. Review the goals of the module, and explain what will be expected of the trainees.**

**Review the modules covered in Level Three and explain how this module fits in.**

**See the Teaching Tip for Sections 1.0.0–4.0.0 at the end of this module.**



MODULE 15305-08 ◆ INSTALLING MECHANICAL SEALS 5.1

Discuss how the Clean Air Act led to the development of the mechanical seal.

Explain the advantages and disadvantages of mechanical seals.

Describe how mechanical seals function.

Discuss the criteria to consider when selecting a mechanical seal.

Explain that a common application for a mechanical seal is a centrifugal pump.

Provide a centrifugal pump for trainees to examine. Explain the role of the mechanical seal in the pump.

## 1.0.0 ◆ INTRODUCTION

Traditionally, shaft seals for pumps have consisted of rope or braided style packing in a **stuffing box** around the shaft. A series of packing rings were installed into the pump stuffing box and they were compressed tightly so that they created a makeshift seal around the shaft. A certain amount of leakage was expected, and the unit could be tightened until the packing failed, and then replacement was necessary. This leakage resulted in loss of product, excessive pollution, and unnecessary hazards to the environment and anyone working around it. The Clean Air Act amendments of 1990 made leaky stuffing boxes unacceptable for use in the United States.

The mechanical seal was developed to replace the compression packing in stuffing boxes and alleviate the problems that it caused. Mechanical seals offer a dramatic reduction in leakage and, when maintained in good condition, cause no wear on the shaft. However, a higher level of skill is required to install the seals, and the initial installation costs may be higher.

This module will introduce you to the basic design of mechanical seals, the classification system for mechanical seals, and the various types of mechanical seals that fall under these classifications. General procedures for removing, inspecting and installing mechanical seals are also presented.

## 2.0.0 ◆ BASIC DESIGN

Although mechanical seals are made in many varieties, they all function the same way, using two or four contacting faces or sealing surfaces. One surface is stationary as the other rotates, and they are lubricated and cooled by a film of fluid. The seal faces are forced together by springs, by bellows, or sometimes by fluid pressure (*Figure 1*).

Mechanical seals consist of four basic sets of parts:

- A set of primary faces, one that rotates and one that is stationary.
- A closing mechanism that provides the load required to keep the seal faces in contact with each other. These are usually multiple springs, a single spring, or metal bellows.
- A set of secondary seals, such as shaft packing and insert mounting, which includes O-rings, V-rings, wedges, or U-cups.
- Mechanical seal hardware, including gland rings, collars, compression rings, pins, and spring.

The primary seal is achieved by two very flat, lapped faces that are held tightly together by springs, elastomers, or metal bellows. Seal design and composition are selected based on a number of operating conditions including:

- Chemical exposure
- Cold temperatures
- Heat exposure
- Lubrication
- Shaft speed
- Availability

Each seal application determines the best type of mechanical seal to use. A very common application for a mechanical seal is a centrifugal pump for liquids (*Figure 2*).

In the simplest terms, a centrifugal pump is a shaft suspended on bearings, with an impeller attached to one end. The impeller is encased in a housing that is filled with a liquid. As the shaft rotates the impeller, centrifugal force pushes the liquid out through an opening, where it is typically piped into a process or to another collection point. As the liquid exits the case, additional liquid is added to the case so that a flow develops.

305F01.EPS

*Figure 1* ◆ Typical mechanical seal.

305F02.EPS

*Figure 2* ◆ Pump with mechanical seal.



Instructor's Notes:

5.2 MILLWRIGHT ◆ LEVEL THREE

The force of the moving liquid creates pressure. This liquid under pressure will naturally try to move to areas of lower pressure. This is the physical principle of hydraulics. A mechanical seal is used between the rotating shaft and the point where the shaft enters the impeller housing. This seal prevents the liquid from leaking around the shaft and escaping from the impeller housing.

## 3.0.0 ◆ MECHANICAL SEAL CLASSIFICATIONS

Mechanical seals are classified by arrangement and design. Seal classification is very important when selecting a seal for a particular application. Although seal selection is usually performed by the design engineer, millwrights should be familiar with seal design classification. The arrangements deal with single or double seals, inside or outside seals, and other external arrangements of the seal type. By design, seals are classified by internal construction according to their function and application, such as O-rings, V-rings, or U-cup; closing mechanism; or whether it is balanced or unbalanced. The seal features of both design and arrangement are combined to provide specific characteristics for each application (*Figure 3*).

### 3.1.0 Classifying Mechanical Seals by Arrangement

Seals classified by arrangement include single sealing units and multiple sealing units. The single sealing units are designed to be installed either inside or outside of the stuffing box. The multiple sealing units can be classified as double back-to-back seals, double face-to-face seals, or tandem seals. Cartridge seals are also available in single, double and tandem arrangements. These are self-contained units consisting of a shaft sleeve, seal, and gland plate. The unit is fitted onto the pump shaft as a complete assembly, and no further fitting is required.

#### *3.1.1 Single Inside Mechanical Seals*

Inside seals (*Figure 4*) are the most common type of mechanical seal. They are installed inside the stuffing box where fluid pressure assists in sealing. In this type of seal, the **stationary face** is held in place by the stuffing box gland. The rotating seal face fastens to the shaft by setscrews on a collar. The sealing point is between the two highly polished faces, or rings. One ring is usually made of carbon, and the other is usually ceramic, tung-

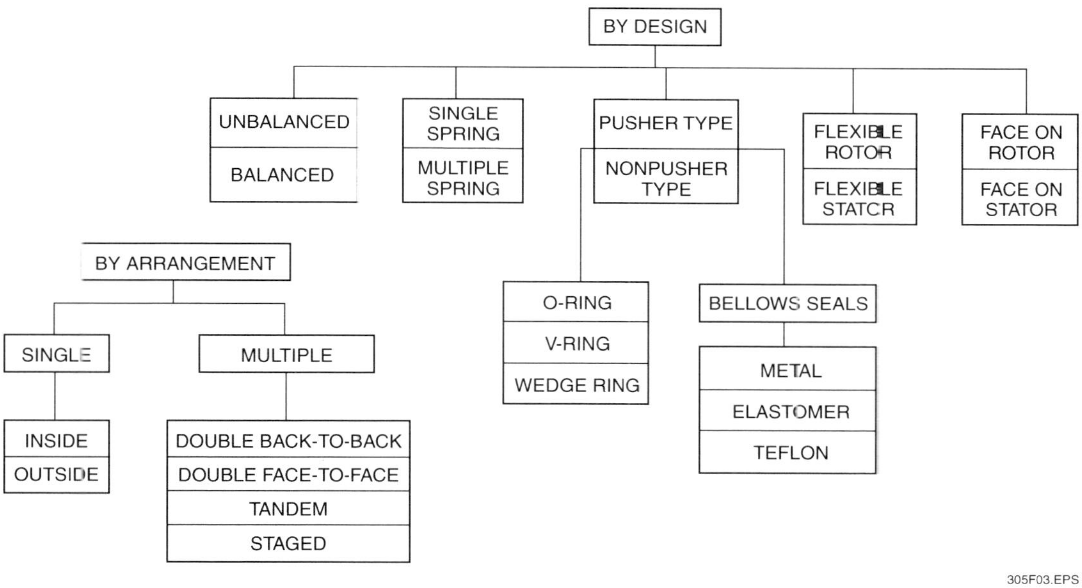


*Figure 3* ◆ Mechanical seal classifications.



**Have trainees review Sections 3.0.0–3.2.8.**

**Ensure that you have everything required to teach this session.**

**Explain that seal classification is based on arrangement and design.**

**Show Transparency 3 (Figure 3). Use the diagram to explain how mechanical seals are classified.**

**Explain that seals classified by arrangement include single sealing units and multiple sealing units.**

**Describe cartridge seals.**

**Explain how inside seals function.**

**Provide inside seals for trainees to examine.**

MODULE 15305-08 ◆ INSTALLING MECHANICAL SEALS 5.3

**Discuss common applications of outside mechanical seals.**

**Provide outside seals for trainees to examine.**

**Discuss applications where double mechanical seals are commonly used.**

**Explain how back-to-back and face-to-face seals function.**

**Discuss tandem mechanical seals and how they are used.**

**Provide tandem seals for trainees to examine.**

sten carbide, or silicon carbide. The two faces are held together by a single spring, or multiple springs, and a collar. O-rings and a gland gasket provide a secondary seal. In inside seals, the liquid being sealed is also used to lubricate the seal. Adjustments require dismantling the equipment unless the seal is **cartridge mounted**.

### *3.1.2 Single Outside Mechanical Seals*

Outside mechanical seals (*Figure 5*) are single seals that are installed outside the stuffing box. They are commonly used in systems that contain highly corrosive fluid because only the insert, seal ring, and secondary seals are exposed to the fluid. Outside seals are used when the stuffing box is too small for an inside seal. They are also easily accessible for troubleshooting and adjustment. Because the pressure of the fluid in the stuffing box tends to open the seal faces rather than close them as in an inside seal, an outside seal is limited to applications of low to moderate system pressures.

### *3.1.3 Double Mechanical Seals*

Double mechanical seals are used for applications in which double protection against leakage is needed, such as with **toxic** chemicals where leakage would be dangerous to the environment, liquids having abrasive properties that would cause rapid wear of the faces, or corrosive liquids that require seals made of expensive materials. The two types of double seals are back-to-back seals and face-to-face seals.

A back-to-back double seal consists of an inner seal and an outer seal, both using a common collar. This type of seal uses a sealing liquid, usually water, injected between the seals to prevent leakage from the system. This liquid, called a buffer fluid, must be maintained at a pressure that is higher than the pressure of the fluid in the system. The buffer fluid prevents the system fluid from coming in contact with the inner part of the seals and provides lubrication to the seal faces. The inner seal prevents the buffer fluid from entering the system, and the outer seal prevents the buffer fluid from leaking to the atmosphere. The life of a double back-to-back seal is many times more than a single seal in extreme conditions.

Face-to-face double seals are used when the stuffing box is too shallow to accommodate a back-to-back seal. This seal is usually cartridge-mounted, with one seal inside the stuffing box and the other seal outside the stuffing box. If the buffer fluid in a double seal is maintained at a pressure higher than system pressure, the buffer fluid acts as a lubricant for the inner seal. If the buffer fluid pressure is lower than system pressure, the purpose of the inner seal is the same as any single seal, and the outside seal acts as a backup seal.

### *3.1.4 Tandem Mechanical Seals*

Tandem mechanical seals (*Figure 6*) are used for extra leak protection in systems that contain products such as vinyl chloride, carbon monoxide, light hydrocarbons, or other **volatile**, toxic, **carcinogenic**, or otherwise hazardous fluids. A tandem seal is actually two seals, an inner seal and an outer seal, installed in tandem. The inner, or primary, seal is designed to handle full system pressure. The outer, or secondary, seal acts as a backup in case the inner seal fails. This type of seal is often installed with an alarm system that produces a warning if the inner seal fails. The outer seal takes over the

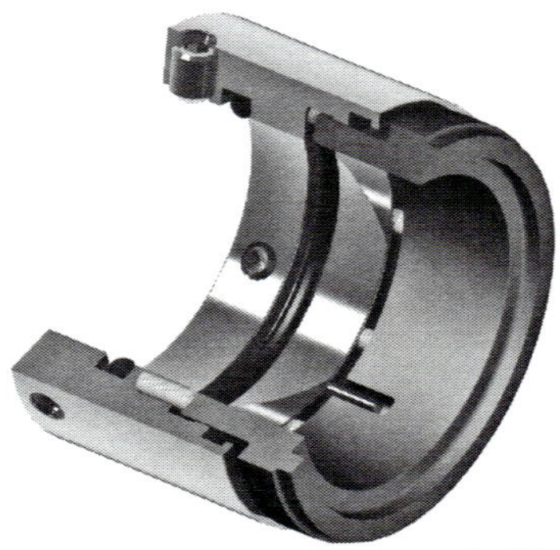

305F04.EPS

*Figure 4* ◆ Inside seal.

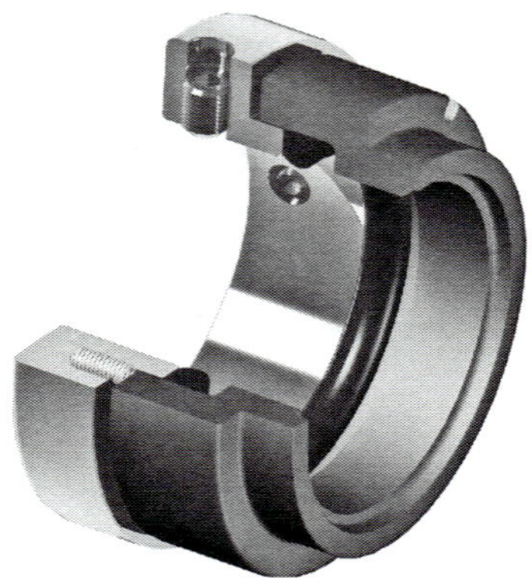

305F05.EPS

*Figure 5* ◆ Outside seal.



Instructor's Notes:

5.4 MILLWRIGHT ◆ LEVEL THREE

sealing function until the equipment can be shut down and the inner seal repaired. Tandem seals increase on-line reliability and leak protection.

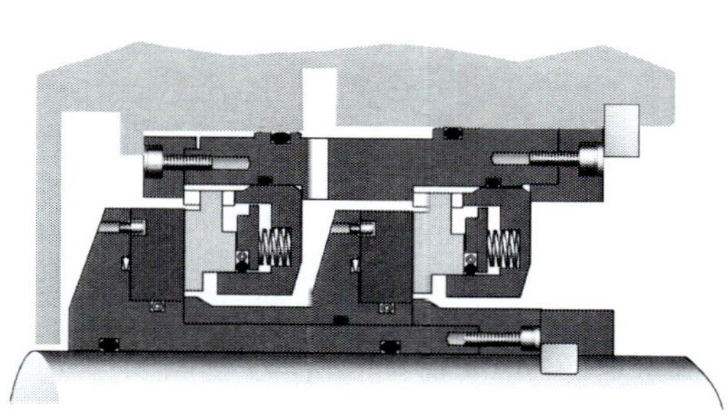


*Figure 6* ◆ Tandem seals.

### 3.1.5 Cartridge Mechanical Seals

Cartridge seals (*Figure 7*) are set at the factory and are installed as complete assemblies. The seal head, along with a stationary seat mounted in a gland plate, is assembled in its set operating length on a sleeve so that no setting measurements are needed. The assembly is mounted on the pump shaft or sleeve and then bolted to the pump body.

**Discuss cartridge seals. Explain that they are installed as complete assemblies.**

**Provide cartridge seals for trainees to examine.**

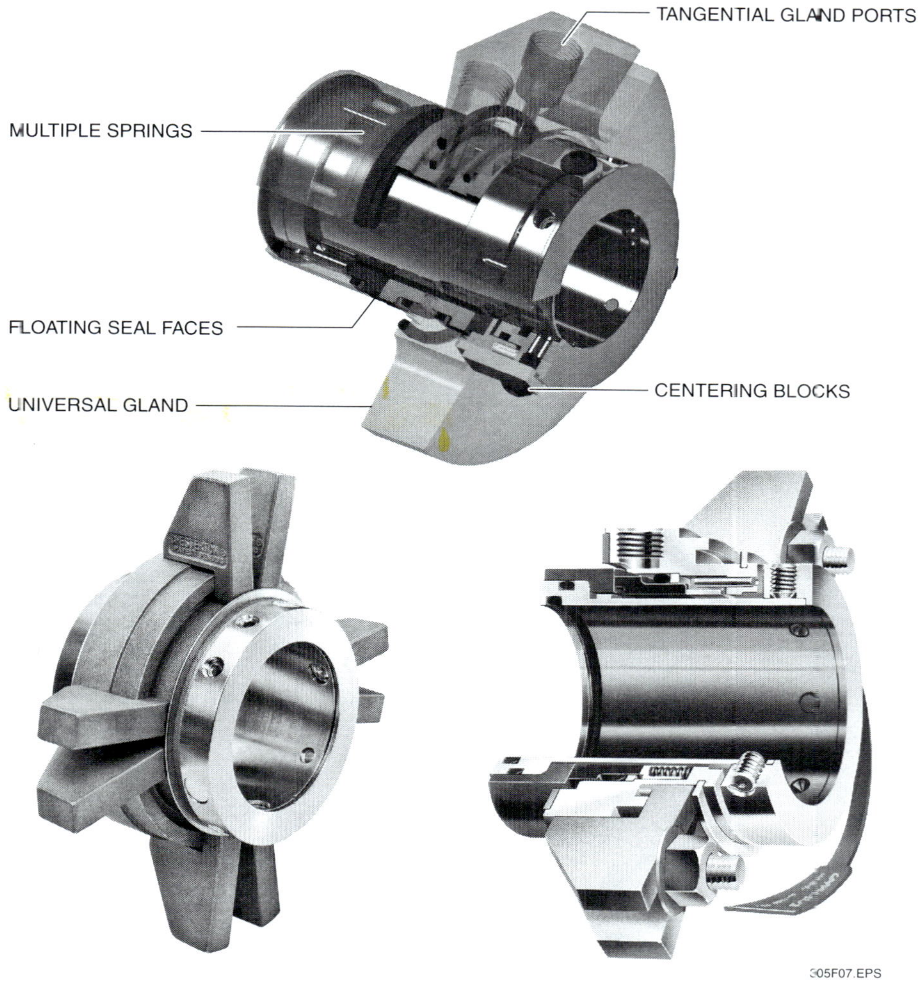


*Figure 7* ◆ Cartridge seal.



MODULE 15305-08 ◆ INSTALLING MECHANICAL SEALS 5.5

Discuss the advantages and disadvantages of cartridge seals.

Review the list of common seal classifications by design.

Discuss the function of balanced seals.

Compare unbalanced seals to balanced seals.

Discuss single-spring seals and their limitations.

Show Transparency 4 (Figure 9).

Provide balanced, unbalanced, and single-spring seals for trainees to examine.

Cartridge seal assemblies come with centering tabs or spacers that are used to center the seal during installation. The tabs are removed after the seal assembly is bolted in place, but should be kept so that they can be reused when the seal is reinstalled after repairs or adjustments.

Another advantage of using cartridge seals is the ability to standardize the seals used since almost all mechanical seals are available in cartridge design. This way, so many different types of seals do not have to be kept in stock. The main drawback of cartridge seals at this time is that older seal chambers designed for compression packing are too small for the cartridge seal units. The cartridge seal has a sleeve, which adds ⅛ to ¼ inch to the diameter; therefore, the chamber must be measured before the seal is installed to ensure that the inside diameter of the stuffing box is large enough.

## 3.2.0 Classifying Mechanical Seals by Design

By design, seals are classified by construction according to function and application, such as type of seal, closing mechanism, or whether it is balanced or unbalanced. The following are some of the most common seal classifications by design:

- Balanced
- Unbalanced
- Single-spring
- Multiple-spring
- Welded metal bellows
- Elastomer bellows
- Rotating
- Stationary

### *3.2.1 Balanced Seals*

The purpose of the balanced seal design (*Figure 8*) is to optimize **seal face loading** so that the faces stay closed without excessive force and so that a thin film of pumped product remains between the face for lubrication. Without this lubrication, friction increases and causes excessive heat, wear, and eventual seal failure. The balance is accomplished by decreasing the area exposed to the fluid, which changes the pressure. Since the system pressure and the spring load remain constant, the face area must be changed to balance the seal. This is done by machining a step in the pump shaft or sleeve to reduce the area of the seal face that is exposed to the pump system pressure. Refer to the manufacturer's manual for instructions on machining a step. The advantage of the balanced seal is that it can be used with higher pressures and velocities than unbalanced seals that have greater seal force.

### *3.2.2 Unbalanced Seals*

In mechanical seals, the amount of seal face leakage is inversely proportional to the amount of seal face loading. The higher the loading of the face, the lower the leakage. Unbalanced seals have a higher face loading than balanced seals and therefore leak less and are more stable than balanced seals when subjected to cavitation, misalignment, or vibration. Unbalanced seals cost less and are more adaptable to standard stuffing boxes than balanced seals.

One disadvantage of an unbalanced seal is that it has a low pressure and velocity limit due to the excessive closing force. If the closing force on the seal exceeds the pressure limit, the lubrication film between the faces is pushed out, causing the faces to wear rapidly.

### *3.2.3 Single-Spring Seals*

Springs provide the force that holds the seal faces together in a mechanical seal. Many seals have a single-coil spring (*Figure 9*) that fits around the shaft against a collar and the rotating seal face. This type of seal is simply designed and easy to use. Its main limitations are the radial space it requires, its tendency to distort at high speeds, and the need for a different spring for each size seal.

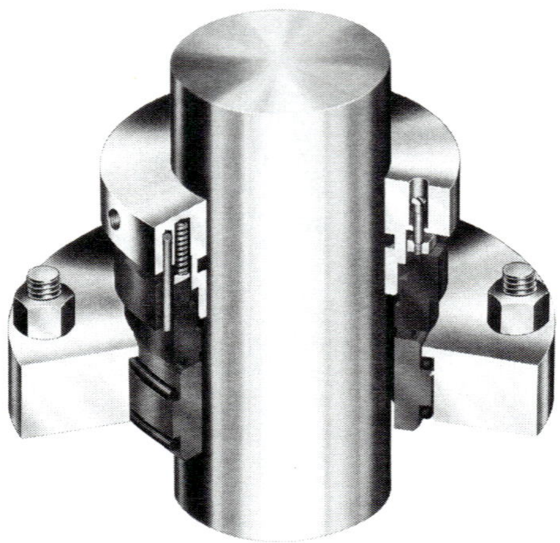

305F08.EPS

*Figure 8* ◆ Balanced seal.



Instructor's Notes:

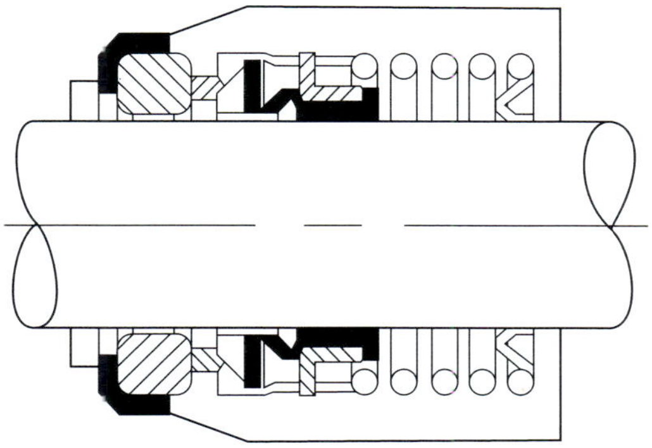

305F09.EPS

*Figure 9* ◆ Single-spring seal.

### 3.2.4 Multiple-Spring Seals

Multiple-spring mechanical seals are used on many high-speed applications because they are not as susceptible to distortion at high speeds. Instead of a single spring that fits around the shaft, the multiple-spring seal has a number of small-coil springs spaced around the **perimeter** of the seal. The multiple springs exert even pressure on the seal face at all times. The same size springs can be used with a variety of shaft size, so it is not necessary to stock a large number of different-sized springs for different-sized seals.

### 3.2.5 Welded Metal Bellows Seals

The welded metal bellows seal has a bellows instead of springs to provide the force that holds the seal faces together. Welded metal bellows seals provide even pressure to all points on the seal ring. They do not require a secondary seal on the shaft, so problems with shaft packing hang-up are eliminated. Welded metal bellows seals are balanced seals and have fewer parts than spring seals.

### 3.2.6 Elastomer Bellows Seals

The elastomer bellows seal has a single-coil spring that fits over the shaft and pushes against an elastomer bellows. Like the welded metal bellows seal, the elastomer bellows seal does not need a secondary seal on the shaft. Elastomer bellows seals are low in cost and are available in standard sizes and in balanced and unbalanced designs.

The disadvantages of the elastomer bellows seal are that it has the same speed limitations of any single-spring seal, and it is not available in special sizes. It also presents some additional repair problems because the elastomer tends to stick to the shaft.

### 3.2.7 Rotating Seals

Mechanical seals are sometimes classified by the position of the springs in the seal. In a rotating seal (*Figure 10*), the springs are located on the rotating part of the seal inside the stuffing box. In this type of seal, the springs apply pressure to the **rotating face** and push it against the stationary face.

### 3.2.8 Stationary Seals

In a stationary seal (*Figure 11*), the springs apply pressure to the stationary face of the seal and push it against the rotating face. Stationary seals are outside mechanical seals.

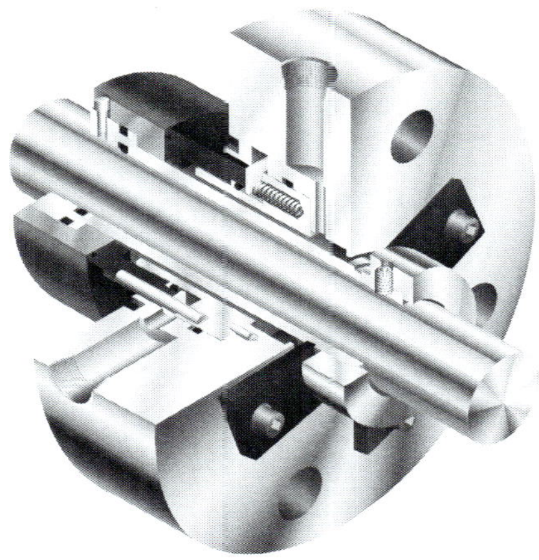

305F10.EPS

*Figure 10* ◆ Rotating seal.

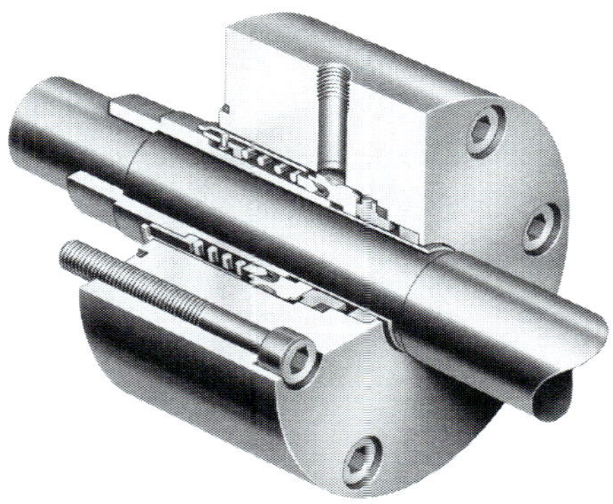

305F11.EPS

*Figure 11* ◆ Stationary seal.

**Discuss the use of multiple-spring, welded metal bellows, and elastomer bellows mechanical seals.**

**Provide samples of these seals for trainees to examine.**

**Discuss rotating and stationary seals. Provide samples of these seals for trainees to examine.**

**See the Teaching Tip for Section 3.0.0 at the end of this module.**

**Have trainees practice identifying types of mechanical seals. Note the proficiency of each trainee. This laboratory corresponds to Performance Task 1.**



MODULE 15305-08 ◆ INSTALLING MECHANICAL SEALS 5.7

**Have trainees review Sections 4.0.0–4.3.0.**

**Ensure that you have everything required to teach this session.**

**Explain that it is necessary to determine the cause of failure when replacing a seal so that the problem can be fixed.**

**Stress the importance of following the manufacturer's repair manual when removing a mechanical seal.**

**Review the steps for removing a mechanical seal from a pump.**

**Explain how to inspect a failed seal to determine the cause of failure.**

**Review the list of problems often found when inspecting seals, and the likely causes of these problems.**

## 4.0.0 ◆ REPLACING MECHANICAL SEALS

A mechanical seal should last until the carbon face has worn away, but approximately 85 percent of the time, this is not the case. It is important to try to determine if a seal has failed prematurely and to determine the cause of the failure. Determining the reason for seal failure is a major objective when removing the seal. Replacing mechanical seals involves removing and inspecting the old seal and installing a new seal.

### 4.1.0 Removing Mechanical Seals

It is very important to follow the manufacturer's repair manual when removing a mechanical seal to avoid damaging the shaft or stuffing box. Follow these steps to remove a mechanical seal from a pump:

*Step 1* Remove all safety covers from the pump and motor.

*Step 2* Examine the coupling to ensure that it has not been damaged.

*Step 3* Gauge the coupling gap to ensure that it is within manufacturer's specifications.

*Step 4* Inspect the motor-to-pump alignment to ensure that it is correct.

*Step 5* Disconnect the pump-to-motor coupling.

*Step 6* Turn the motor shaft by hand to determine if there are rough spots or looseness that indicate motor bearing problems.

*Step 7* Examine barrier fluid piping to the seal to ensure that the piping is not bent or damaged and that it is properly connected.

Step 8 Take precautions to ensure that the barrier fluid gauges are not damaged.

*Step 9* Disassemble the pump to the point at which the seal can be removed.

*Step 10* Remove the seal from the pump shaft.

### 4.2.0 Inspecting Mechanical Seals

All parts and surfaces of a seal should be inspected closely to determine the reason for premature seal failure when it is removed. If the cause of the premature failure can be understood, it can be eliminated. All parts of the seal should be inspected.

Follow these guidelines to inspect mechanical seals:

- Check the seal faces for chipped edges. Chipped edges on the seal faces are commonly caused by the faces separating and slamming back together due to **flashing** of the liquid in a system. When water flashes from a liquid to a gas, it expands and can cause the faces to separate. The remedy for flashing is to reduce face heat by changing the seal type or some other variable.
- Check the hard, stationary seal face for flaking or peeling, which indicate a defective seal face coating or a chemical attack at the face bond. The chemical attack was probably caused by excessive heat at the face.
- Check the carbon face for signs of pitting, blistering, or corrosion, which indicate that the wrong grade or type of carbon was used in the seal.
- Check the stationary seal ring for worn spots, which indicate that the seal flush line is directed directly at the seal. The flush should come in at an angle, causing the fluid in the stuffing box to circulate.
- Check the hard face for deep wear, which indicates that particles are embedded in the carbon face and are grinding on the hard face. This occurs in outside seals, seals in misaligned pumps, and seals in severe abrasive service. It is caused by the faces separating and large particles getting between the faces.
- Check the hard face for a widened wear track, which indicates pump misalignment. This can be caused by bad bearings, shaft **deflection**, shaft whip, a bent shaft, bad coupling alignment, vibrations, severe pipe strain, or a tilted stationary seat.
- Check the hard face for a narrow wear track, which indicates that the seal has been overpressurized and has bowed away from the pressure. This causes the seal to be limited to only a portion of the face width.
- Check the hard seal for the absence of a wear track, which indicates that the rotating head is remaining stationary or is running against a surface other than the stationary seal. This can be caused by misalignment of the gland, an undersized shaft, or the sleeve not being properly attached to the shaft.
- Check the hard face for shiny spots, which indicate warping of the faces. Warping can be caused by excessive pressure, improper gland bolt tightening, or improper clamping of the seat.

5.8 MILLWRIGHT ◆ LEVEL THREE

Instructor's Notes:

- Check the hard face for an off-center wear track, which indicates that the stationary part of the seal is not properly aligned with the rotating part.
- Check the hard face for circular grooves, which indicate that particles between the seal faces have scratched the face.
- Check the hard face for heat cracking, which indicates incorrect plating of the hard face. These cracks will cut the softer carbon face.
- Check for broken springs or bellows.
- Check for clogged springs. Springs clog when the system fluid is dirty and the seal is not moving axially. Some multiple-spring seals become clogged very easily and should not be used in a dirty fluid without a clean flushing fluid.
- Check the shaft, the seal, the gland, and the stuffing box for signs of rubbing. Some situations that cause rubbing are the following:
  - Flushing lines protruding into the stuffing box
  - The **nonpiloting** gland slipping down and hitting the seal
  - Gaskets protruding into the seal cavity
  - Nonpiloting stationary rings coming in contact with the rotating shaft
  - Scale building up in the stuffing box
  - The stuffing box not being concentric to the shaft
  - Setscrews backed out and hitting the stuffing box
- Check for seals that look blistered, swollen, broken, or crumbled. These conditions can be caused by chemical attack or excessive temperature. This is the result of incorrect material selection, loss of barrier fluid, or contamination of barrier fluid. Excessive pressure can extrude and cut O-rings. In these cases, the pumped fluid should be chemically analyzed, the operating temperature checked, the **trace elements** analyzed, and the seal manufacturers' specifications compared to the collected data.
- Check for **leaching** through the seal faces, which causes some increase in leakage but a large increase in face wear. A good remedy is to upgrade to better face materials or to change to a design that allows the seal to be flushed with a suitable barrier fluid.
- Check for face distortion, which causes excessive leakage and shows an uneven wear pattern on the sealing faces. The usual cause is uneven torque on the gland nut, improper assembly, or **thermal spikes** that warp the sealing faces. The sealing faces should be replaced, or at least lapped, to remove distortions. It is good to check shaft and pump alignment and look for mistakes in equipment mounting.
- Check for seal face deflection, which is evident by a uneven or narrow wear band on the wide face. This is usually concave or convex. Causes may be improper stationary seal face support, swelling of seals, high pressure, or an imbalance of hydraulic and mechanical loads on the primary seal faces.
- Check for pump-to-motor misalignment. This misalignment is always present to some degree because of mounting imperfections, shaft runout, bearing problems, or shaft deflection. All seals are designed to compensate for this misalignment by allowing some back-and-forth movement, but in extreme cases, **fretting** corrosion will occur and cause leakage at secondary seals and damage the sleeve directly under the secondary seal.
- Check for heat checking on the seal ring, which appears as fine to large cracks that radiate from the center of the seal ring. These cracks cut and scrape the opposing seal ring and cause leakage. This can be caused by vaporization at the seal faces or excessive heat, pressure, or speed. Cooling the seal faces or decreasing the load can prevent heat checking.
- Check for blistering on the seal faces, which looks like small, circular, raised sections on the seal faces. These blisters separate the seal faces and cause leakage. Blistering occurs in seals that are stopped and started frequently in high-viscosity fluid applications. Changing the viscosity of the fluid, eliminating the frequent stops and starts, or changing seal face materials to tungsten carbide or bronze may help.

In all cases, it is good to check the seal design operating limits, consider other seal designs, consider a flexible mount for the stationary seal face, or replace the seal faces with a material more suited to the application.

Seals can be rebuilt. Not all companies do the rebuilding in house, and cartridge seals are usually either discarded or returned to the manufacturer. However, if necessary, seals can be rebuilt. Due to the requirements in relation to downtime, it is common to keep a spare on hand and to install it when the old seal is removed. Then the old seal is taken to the shop for rebuilding.

If you are rebuilding an old seal, it is common to replace the O-rings and, if possible, the springs, and sometimes to lap the face, to ensure that the seal will hold pressure. Certainly, the face must be checked for flatness.

**Provide examples of damaged seals and have trainees troubleshoot the probable cause of failure for each.**

**Discuss rebuilding damaged seals.**



MODULE 15305-08 ◆ INSTALLING MECHANICAL SEALS 5.9

**Define flatness. Explain how flatness can be measured using beams of light.**

**Stress the importance of keeping seals clean.**

**List the signs of excessive heat buildup between seals. Describe how these problems can be rectified.**

**Discuss conditions of mechanical seals that indicate misalignment, and probable causes of these conditions.**

### *4.2.1 Flatness*

Flatness, as it applies to seal faces, is the property of having a plane surface without elevations or depressions from that flat surface. In the case of seals that may have to prevent leakage of possibly toxic fluids, the mating surfaces must be flat enough to maintain a vacuum of one milligram of mercury.

The problem is that measuring flatness requires the ability to see deviations far beyond the capacity of the human eye. The solution is to use a property of light. If two light beams are at the same frequency but synchronized so that they are half a wavelength apart, the beams cancel each other out. The result is a dark line because neither beam is visible. This property, of the two waves interfering with each other's light, is called interference.

If an optically flat lens (a lens that is flat on one side and as accurately made as a lens from a pair of binoculars) is laid on a flat surface and a light that has one color (monochromatic) is shined upon it, the reflection from a flat surface will produce interference patterns, depending on the flatness of the surface on which the lens is resting. The goal is to see not more than three interference lines.

The light can be of any single wavelength, which means it will be one color. The most commonly used color is a pink color from a helium gas light source. This has a wavelength of approximately 0.6 microns. If the distance between the optically flat glass and the surface beneath is 0.3 microns, the waves would interfere, producing a line. To understand how small a distance this is, the smallest object the human eye can see is approximately 40 microns. An ordinary piece of newsprint, such as is used in a newspaper, might be 70 microns thick.

It is very important that the seal be kept very clean, since any dirt that may intrude will cause wear and loss of sealing. Any flexible members of the seal must be examined and considered for replacement, especially if there are any signs of fatigue or wear.

### *4.2.2 Hard Face Area*

Inspect the hard face of stationary seals for the following indicators of problems that may have caused the seal to fail at an early date. Excessive heat buildup between the seals may cause one or more of the following signs:

- Chipped edges
- Flaking
- Peeling
- Cracking (from heat)

These problems may sometimes be solved by changing the grade or type of seal being used to one of a different composition. Other changes can sometimes be made to cause a change to some other variable in the seal area.

Leaching through seal faces causes a negligible increase in leakage but a large increase in the amount of wear on the face. Selecting a better face material or changing to a design that allows the seal to be flushed with a suitable barrier fluid normally alleviates this condition.

Heat checking on the seal ring usually appears as fine to large cracks that radiate from the center of the seal ring outward. These cracks cut and scrape the opposing seal ring and cause leakage. This can be caused by vaporization at the seal faces due to excessive heat, pressure, or speed. Cooling the seal faces or decreasing the load can prevent heat checking.

Blistering on seal faces appears as small, circular, raised areas on the seal faces. These raised blister areas keep the seal faces separated and cause leakage. Blistering occurs in seals that are stopped and started frequently in high-viscosity fluid applications. Changing the viscosity of the fluid, eliminating the frequent stops and starts, or changing seal face materials to tungsten carbide or bronze should be considered.

In all cases, the manufacturer's recommendations and seal design limits should be studied. Other seal designs should be considered in order to find the best replacement.

### *4.2.3 Mechanical Failures*

The following group of conditions reveals problems that are of a more mechanical nature. The indications show that the seal has not been in proper alignment for some reason. The major indication is that clearance is not set properly in some area and the severe wear (or lack thereof) displayed has been built into the setup. If the seal was not properly aligned or there is excessive wear in some bearing or shaft, unusual wear tracks will be the indicator. If the seal was incorrectly tightened when installed, the seal was deformed, causing the unusual wear patterns. A very narrow wear track is indicative of the unit being overpressured, causing it to flex away from the face. Deep wear rings are probably due to the mating surfaces allowing abrasive material to enter the face areas. A piece of grit or metal can become imbedded in the carbon face of the seal and cut the ring deeply as it turns. A simplified list of signs of a problem with the installation would include:

- Areas or rings of deep wear



Instructor's Notes:

5.10 MILLWRIGHT ◆ LEVEL THREE

- Uneven or widened areas in the wear track
- Narrow wear track
- Absence of a wear track
- Shiny spots in the wear track

Worn spots on the stationary seal ring are an indication that the seal flush line is not adjusted correctly. The flush line should be directed toward the seal at an angle, causing the fluid in the stuffing box to circulate. Without this angle, the force of the flushing fluid will cause accelerated seal wear in one area. Manufacturer's recommendations must be followed.

The absence of a wear track indicates that the rotating head is remaining stationary or is running against a surface other than the stationary seal. This can be caused by misalignment of the gland, an undersized shaft, or the sleeve not being properly attached to the shaft.

Some units employ springs or bellows-type parts to assist in maintaining a mechanical seal. These parts must be inspected closely. Springs and bellows do break. However, most of the time, they become clogged with dirt or sludge because the system fluid has become dirty. The seal does not move axially when the springs become inactive. Some multiple-spring seals become clogged very easily. They should not be used in a system pumping a dirty fluid without using a clean flushing fluid.

### *4.2.4 Incorrect Seal Composition*

Carbon seals are available in a number of different grades and compositions. Damages indicating that an incorrect seal grade or composition may have been used include pitting, blistering, and corrosion. However, the type of service the pump was being used for may have changed and the fluid now being pumped might be incompatible with the composition of the original seals. It may be possible to select a more compatible seal material for the replacement.

### *4.2.5 Improperly Installed Components*

An incorrectly installed shaft, seal, gland, and/or stuffing box will result in components rubbing against other components during operation. If inspection reveals signs of rubbing, one or more of the following conditions could be the cause:

- Flushing lines protruding into the stuffing box
- The nonpiloting gland slipping down and hitting the seal
- Gaskets protruding into the seal cavity
- Nonpiloting stationary rings coming into contact with the rotating shaft
- Scale buildup in the stuffing box
- Stuffing box not being concentric to the shaft
- Setscrews backed out and hitting the stuffing box

### *4.2.6 Chemical Damage*

If inspection reveals that the seals appear to be blistered, swollen, broken, or crumbled, the conditions can indicate a chemical reaction and/or excessive temperature. The probable cause is that a seal of incorrect material was selected. The condition could also be the result of a loss of barrier fluid or contamination of barrier fluid. Excessive pressure could have extruded the O-rings, allowing them to be cut, releasing contamination into the barrier fluid.

In order to identify the cause, the fluid being pumped should be chemically analyzed, the operating temperature determined, and the trace elements analyzed. When this data has been collected, it should be compared to the seal manufacturer's specifications. Another type of seal should be selected for the repair.

### *4.2.7 Shaft Alignment*

A small amount of pump-to-motor misalignment is always present because of mounting imperfections, shaft runout, bearing problems, and shaft deflection. All seals are designed to compensate for this misalignment by allowing some back-and-forth movement. In extreme cases, fretting corrosion will occur and cause leakage at secondary seals and damage the sleeve directly under the secondary seal. A unit experiencing this condition may require inspection by an engineer if the cause is not immediately obvious Some major actions may be required to repair this condition.

A related condition that causes excessive leakage is revealed by an uneven wear pattern on the sealing faces. Seal face distortion can cause excessive leakage. Seal face distortion is revealed by an uneven wear pattern on the sealing faces. The usual cause is uneven torque on the gland nut, an improper assembly procedure, or a thermal spike that warps the seal face. The seal faces should be replaced, or at least lapped, to remove distortions. Examining shaft and pump alignment for possible mistakes in equipment mounting, and correcting them, could be beneficial in correcting this problem and preventing others in the entire assembly.

**Discuss mechanical failures due to misalignment, excessive wear, and clogging.**

**Discuss symptoms that indicate that an incorrect seal composition may have been used.**

**Discuss indications of improperly installed components.**

**Describe signs of chemical damage and excessive temperature.**

**Discuss causes and indications of shaft misalignment.**



MODULE 15305-08 ◆ INSTALLING MECHANICAL SEALS 5.11

**Review the rules for installing mechanical seals.**

**Explain that drawings and instructions must be studied before installing a mechanical seal.**

**Stress the importance of using required safety devices to prevent injury.**

**Review the steps for installing a mechanical seal on a pump.**

**Show Transparency 5 (Figure 12). Explain how to check end play.**

**Show Transparency 6 (Figure 13). Explain how to check shaft whip and deflection.**

**Show Transparency 7 (Figure 14). Explain how to check stuffing box shaft runout.**

**Show Transparency 8 (Figure 15). Explain how to check stuffing box face runout.**

**Show Transparency 9 (Figure 16). Explain how to check stuffing box concentricity.**

## 4.3.0 Installing Mechanical Seals

Following proper installation procedures for mechanical seals is critical to the life expectancy and efficiency of the seal. Mechanical seals are designed to provide many years of efficient service. There are four rules that apply when installing mechanical seals. These rules are:

- The equipment must be properly prepared to receive a mechanical seal.
- The seal installed must be right for the application.
- The seal must be properly installed according to the manufacturer's specifications.
- The necessary environmental controls must be installed and maintained.

There are many types and designs of mechanical seals, and the installation procedure for each type of seal is different. Each seal should come with manufacturer's drawings and installation instructions. Study these drawings and instructions until you are very familiar with them before installing the seal. Follow these steps to install a mechanical seal on a pump:

> **WARNING!**
> Ensure that all potential hazards are protected against. Install safety devices, tags, and/or restraints to prevent accidental startup of equipment to protect personnel and equipment from injury or damage.

*Step 1* Ensure that the pump stuffing box and the shaft have been thoroughly cleaned.

*Step 2* Inspect the stuffing box and the shaft to ensure that there are no burrs, nicks, scratches, or excessive wear. Any burrs or nicks found in the stuffing box or on the shaft must be dressed using an appropriate hone. If there are deep scratches or excessive wear, an evaluation must be made by an engineer as to the need for repairs before the seal is installed.

*Step 3* Measure the shaft or sleeve diameter using a micrometer and ensure that it is the proper size according to the seal installation specifications.

*Step 4* Measure the inside diameter of the stuffing box using a micrometer and ensure that it is within the seal installation specifications.

*Step 5* Attach a dial indicator to the pump so that its stem touches the shoulder of the shaft.

*Step 6* Tap the shaft on one end and then the other, using a soft mallet, to check the shaft end play (*Figure 12*). End play should be within the manufacturer's specifications.

*Step 7* Attach the dial indicator to the pump so that the stem touches the side of the shaft near the back of the stuffing box.

*Step 8* Push down lightly on the impeller end of the shaft, and note any radial movement that registers on the indicator. Radial movement should be no more than 0.002 to 0.003 inch. Radial movement of the shaft will cause shaft whip and deflection (*Figure 13*).

*Step 9* Attach the dial indicator to the pump housing so that the stem touches the shaft.

*Step 10* Turn the shaft by hand, and note the shaft runout that registers on the dial indicator (see *Figure 14*). Shaft runout should be checked at two or more points on the shaft. Shaft runout indicates a bent shaft and should be no more than 0.0015 to 0.003 inch.

*Step 11* Ensure that the stuffing box is bolted in place, and attach the dial indicator to the shaft so that the stem touches the face of the stuffing box.

*Step 12* Turn the shaft by hand, and note any runout that registers on the dial indicator (see *Figure 15*). This measurement indicates stuffing box face runout. Runout should be no more than 0.003 inch.

*Step 13* Attach the dial indicator to the shaft so that the stem touches the inside of the stuffing box.

*Step 14* Turn the shaft by hand, and note any deviation in the dial indicator reading. This reading indicates the concentricity of the stuffing box bore. The stuffing box should be concentric to the shaft within 0.005 inch (*Figure 16*).

*Step 15* Remove all burrs and sharp edges from the shaft, including edges of keyways and threads, using an emery cloth or a hone.



Instructor's Notes:

5.12 MILLWRIGHT ◆ LEVEL THREE

*Step 16* Check the stuffing box bore and face to ensure that they are free of burrs.

*Step 17* Make any required adjustments to the pump. On some pumps, like the open impeller pump, initial impeller adjustments must be made before the seal is installed.

*Step 18* Oil the shaft lightly.

*Step 19* Install the seal according to the manufacturer's drawings and instructions.

- Keep the seal and all of its components clean and protected from damage.
- Do not touch the seal faces with your hands. Handle the seals by the edges.
- Ensure that the seal is compressed to the proper face load.
- Maintain concentricity of the seal to the shaft.
- Follow the proper bolt-tightening procedure.

**NOTE**

It is very important to follow the manufacturer's drawings and instructions exactly to ensure a successful installation.

*Step 20* Reassemble the pump.

*Step 21* Connect all seal piping to the pump.

*Step 22* Install the pump on its base.

*Step 23* Align the pump to the motor.

**CAUTION**

Proper alignment is critical to the operation, life expectancy, and efficiency of the seal.

*Step 24* Replace all protective covers.

*Step 25* Recommission pump per company procedures. If barrier buffer fluid is used, turn it on and adjust it to the proper pressure and flow.

*Step 26* Test-run the pump and monitor it for noise, vibration, and excessive heat.

**CAUTION**

Follow standard operating procedures when running the equipment to avoid damaging the equipment or the seal.

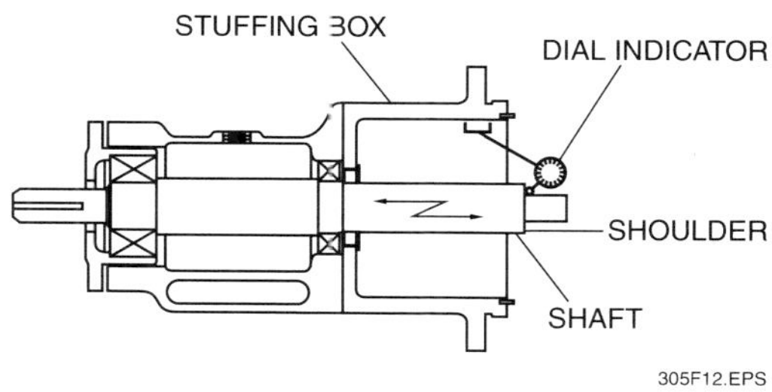


*Figure 12* ◆ Checking end play.

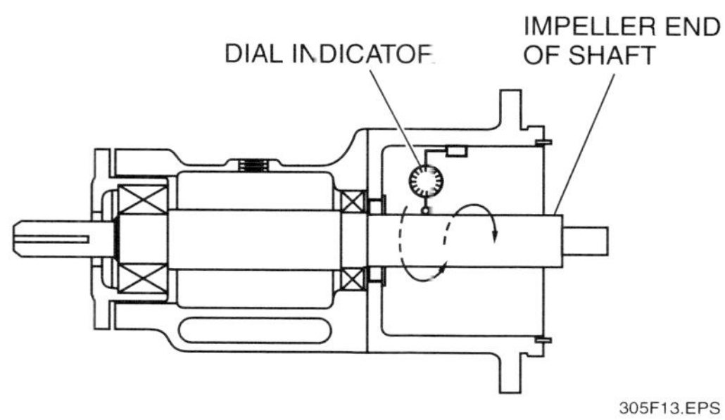


*Figure 13* ◆ Checking shaft whip and deflection.

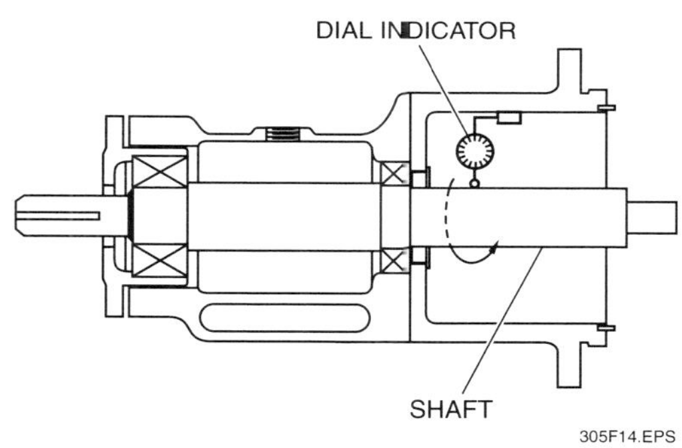


*Figure 14* ◆ Checking stuffing box shaft runout.

**Review and demonstrate how to remove, inspect, and install mechanical seals.**

**Have trainees practice removing, inspecting, and installing mechanical seals. Note the proficiency of each trainee. This laboratory corresponds to Performance Task 2.**

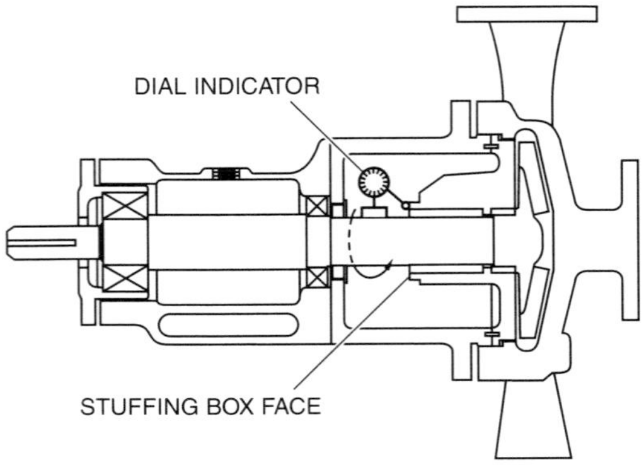


*Figure 15* ◆ Checking stuffing box face runout.

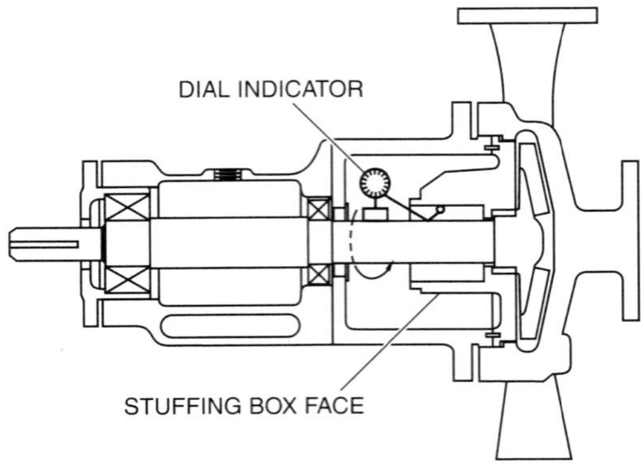


*Figure 16* ◆ Checking the stuffing box concentricity.



Instructor's Notes:

5.14 MILLWRIGHT ◆ LEVEL THREE

## Review Questions

1. Mechanical seals are often used in place of _____ in a stuffing box.
   a. O-rings
   b. gaskets
   c. packing
   d. glands

2. A higher level of skill is required to install mechanical seals, and the initial installation costs may be higher than using compression packing.
   a. True
   b. False

3. The most common type of mechanical seal is the _____ mechanical seal.
   a. inside
   b. outside
   c. double
   d. tandem

4. In an inside mechanical seal, the rotating face fastens to the _____.
   a. gland
   b. stuffing box housing
   c. shaft
   d. collar

5. In an inside mechanical seal, _____ provide a secondary seal.
   a. O-rings and a gland gasket
   b. glands
   c. collars
   d. springs

6. The seal that is commonly used in systems that contain highly corrosive fluid is the _____ mechanical seal.
   a. inside
   b. outside
   c. double
   d. tandem

7. The seal that is limited to applications of low to moderate pressures is the _____ mechanical seal.
   a. inside
   b. outside
   c. double
   d. tandem

8. The seal that is used for extra leak protection with toxic chemicals is the _____ mechanical seal.
   a. inside
   b. outside
   c. double
   d. single-spring

9. The double seal in which two seals use a common collar is the _____ seal.
   a. back-to-back
   b. face-to-face
   c. side-to-side
   d. face-to-back

10. The type of seal that is often installed with an alarm in case the inner seal fails is the _____ seal.
    a. inside
    b. outside
    c. double
    d. tandem

11. The seal that does not require the usual setting measurements for installation is the _____ seal.
    a. double
    b. spacer
    c. cartridge
    d. tandem

12. A seal that can be used with higher pressure and velocity due to reducing the area of face contact is the _____ type.
    a. rotating
    b. balanced
    c. multiple spring
    d. elastomer bellows

13. One disadvantage of an unbalanced seal is that it has a _____.
    a. low pressure limit
    b. low face loading
    c. high cost
    d. low life expectancy



**Have trainees complete the Review Questions, and go over the answers prior to administering the Module Examination.**

## Review Questions

14. One limitation of a single-spring seal is its tendency to _____.
    a. stick at low speeds
    b. distort at high speeds
    c. leak excessively
    d. apply too much pressure

15. Welded metal bellows seals are unbalanced seals that have more parts than spring seals.
    a. True
    b. False

16. The elastomer bellows seal presents some additional repair problems because the elastomer tends to stick to the _____.
    a. gland
    b. shaft
    c. seal
    d. bellows

17. Seals that are blistered, swollen, broken, or crumbled may have been damaged by chemical attack or by _____.
    a. large motors
    b. start-up
    c. excessive temperature
    d. soft foot

18. Seal faces should be replaced or _____ to remove distortions.
    a. rotated
    b. turned
    c. lapped
    d. polished

19. Seal face distortion is revealed by an uneven wear pattern on the seal faces.
    a. True
    b. False

20. Radial movement of the shaft will cause _____.
    a. seal face distortion
    b. shaft whip and deflection
    c. pitting and blistering of the seal face
    d. fretting corrosion



Instructor's Notes:

## Summary

Mechanical seals are essential to properly sealing shafts that must operate in fluid environments. Well-maintained mechanical seals protect the shaft and all moving components. They also prevent the loss of product, unnecessary pollution, and the possible exposure to hazards for people who must work around the equipment.

There are various types of mechanical seals manufactured today. Seal design and composition are selected based on a number of operating conditions including chemical exposure, temperature, and shaft speed. Each seal application must be carefully studied to determine the best type of mechanical seal to use.

Mechanical seals are designed to give many years of service; however, they are more expensive and require greater mechanical skill to install than traditional stuffing box compression packing. The most common causes of mechanical seal failure include improper installation, incorrect seal selection, environmental exposure, or a combination of these factors.

The operating life of a seal is determined by how well you follow proper procedures when installing, inspecting, and maintaining mechanical seals. You must develop skills through training and experience in order to properly perform these tasks.

## Notes



**Summarize the major concepts presented in the module.**

**Administer the Module Examination. Record the results on Craft Training Report Form 200, and submit the results to the Training Program Sponsor.**

**Administer the Performance Test, and fill out Performance Profile Sheets for each trainee. If desired, trainee proficiency noted during laboratory sessions may be used to complete the Performance Test. Record the results on Craft Training Report Form 200, and submit the results to the Training Program Sponsor.**

## Trade Terms Introduced in This Module

*Carcinogenic:* Cancer-causing.

*Cartridge mount:* A complete seal assembly that is installed in one piece.

*Deflection:* The deviation from straight and true as shown by a measuring device, such as a dial indicator.

*Flashing:* The almost instantaneous vaporization of a liquid to a vapor.

*Fretting:* Losing material due to excessive vibration and rubbing.

*Leaching:* Causing a liquid to filter through a material.

*Nonpiloting:* Not being automatically centered.

*Perimeter:* The outer boundary of an object or an area.

*Rotating face:* The sealing face of a mechanical seal that rotates with the pump shaft and presses against the stationary face.

*Seal face loading:* The amount of pressure or force that is applied to the seal face and that acts to close it.

*Stationary face:* The nonmoving sealing face of a mechanical seal.

*Stuffing box:* The housing that holds the packing that controls leakage along a shaft or rod.

*Thermal spike:* A very rapid increase in temperature, which looks like a spike on a temperature chart graph.

*Toxic:* Harmful, destructive, or deadly.

*Trace element:* A very minute amount of a chemical or particle in a fluid, which can only be detected by lab instruments.

*Volatile:* Able to rapidly change to vapor.



Instructor's Notes:

5.18 MILLWRIGHT ◆ LEVEL THREE

# Resources & Acknowledgments

## Additional Resources

This module is intended to present thorough resources for task training. The following reference works are suggested for further study. These are optional materials for continued education rather than for task training.

Flowserve Corporation, www.flowserve.com/eim/EducationalServices

Mechanical Seals.Net, www.mechanicalseals.net/Mechanical_Seal_Repair.htm

McNally Institute, www.mcnallyinstitute.com/10-html/10-2.html

## Figure Credits

Burgmann Industries GmbH & Co. KG, 305F01

UTEX Industries, Inc., 305F02

Advanced Sealing Technology, 305F04, 305F05, 305F07 (top)

Flowserve Corporation, 305F06, 305F08, 305F10, 305F11

Provided by Flowserve Corporation, a global leader in fluid motion and control. More information on Flowserve and its products can be found at www.flowserve.com

A.W. Chesterton, 305F07 (bottom left, bottom right)

# MODULE 15305-08 — TEACHING TIPS

The following are suggested activities or instructional methods to help you teach the material in this module.

## General

When you call on someone to answer a question, the rest of the class relaxes or even tunes out because they expect that the question and answer will take place only between you and the trainee you called on. Instead, use this technique to involve more trainees in answering questions and to keep them on their toes.

1. Ask trainees to define a term or explain a concept.
2. After one trainee has answered, ask a trainee seated nearby if the answer is right. Then ask whether a trainee in the back of the room agrees.
3. Ask trainees to explain why they think an answer is right or wrong.
4. Use the session to clear up incorrect ideas and encourage trainees to learn from their mistakes.

## Sections 1.0.0 – 4.0.0

### *Trade Terms Quiz*

This Quick Quiz will familiarize trainees with the trade terms commonly used when installing mechanical seals. You will need photocopies of the quiz provided on the following page. Trainees will need pencils. If you allow trainees to use the Trainee Guide, decrease the amount of time you give them to complete the quiz.

1. Make a photocopy of the quiz for each trainee.
2. Give trainees between 5 and 10 minutes to complete the quiz.
3. Go over the answers to the quiz.
4. Ask trainees if they have questions.

### *Answers to Quick Quiz*

1. b
2. e
3. c
4. i
5. h
6. d
7. g
8. a
9. f
10. j

## Quick Quiz *Trade Terms*

For each description listed, identify the term that the text best describes. Write the corresponding letter in the blank provided.

_____ 1. The deviation from straight and true, as shown by a measuring device such as a dial indicator, is called _____.

_____ 2. When a liquid is made to filter through a material, it is called _____.

_____ 3. The almost instantaneous vaporization of a liquid to a vapor is called _____.

_____ 4. A very rapid increase in temperature is called a _____.

_____ 5. In a mechanical seal, the _____ is held in place by the stuffing box gland.

_____ 6. Losing material due to excessive vibration and rubbing is called _____.

_____ 7. The purpose of the balanced seal design is to optimize _____ so that the faces stay closed without excessive force.

_____ 8. A complete seal assembly that is installed in one piece is a _____.

_____ 9. One situation that can cause rubbing is when the _____ gland is slipping down and hitting the seal.

_____ 10. A very minute amount of a chemical or particle in a fluid, which can only be detected by lab instruments, is called a _____.

a. cartridge mount
b. deflection
c. flashing
d. fretting
e. leaching
f. nonpiloting
g. seal face loading
h. stationary face
i. thermal spike
j. trace element

## Section 3.0.0 *Mechanical Seal Classifications*

This exercise provides practice in identifying types of mechanical seals. Trainees will need pencils and paper. You will need an overhead projector, acetate sheets, and transparency markers. Allow 20 to 30 minutes for this exercise. This exercise corresponds to Performance Task 1.

1. Divide the class in half. On one sheet write "Team One" on one half, and "Team Two" on the other half.
2. Name and or hold up examples of various types of mechanical seals. Have trainees identify the seal. Each correct answer gets a check on the acetate sheet. An incorrect answer means the other team has the opportunity to answer.
3. Keep score to encourage class participation. To ensure complete participation, each team member can answer no more than two times (adjust according to the number of participants and items to be identified).
4. At the end of the game, answer any questions.

## MODULE 15305-08 — ANSWERS TO REVIEW QUESTIONS

| Answer | Section |
|---|---|
| 1. c | 1.0.0 |
| 2. a | 1.0.0 |
| 3. a | 3.1.1 |
| 4. c | 3.1.1 |
| 5. a | 3.1.1 |
| 6. b | 3.1.2 |
| 7. b | 3.1.2 |
| 8. c | 3.1.3 |
| 9. a | 3.1.3 |
| 10. d | 3.1.4 |
| 11. c | 3.1.5 |
| 12. b | 3.2.1 |
| 13. a | 3.2.2 |
| 14. b | 3.2.3 |
| 15. b | 3.2.5 |
| 16. b | 3.2.6 |
| 17. c | 4.2.5 |
| 18. c | 4.2.6 |
| 19. a | 4.2.6 |
| 20. b | 4.3.0 |

## CONTREN® LEARNING SERIES — USER UPDATE

The NCCER makes every effort to keep these textbooks up-to-date and free of technical errors. We appreciate your help in this process. If you have an idea for improving this textbook, or if you find an error, a typographical mistake, or an inaccuracy in NCCER's Contren® textbooks, please write us, using this form or a photocopy. Be sure to include the exact module number, page number, a detailed description, and the correction, if applicable. Your input will be brought to the attention of the Technical Review Committee. Thank you for your assistance.

*Instructors* – If you found that additional materials were necessary in order to teach this module effectively, please let us know so that we may include them in the Equipment/Materials list in the Annotated Instructor's Guide.

**Write:** Product Development and Revision
National Center for Construction Education and Research
3600 NW 43rd St., Bldg. G, Gainesville, FL 32606

**Fax:** 352-334-0932

**E-mail:** curriculum@nccer.org

Craft ______ Module Name ______

Copyright Date ______ Module Number ______ Page Number(s) ______

Description

______

______

______

(Optional) Correction

______

______

(Optional) Your Name and Address

______

______

Module 15306-08

# *Removing and Installing Bearings*

## NCCER STANDARDIZED CRAFT TRAINING PROGRAM

The National Center for Construction Education and Research (NCCER) provides a standardized national program of accredited craft training. Key features of the program include instructor certification, competency-based training, and performance testing. The program provides trainees, instructors, and companies with a standard form of recognition through a National Craft Training Registry. The program is described in full in the *Guidelines for Accreditation*, published by NCCER. For more information on standardized craft training, contact the NCCER by writing us at 3600 NW 43rd St., Bldg. G, Gainesville, FL 32606; calling 352-334-0911; or emailing info@nccer.org. More information may be found at our website, www.nccer.org.

## HOW TO USE THIS ANNOTATED INSTRUCTOR'S GUIDE

Each page presents two sections of information. The larger section displays each page exactly as it appears in the Trainee Module. The narrow column ties suggested trainee and instructor actions to each page and provides icons (detailed below) to call your attention to material, safety, audiovisual, or testing requirements. The bottom of each page includes space for your notes.

The **Audiovisual** icon indicates an appropriate time to show a transparency or other audiovisual aid.

The **Classroom** icon prompts you to define a term, stress a point, ask trainees to explain a concept, or give examples.

The **Demonstration** icon directs you to show trainees how to perform tasks.

The **Examination** icon tells you to administer the written module examination.

The **Homework** icon is placed where you may wish to assign reading for the next class, assign a project, or advise trainees to prepare for an examination.

The **Laboratory** icon is used when trainees are to practice performing tasks.

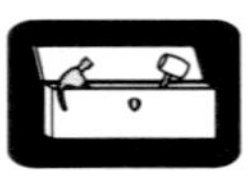
The **Materials** icon is a reminder for you to gather materials needed for classes, labs, and testing.

The **Performance Testing** icon tells you to administer a performance test or a portion thereof.

The **Safety** icon is used to emphasize safety issues. It is often keyed to *Caution* and *Warning!* statements in the Trainee Module.

The **Teaching Tip** icon indicates additional guidance is available, such as how to conduct an exercise, get the most educational value from a field trip, or encourage class participation. Teaching Tips may expand on a feature (*Think About It*, *Did You Know?*) or provide *Quick Quizzes* or similar exercises. You will be referred to the Teaching Tips section at the back of the module if there is additional material.

The **Combination** icon indicates that the laboratory listed corresponds with a performance task. If desired, you can note the proficiency of the trainees during the laboratory, and use it to satisfy performance testing requirements.

## PREPARATION

Before teaching this module, you should review the Objectives, Performance Tasks, Materials and Equipment List, and Module Outline. Be sure to allow ample time to prepare your own training or lesson plan and gather all required materials and equipment.

## MODULE OVERVIEW

This module explains how to remove, install, and maintain different types of bearings.

## PREREQUISITES

Prior to training with this module, it is recommended that the trainee shall have successfully completed the following: *Core Curriculum; Millwright Level One; Millwright Level Two;* and *Millwright Level Three,* Modules 15301-08 through 15305-08.

## OBJECTIVES

Upon completion of this module, the trainee will be able to do the following:

1. Identify common bearing failure modes and describe the conditions that cause them.
2. Describe the safety precautions that must be followed and the personal protective equipment that must be worn when removing and installing bearings.
3. Remove defective bearings using manual pullers and/or a press.
4. Describe using heat to remove a defective bearing.
5. Install new bearings by heating the bearing or applying pressure.
6. Install pillow block bearings.

## PERFORMANCE TASKS

Under the supervision of the instructor, the trainee should be able to do the following:

1. Remove a defective bearing using manual pullers and/or a press.
2. Install a new bearing by heating the bearing or applying pressure.
3. Install a pillow block bearing.

## MATERIALS AND EQUIPMENT LIST

Overhead projector and screen
Transparencies
Blank acetate sheets
Transparency pens
Whiteboard/chalkboard
Markers/chalk
Pencils and scratch paper
Appropriate personal protective equipment
Manual puller
Hydraulic press (optional)
Manual (arbor) press (optional)
Aluminum heating ring
Pillow block bearing
Split housing pillow block bearing
Angular-contact ball bearing

Bearings with the following conditions:
- Flaking
- Spalling
- Brinelling
- Misalignment damage
- Thrust failure
- Broken cam
- Electric arcing damage
- Fluting
- Lubrication failure
- Contamination failure

Induction-type bearing heater
Module Examinations*
Performance Profile Sheets*

* Located in the Test Booklet

## SAFETY CONSIDERATIONS

Ensure that the trainees are equipped with appropriate personal protective equipment and know how to use it properly. This module may require that the trainees visit job sites. Ensure that trainees are briefed on site safety policies prior to any site visits. This module requires trainees to use heat removal methods. Make sure trainees are briefed on appropriate safety procedures for using heat and cutting torches to remove bearings.

## ADDITIONAL RESOURCES

This module is intended to present thorough resources for task training. The following reference works are suggested for both instructors and motivated trainees interested in further study. These are optional materials for continued education rather than for task training.

*Installing and Replacing Bearings*, TPC Training Systems; 310 S. Michigan Avenue, Chicago, IL 60604, (312) 987-4100.

*Care and Maintenance of Bearings, Cat. No 3017/E,* NTN Corporation; 1-3-17 Kyomachibori Nishi-ku, Osaka-shi, Japan.

## TEACHING TIME FOR THIS MODULE

An outline for use in developing your lesson plan is presented below. Note that each Roman numeral in the outline equates to one session of instruction. Each session has a suggested time period of 2½ hours. This includes 10 minutes at the beginning of each session for administrative tasks and one 10-minute break during the session. Approximately 20 hours are suggested to cover *Removing and Installing Bearings.* You will need to adjust the time required for hands-on activity and testing based on your class size and resources. Because laboratories often correspond to Performance Tasks, the proficiency of the trainees may be noted during these exercises for Performance Testing purposes.

| Topic | Planned Time |
|---|---|
| **Session I. Introduction** | |
| A. Introduction | ____________ |
| **Session II. Removing Bearings** | |
| A. Removing Bearings | ____________ |
| B. Removal Methods | ____________ |
| C. Laboratory<br>Trainees practice removing a defective bearing using manual pullers and/or a press. This laboratory corresponds to Performance Task 1. | ____________ |
| **Session III. Troubleshooting Antifriction Bearings** | |
| A. Troubleshooting Antifriction Bearings | ____________ |
| B. Types of Failures | ____________ |
| **Session IV. Installing Bearings** | |
| A. Installing Tapered Roller Bearings, Thrust Bearings, Spherical Roller Bearings | ____________ |
| **Session V. Installing Bearings** | |
| A. Installing Pillow Block Bearings and Angular-Contact Ball Bearings | ____________ |
| **Session VI. Installing Bearings** | |
| A. Laboratory<br>Trainees practice installing a new bearing by heating the bearing or applying pressure. This laboratory corresponds to Performance Task 2. | ____________ |
| B. Laboratory<br>Trainees practice installing a pillow block bearing. This laboratory corresponds to Performance Task 3. | ____________ |

**Session VII. Module Review**

A. Module Review ____________

**Session VIII. Testing**

A. Module Examination ____________

1. Trainees must score 70% or higher to receive recognition from NCCER.
2. Record the testing results on Craft Training Report Form 200, and submit the results to the Training Program Sponsor.

B. Performance Testing ____________

1. Trainees must perform each task to the satisfaction of the instructor to receive recognition from NCCER. If applicable, proficiency noted during laboratory exercises can be used to satisfy the Performance Testing requirements.
2. Record the testing results on Craft Training Report Form 200, and submit the results to the Training Program Sponsor.

15306-08

# Removing and Installing Bearings

Assign reading of Module 15306-08.

# 15306-08
# *Removing and Installing Bearings*

*Topics to be presented in this module include:*



## Overview

Much of the millwright's work involves rotating equipment. The operation of bearings is a key function of rotating equipment. In this module, you will learn how to remove, install, and maintain different types of bearings.

*Instructor's Notes:*

## Objectives

When you have completed this module, you will be able to do the following:

1. Identify common bearing failure modes and describe the conditions that cause them.
2. Describe the safety precautions that must be followed and the personal protective equipment that must be worn when removing and installing bearings.
3. Remove defective bearings using manual pullers and/or a press.
4. Describe using heat to remove a defective bearing.
5. Install new bearings by heating the bearing or applying pressure.
6. Install pillow block bearings.

## Trade Terms

Burr
Cam
Flaking
Fluting
Spalling
Thrust

## Required Trainee Materials

1. Pencil and paper
2. Appropriate personal protective equipment, including safety glasses with side shield, heat-resistant gloves, and a face shield

## Prerequisites

Before you begin this module, it is recommended that you successfully complete *Core Curriculum*; *Millwright Level One*; *Millwright Level Two*; and *Millwright Level Three*, Modules 15301-08 and 15305-08.

This course map shows all of the modules in the third level of the *Millwright* curriculum. The suggested training order begins at the bottom and proceeds up. Skill levels increase as you advance on the course map. The local Training Program Sponsor may adjust the training order.

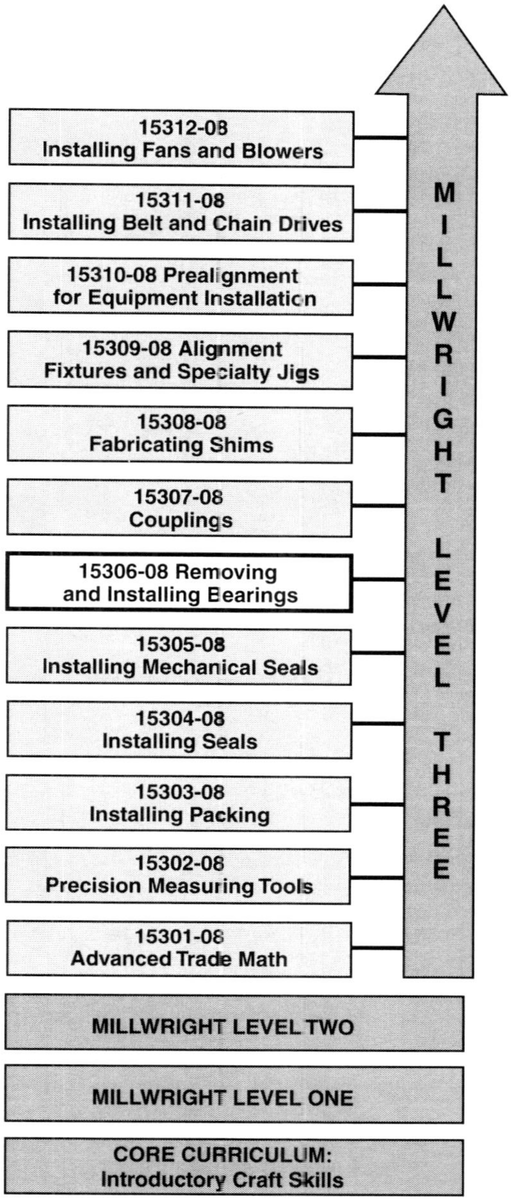


306CMAP.EPS



**Ensure that you have everything required to teach the course. Check the Materials and Equipment list at the front of this module.**

**See the general Teaching Tip at the end of this module.**

**Explain that terms shown in bold are defined in the Glossary at the back of this module.**

**Show Transparency 1, Objectives, and Transparency 2, Performance Tasks. Review the goals of the module, and explain what will be expected of the trainees.**

**Review the modules covered in Level Three and explain how this module fits in.**

MODULE 15306-08 ◆ REMOVING AND INSTALLING BEARINGS 6.1

**Discuss the purpose and function of bearings.**

**Stress the need to use appropriate PPE when working with bearings.**

**Have trainees review Sections 2.0.0–2.5.0.**

**Ensure that you have everything required to teach this session.**

**Emphasize that care must be taken to avoid damaging equipment.**

**Explain that old bearings should be kept and inspected to determine the cause of failure.**

**Introduce the tools and methods used to remove bearings.**

**Stress that bearings should be shielded for protection from injury.**

**Discuss bearing pullers.**

**Provide a manual bearing puller for trainees to examine, and explain how it is used.**

**Review the steps to remove a bearing using a manual bearing puller.**

## 1.0.0 ◆ INTRODUCTION

Bearings must be removed and installed from time to time either because they wear out or because they must be removed to disassemble a piece of equipment for repair or maintenance. Bearings are designed to give the longest and best possible service life for a particular application. The service life of a bearing depends on the proper installation and maintenance of the bearing.

Bearings are used to reduce friction between the moving parts of pieces of equipment that have rotating shafts. There are many types of bearings in use, and there are different methods for removing and installing them. Removing and installing bearings are major parts of a millwright's job. This module explains some of the most common methods for removing and installing bearings.

> **WARNING!**
> When removing, installing, and heating bearings, always wear the proper personal safety equipment, including safety glasses and heat-resistant gloves.

## 2.0.0 ◆ REMOVING BEARINGS

When removing a bearing, it is very important to follow the proper procedures to prevent damaging the bearing or the shaft. Sometimes a bearing is removed for maintenance or inspection and is reused on the equipment. In this case, improper removal techniques can damage the shaft and make it unusable. When a bearing is removed because it is worn out or has failed, it is replaced with a new one. In this case, the old bearing should be kept so that it can be inspected to determine why it failed. The information can be used for predictive and preventive maintenance purposes and to retain historical data of the equipment.

The most common tools for removing bearings are bearing pullers and presses. The most common methods of removing bearings are hydraulic removal, temperature removal, and using a cutting torch.

> **WARNING!**
> Bearings may shatter or disintegrate during removal. To prevent personal injury, shield the bearing with a material strong enough to prevent debris from escaping and causing injury.

### 2.1.0 Using Bearing Pullers

Using bearing pullers is the most common way to remove bearings. One advantage of a puller is that it can be taken in the field to remove the bearing from the shaft while the bearing is still in the machine. Pullers come in various styles and sizes. Most pullers come with different attachments for different applications.

Both manual and hydraulic pullers are available. The manual puller (*Figure 1*) has a bolt that is turned using a wrench to provide the pressure to pull the bearing. The hydraulic puller has a hydraulic cylinder and pump that provide the pulling pressure. Both types have the same attachments.

> **NOTE**
> Smaller bearings can often be pulled with a slide hammer puller.

Follow these steps to remove a bearing using a manual bearing puller.

*Step 1* Ensure that the puller is clean.

*Step 2* Position the puller jaws behind the bearing so that they press against only the inner race of the bearing.

> **CAUTION**
> The puller jaws must apply pressure only to the inner race of the bearing. If pressure is applied to the outer race, the bearing will be damaged and may come apart.

*Step 3* Hold the jaws in place, and screw in the bolt manually until it touches the end of the shaft.

*Step 4* Check the alignment of the puller to ensure that it will pull evenly on the bearing.

> **CAUTION**
> If the puller is misaligned and not pulling straight, the bearing will become cocked and may damage the shaft.



Instructor's Notes:

6.2 MILLWRIGHT ◆ LEVEL THREE

*Step 5* Apply a light coat of oil to the shaft to make the bearing slide off the shaft easily.

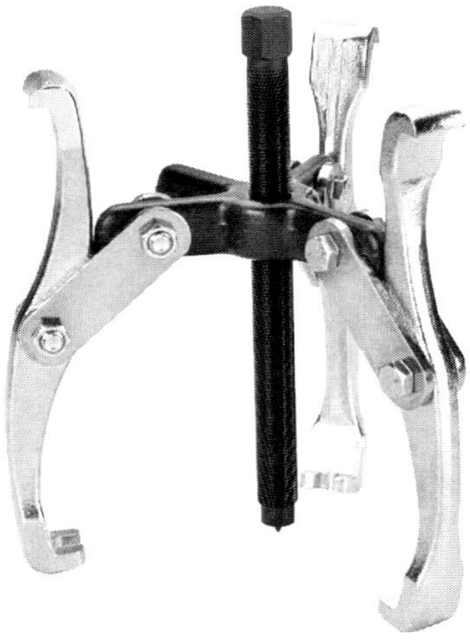

306F01.EPS

*Figure 1* ◆ Manual bearing puller.

*Step 6* Turn the bolt slowly, using a wrench, to apply pressure. *Figure 2* shows how to remove a bearing, using a manual bearing puller.

*Step 7* Continue to turn the bolt until the bearing comes off the shaft.

> **CAUTION**
> Do not let the bearing fall on the floor when it comes off the shaft because it could be damaged and get dirty.

## 2.2.0 Presses

There are two types of presses that are used for removing bearings: the hydraulic press and the manual, or arbor, press (*Figure 3*). The hydraulic press can generate a great amount of force and is used on large and small bearings. The arbor press generates much less force and is used on small bearings. When using presses, the shaft usually must be removed from the equipment and

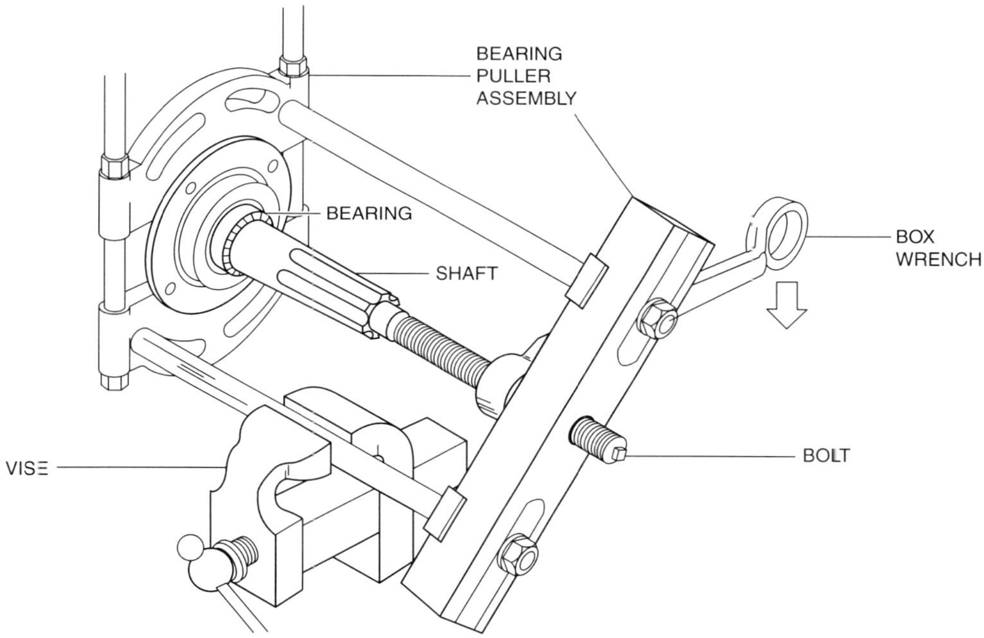


306F02.EPS

*Figure 2* ◆ Removing a bearing using a manual bearing puller.

**Show Transparency 3 (Figure 2).**

**Discuss hydraulic and manual (arbor) presses.**

**If possible, provide a hydraulic and arbor press for trainees to examine. Alternately, arrange to visit a job site where hydraulic and/or arbor presses are in use.**



MODULE 15306-08 ◆ REMOVING AND INSTALLING BEARINGS 6.3

Explain how a press is used to remove bearings.

Review the steps for removing a bearing using a press.

Discuss the hydraulic removal method and describe situations where it can be used.

*Figure 3* ◆ Hydraulic press and arbor press.

brought to the press. When removing a bearing from a shaft, position the bearing so that pressure will be exerted only on the inner ring. Pressure applied to the outer ring will damage the bearing. When removing a bearing from a housing, position the bearing so that pressure will be exerted only on the outer race.

Follow these steps to remove a bearing using a press:

*Step 1* Place the shaft and bearing in the press so that the inner race of the bearing is supported by two blocks of the same size.

*Step 2* Lower the ram of the press so that it touches the end of the shaft.

*Step 3* Ensure that the shaft is in the fully vertical position to prevent the bearing from cocking when the shaft is pressed out of the bearing.

*Step 4* Apply a light coat of oil to the shaft to help the bearing slide off easily.

*Step 5* Apply pressure slowly to press the shaft out of the bearing.

**CAUTION**

Keep the shaft vertical to prevent the bearing from cocking and gouging the shaft. Do not let the shaft fall to the floor when it is released from the bearing.

### 2.3.0 Hydraulic Removal Method

Hydraulic bearing removal is possible only if the shaft has been prepared in advance of the bearing installation. A hole is drilled in the shaft so that pressurized hydraulic fluid can be applied to the inner race of the bearing. The fluid is pumped into the hole through a hydraulic fitting where the fluid expands the inner race of the bearing. This expansion allows the bearing to slip off the shaft. Hydraulic bearing removal can be done while the



Instructor's Notes:

6.4 MILLWRIGHT ◆ LEVEL THREE

shaft is in the machine. *Figure 4* shows how the bearing is set up for removal.

Follow these steps to remove a bearing using the hydraulic removal method:

*Step 1* Clean the shaft, including the hole for the hydraulic fluid.

*Step 2* Install a hydraulic fitting in the hole.

*Step 3* Connect the hydraulic pump hose to the fitting.

*Step 4* Pump the hydraulic pump to apply pressure to the bearing inner race until the bearing slips off the shaft.

**WARNING!**

Some bearings, especially tapered seat bearings, will pop off the shaft with considerable force and can cause personal injury. Do not stand in front of the bearing while applying pressure. Leave the locking nut, with some clearance, on the end of the shaft to prevent the bearing from flying off.

## 2.4.0 Temperature Removal Method

Bearings that are shrink-fitted to the shaft sometimes require the use of heat to expand the bearings so that they will slip off the shaft. Timing is the most important factor when using heat to remove a bearing. The bearing must be slipped off the shaft as soon as the bearing has sufficiently expanded and before the shaft expands. The most common heating tool used for bearing removal (for replacement) is a torch. Another method used for heating bearings is the aluminum heating ring (*Figure 5*).

**CAUTION**

If the bearing is going to be re-used, do not heat the bearing with a torch or heat it above 250°F. The direct heat from the torch heats the bearing unevenly and tends to overheat it. This will ruin the bearing.

Follow these steps to remove a bearing using the aluminum heating ring:

*Step 1* Clean the shaft to remove any dirt or grit.

*Step 2* Dress down any burrs or nicks from the shaft.

*Step 3* Disassemble the bearing, leaving only the inner ring on the shaft.

*Step 4* Select an aluminum heating ring that fits the bearing ring.

*Step 5* Heat the aluminum heating ring to approximately 500°F, using a bearing heater or a similar method.

*Step 6* Slip the heating ring over the bearing ring, and squeeze the two handles on one side together to clamp the bearing ring.

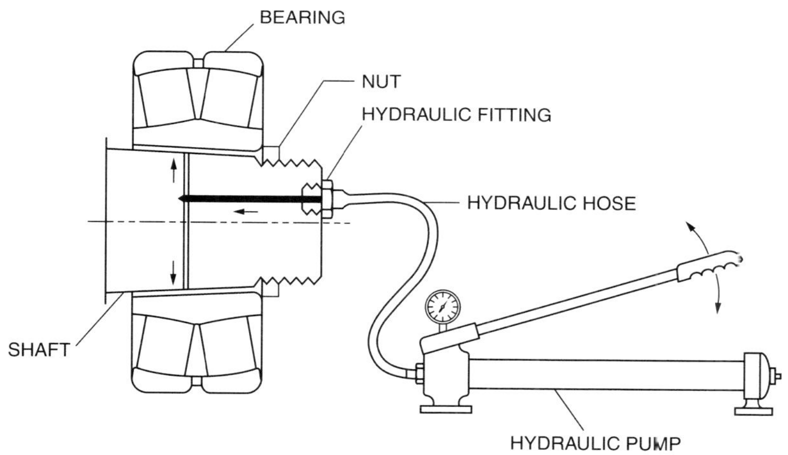


306F04.EPS

*Figure 4* ◆ Hydraulic removal setup.

**Show Transparency 4 (Figure 4). Explain the hydraulic removal setup.**

**Review the steps for the hydraulic removal method.**

**Explain how to avoid injury when performing hydraulic bearing removal.**

**Explain that the temperature removal method uses heat to expand the bearing, causing it to slip off.**

**Show Transparency 5 (Figure 5). Explain how the aluminum heating ring is used for heating bearings.**

**Review the steps to remove a bearing using the aluminum heating ring.**



MODULE 15306-08 ◆ REMOVING AND INSTALLING BEARINGS 6.5

**Point out that gloves and eye protection are required when using an aluminum heating ring.**

**Provide examples of situations that may require the use of a cutting torch to remove a bearing.**

**Stress that the cutting torch method requires that fire safety guidelines be followed. Discuss these guidelines.**

**Review the steps for removing a bearing using a cutting torch.**

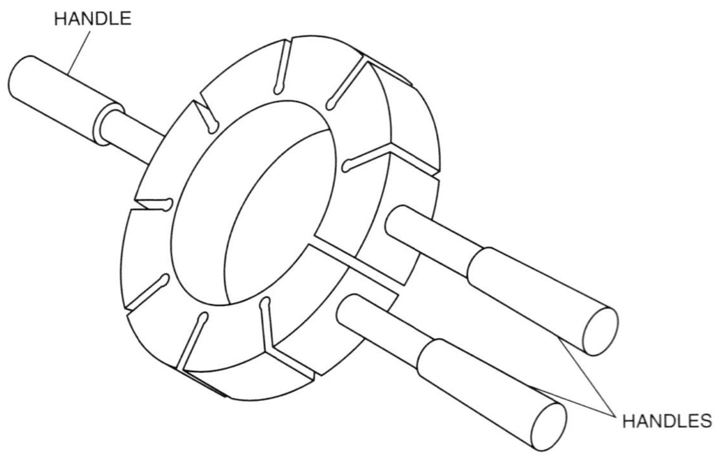


*Figure 5* ◆ Aluminum heating ring.

> **WARNING!**
> Wear protective gloves when handling the heating ring to avoid getting burned.

*Step 7* Grip the handles on both sides of the heating ring, and apply force back and forth to turn the bearing ring. When the bearing ring has sufficiently expanded, it will turn on the shaft.

*Step 8* Slip the ring off the shaft.

> **NOTE**
> Cooling the shaft on which the bearing is fitted while heating the bearing can help remove the bearing faster. This simultaneously shrinks the shaft while expanding the bearing, increasing the clearance between the two. Cooling can be done with liquid nitrogen or dry ice.

> **WARNING!**
> When performing this procedure, wear gloves and eye protection, ensure the area is well ventilated, and handle the coolant according to instructions. Cool only the shaft, not the bearing.

## 2.5.0 Using a Cutting Torch

Sometimes an inner bearing ring will become seized on the shaft, and common bearing removal methods will not work. In this situation, the bearing must be cut off the shaft, using a cutting torch. This will destroy the bearing, but if done correctly, the shaft will not be damaged. Using the cutting torch to remove bearings is a last resort. Follow these steps to remove a bearing using a cutting torch.

> **WARNING!**
> Follow all fire safety guidelines to ensure that the cutting operation will not present a fire hazard. Have a fire extinguisher available.

*Step 1* Disassemble the bearing, leaving only the inner ring on the shaft.

*Step 2* Clean the shaft and ring to remove any oil or grease that may catch fire during the cutting operation.

*Step 3* Determine the best place to cut the ring. If the equipment has a key, cut the ring on top of the key. This will reduce the risk of the shaft being damaged by the heat.



Instructor's Notes:

6.6 MILLWRIGHT ◆ LEVEL THREE

> **WARNING!**
> Only persons properly trained in the setup and use of oxyacetylene torches should attempt to remove bearings with a cutting torch.

*Step 4* Cut the ring.

> **CAUTION**
> Direct the heat across the face of the ring instead of directing it toward the shaft, and cut the ring from side to side. Make the cut as quickly as possible to keep residual heat to a minimum. Use a bent cutting tip to reduce the likelihood of damaging the shaft. Make the cut without depressing the oxygen lever.

*Step 5* Drive a chisel into the cut to pry the ring open.

*Step 6* Slip the ring off the shaft. If the ring will not come off the shaft, make a second cut 180 degrees from the first, and remove the ring in two pieces.

*Step 7* Clean and dress the shaft to remove any slag, burrs, or scratches.

## 3.0.0 ◆ TROUBLESHOOTING ANTIFRICTION BEARINGS

Troubleshooting bearings means identifying the causes of bearing failure. Bearing failures can be divided into the following nine basic types:

- Fatigue failure
- Brinelling
- False brinelling
- Misalignment
- Thrust failure
- Broken cam
- Electric arcing
- Lubrication failure
- Contamination failure

Predictive maintenance records show that 39 percent of all bearing failures are due to lubrication problems. Improper assembly problems account for 31 percent of bearing failures, and other problems, including static brinelling, manufacturing defects, temperature extremes, and shock loading, total 19 percent. Normal fatigue problems that occur within 36 to 88 months total 11 percent.

### 3.1.0 Fatigue Failure

Fatigue failures can be divided into two classes: normal and premature. Normal fatigue failure occurs when a bearing wears out. The bearing has lasted its expected life and has failed. This failure is usually evidenced by **flaking** or **spalling** of parts of the inner race. This type of failure is indicated by a noisy bearing accompanied by excessive vibration. The solution is to replace the bearing.

Premature fatigue failure is caused by some form of an overload in the bearing. The overload may be caused by an actual overload or by a parasitic load. A parasitic load is one that is applied to a bearing in addition to its normal load. When the two loads are combined, they overload the bearing.

Typical parasitic loads are caused by undersized housings, oversized shafts, out-of-round shafts or housings, burrs, nicks, and defective shaft shoulders. While the bearing will not immediately fail, it will wear more rapidly than it should. For example, the bearing life may be decreased from 6 years to 6 months or less. When a bearing fails in a short time, check for any of the above causes of parasitic loads.

### 3.2.0 Brinelling

Brinelling (*Figure 6*) is a denting of the raceways by the rolling element. This condition is either caused by a mounting error or by shock load while the bearing is stationary.

The most common cause of brinelling is a mounting error. If the pressure is applied to the wrong race during mounting or dismounting, the force is transmitted through the rolling elements.

306F06.EPS

*Figure 6* ◆ Brinelling.

Explain that only trained personnel are allowed to use a cutting torch.

Show trainees how to remove a defective bearing using manual pullers and/or a press.

Have trainees practice removing a defective bearing using manual pullers and/or a press. Note the proficiency of each trainee. This laboratory corresponds to Performance Task 1.

Have trainees review Sections 3.0.0–3.9.0.

Ensure that you have everything required to teach this session.

Discuss the basic types of antifriction bearing failures.

Explain normal and premature fatigue failure and the causes of each.

Discuss the causes of brinelling.

Provide samples of bearings with flaking, spalling, and brinelling damage for trainees to examine.

See the Teaching Tip for Section 3.0.0 at the end of this module.



MODULE 15306-08 ◆ REMOVING AND INSTALLING BEARINGS 6.7

Discuss the conditions that will cause false brinelling. Explain that it can be prevented by eliminating one of these conditions.

Discuss how to detect a misalignment failure.

Describe the three types of misalignment.

Provide samples of bearings with damage caused by misalignment.

Describe improper installation and application problems that can result in thrust failure.

Show Transparency 6 (Figure 8).

Discuss the cause of broken cams and explain where they occur.

Show Transparency 7 (Figure 9).

This force is so concentrated that it dents the race. This dent causes noisy vibration during operation. The rolling element will usually have a smaller dent. When the bearing is put in operation, spalling and flaking result. These usually occur at the spacing of the rolling elements.

A shock load to the bearing also causes brinelling. A shock load is caused by excessive bearing loading. The bearing races usually fracture under the shock load. This possibility should be considered if brinelling is a problem.

## 3.3.0 False Brinelling

False brinelling looks similar to brinelling. It is caused by vibration between the rollers in the bearing raceways. There will be marks in the raceway of the same spacing as the rolling elements. The difference is that a true brinell is a dent, while false brinelling is actual removal of metal from the bearing race.

False brinelling actually removes microscopic pieces from the bearing race. The following conditions must exist for false brinelling to occur:

- Stationary bearing
- Bearing under load
- Vibration

If these three conditions exist, false brinelling will occur. An electric motor in storage is likely to experience false brinelling. The bearing is mounted, stationary, and under load from the weight of the armature of the motor. Any vibration that occurs will cause false brinelling. False brinelling is also common in down equipment, such as standby pumps and motors.

False brinelling can be prevented by eliminating any one of the three needed conditions. If the armature in the electric motor is supported, taking the load off the bearings, false brinelling will not occur. When shipping machinery, care must be taken to prevent false brinelling.

## 3.4.0 Misalignment

A misalignment failure (*Figure 7*) is usually evidenced by a rolling element track in the raceway. If the failure is due to misalignment, the track usually runs from side to side in the stationary, or inner race. The rotating, or outer, race has a wide rolling element path.

The retainer may be damaged from coming in contact with the race. It is possible for the retainer to break, releasing the rolling elements in the bearing. The rolling elements, raceways, and retainers may be discolored from the excessive heat.

The types of misalignment are shaft misalignment, housing misalignment, and shaft deflection. Shaft misalignment results when the shaft is out of line with the housing. This may be due to misalignment of the drive and driver in a system. Housing misalignment is usually caused by settling or shifting of a foundation, and shaft deflection may be caused by overloads or improper installation.

## 3.5.0 Thrust Failure

**Thrust** failures are caused by improper installation and improper application. Improper installation includes mounting the bearing backward, improperly positioning the bearing on the shaft, or failing to lock a hot bearing on the shaft before the bearing cools. Improper application results from using a bearing that is not meant to take thrust load. The thrust load will destroy the bearing. *Figure 8* shows thrust failure of a maximum-capacity bearing. In this example, the maximum-capacity bearing was subjected to a thrust load. Since the bearing could not handle thrust loads because of the loading slots, the raceway was damaged.

## 3.6.0 Broken Cam

Broken **cams** (*Figure 9*) occur on pillow block bearings that use locking rings to hold the inner ring on the shaft. These bearings have a special extension on the inner race.

Broken cams are usually caused by the use of an undersized shaft in the installation. The shaft and ring are locked together off center, which puts excessive force on the locking ring. As the bearing rotates, the ring is forced to flex. This flexing

306F07.EPS

*Figure 7* ◆ Damage caused by misalignment.



Instructor's Notes:

6.8 MILLWRIGHT ◆ LEVEL THREE

quickly fatigues the area of the locking screw. Once the ring breaks, it is no longer fastened to the shaft. This causes rapid wear of the bearing race. The best prevention of this failure is to use the correct size shaft.

### 3.7.0 Electric Arcing

Electric arcing is caused by electrical current passing through a bearing. As the rolling elements travel, the current makes microscopic welds to the raceway. As the rolling element breaks the weld, small particles are pulled from the raceway and the rolling element. This does not destroy the bearing immediately, but it does cause excessive noise and vibration.

The most common cause of electric arcing is welding on equipment during installation or repair. When welding around equipment with bearings, the welding machine should be directly grounded as close as possible to the work to be welded. If the path to ground is through a bearing, some electric arcing will occur. Another cause is stray current in electrical motors and generators. The bearings in these units become a path to ground, causing electric arcing. Also, installations that develop static electricity may develop enough static electricity to cause electric arcing. If conditions are right, electric arcing may develop into a condition called **fluting** (*Figure 10*). This is the presence of long, rounded grooves in the raceway. The grooves develop and deepen, causing destruction of the bearing.

### 3.8.0 Lubrication Failure

Lubrication failures are indicated by a change in the color of the lubricant. The grease is dark and caked and usually has a burnt odor. The bearing itself may be brown or blue, indicating an extremely high operating temperature. The failure is usually accompanied by a high-pitched whistling noise. The most common causes of lubricant failure are too much or too little lubricant, the wrong type of lubricant, and bearing overload. The overload condition will rupture the lubricant film barrier, generating intense heat. This heat breaks down the lubricant, which in turn will destroy the bearing. The best prevention of this failure is to always use the correct type and amount of lubricant.

### 3.9.0 Contamination Failure

Contamination in a bearing is evidenced by scratches, pits, or scoring. In some idle bearings, rust may form. The cause of contamination is the entrance of dirt or some foreign liquid into the bearing. This may be caused by improper handling, poor sealing, or contaminated environ-

**Discuss the problem of electric arcing and how it is caused.**

**Explain how electric arcing can develop into fluting.**

**Describe indications of lubrication failure.**

**Describe indications of contamination failure.**

**Provide samples of bearings with damage caused by thrust failure, broken cam, electric arcing, fluting, lubrication failure, and contamination failure for trainees to examine.**

306F08.EPS

*Figure 8* ◆ Thrust failure of maximum-capacity bearing.

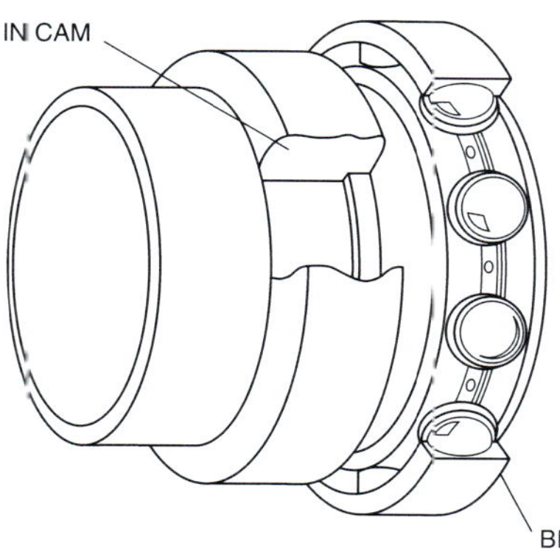


306F09.EPS

*Figure 9* ◆ Broken cam.

306F10.EPS

*Figure 10* ◆ Fluting.



MODULE 15306-08 ◆ REMOVING AND INSTALLING BEARINGS 6.9

Have trainees review Sections 4.0.0–4.3.0.

Ensure that you have everything required to teach this session.

Discuss the different ways that bearings are mounted, depending on the application.

Stress that bearings must be installed precisely.

Explain the need for proper bearing fit. Define slip fit and interference fit.

Discuss how tapered roller bearings are mounted.

Review the steps for installing a tapered roller bearing using the temperature mounting method.

ments. The best prevention of this failure is to keep all dirt and debris out of the bearing. Cleanliness at all times will prevent contamination failure.

> **NOTE**
> Worn or damaged grease seals allow contaminants to enter the bearing and allow lubricant to escape. When removing or replacing a bearing, always inspect the seal and replace it if any wear or damage is noticed.

## 4.0.0 ◆ INSTALLING BEARINGS

The method used to install a bearing depends the type of bearing being installed and its application. Bearings are mounted in a variety of ways. Some bearings are installed on a shaft and then mounted in a specially made bore in a machine. Shaft bearings in a transmission or gear reducer are mounted in bores in the housing. Some bearings are mounted in pillow block housings or flanged housings and support the shaft of a piece of equipment. Bearings are precision parts that must be installed as precisely as possible.

Two kinds of fits for bearings are the slip fit and the interference fit. The slip fit is the simplest to install because it fits fairly loosely, and can usually be pushed into place by hand. The interference fit is much tighter and requires more effort to press the bearing into place. Bearings usually have a slip fit on one ring and a press fit on the other. The ring that rotates is usually interference-fitted. In most applications, the inner ring of the bearing rotates. However, in some applications the outer ring rotates and, therefore, gets the interference fit.

Proper bearing fit is very important because a bad fit can result in premature bearing failure. A fit that is too loose causes the bearing and the shaft to wear rapidly due to the shaft sliding in the ring. A fit that is too tight can cause increased friction due to decreased clearances in the bearing. This friction results in high operating temperatures. A bearing with a fit that is too tight will fail early.

Properly sizing bearings is a major part of bearing installation. Installation methods are the same for different types of bearings. Once you have learned the proper installation procedures for a few types, you should be able to install most types of bearings. The following sections explain how to install the following types of bearings using various methods:

- Tapered roller
- Thrust
- Spherical roller
- Pillow block
- Angular-contact ball

### 4.1.0 Installing Tapered Roller Bearings Using the Temperature Mounting Method

Tapered roller bearings and angular-contact ball bearings are always mounted in pairs in opposition to each other. The clearances are adjusted at mounting. Tapered roller bearings and angular-contact ball bearings are installed in basically the same way. They can be mounted by press mounting or temperature mounting. This section describes installing a tapered roller bearing by the temperature mounting method. Temperature mounting can be performed in the shop or in the field while the shaft is still in the equipment. Follow these steps to install a tapered roller bearing using the temperature mounting method.

> **NOTE**
> Bearings are often heated by induction. Smaller bearings can be heated on an electric bearing heater that resembles a kitchen hotplate. Another method involves immersing the bearing in a bath of heated and temperature-controlled oil.

*Step 1* Clean the shaft thoroughly. Use environmentally friendly solvents when cleaning bearings and shafts prior to installation.

> **CAUTION**
> Keep the bearing in its protective wrapping until you are ready to install it to prevent contamination from dirt, dust, and grit. For the same reason, handle the bearing only with clean gloves, and keep the work area and all tools clean.

*Step 2* Inspect the shaft, and dress any burrs or nicks that would interfere with the bearing installation.

*Step 3* Polish the shaft lightly, using an emery cloth.

*Step 4* Measure the shaft and housing in several places, using a micrometer, to ensure that the diameter of the shaft and other parts are within specifications for the bearing being installed and to ensure that the shaft is not out of round.



Instructor's Notes:

6.10 MILLWRIGHT ◆ LEVEL THREE

**NOTE**

The shaft size and all other critical dimensions can be found in the manufacturer's specifications that come with the bearing. Follow the manufacturer's specifications and instructions when installing any bearing.

*Step 5* Remove the new bearing from its protective wrapping.

*Step 6* Place the bearing on an induction-type bearing heater (*Figure 11*).

**CAUTION**

The hot oil constitutes a hazard, both of dropping the bearing and of burns. Use personal protective equipment appropriate to the hazards.

*Step 7* Set the bearing heater to the temperature required by the manufacturer.

**CAUTION**

The maximum temperature to which a bearing should be heated is 250°F. Overheating a bearing can adversely affect the hardness of the bearing steel.

**WARNING!**

Wear insulated gloves when handling heated bearings.

*Step 8* Using an inside micrometer or similar method, check the bearing bore size periodically as the bearing is being heated, until it has expanded to the proper size to slip onto the shaft.

**NOTE**

There may be no need to heat the bearing to the maximum temperature. Heat it only until it has expanded enough to slip onto the shaft. Install the backing plate and other parts at this point if necessary.

*Step 9* Remove the bearing from the heater, and slip it onto the shaft (*Figure 12*). The bearing must be slipped onto the shaft, quickly moved to its proper position, and held there to prevent it from moving off the shoulder. When the bearing cools, it shrinks to fit the shaft. If it shrinks in the wrong position on the shaft, the bearing will have to be reheated, removed, and remounted.

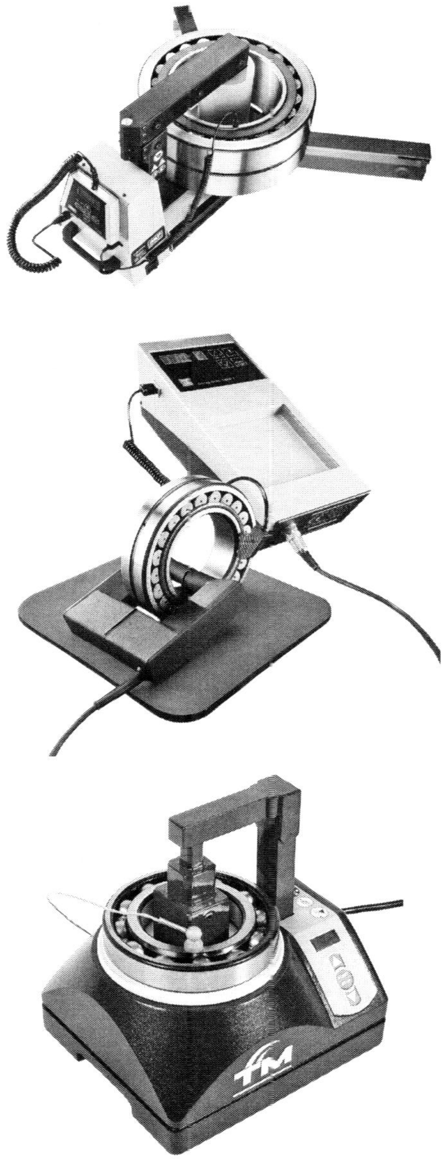

306F11.EPS

*Figure 11* ◆ Portable induction-type bearing heater.

**Explain how induction-type bearing heaters are used.**

**Provide an induction-type bearing heater for trainees to examine.**

**Stress the dangers of hot oil and overheated bearings. Remind trainees they should always wear insulated gloves when handling heated bearings.**

**Show Transparency 8 (Figure 12). Explain how to move the bearing into proper position.**



MODULE 15306-08 ◆ REMOVING AND INSTALLING BEARINGS 6.11

**Show Transparency 9 (Figure 13). Explain how to lock the bearing into position.**

**Explain how thrust bearings are mounted.**

**Review the steps for installing a thrust bearing using the press mounting method.**

*Step 10* Lock the bearing in place, using a locknut (*Figure 13*). The bearing should be locked in position with the locknut to prevent it from moving as it shrinks. When the bearing has shrunk enough to sieze the shaft, gently loosen the locknut. There may or may not be a lock washer, also.

*Step 11* Allow the bearing to cool to ambient temperature.

*Step 12* Loosen the locknut.

*Step 13* Turn the bearing with an ungloved hand, and feel for rough spots.

> **WARNING!**
> To avoid being burned, wait until the bearing has cooled before touching it with your hand.

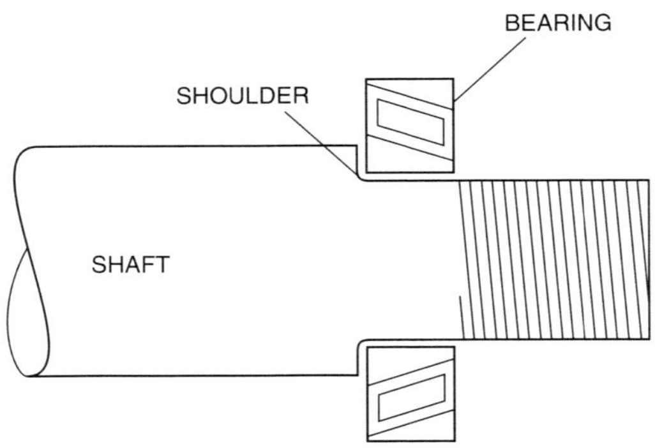


306F12.EPS

*Figure 12* ◆ Bearing in installed position.

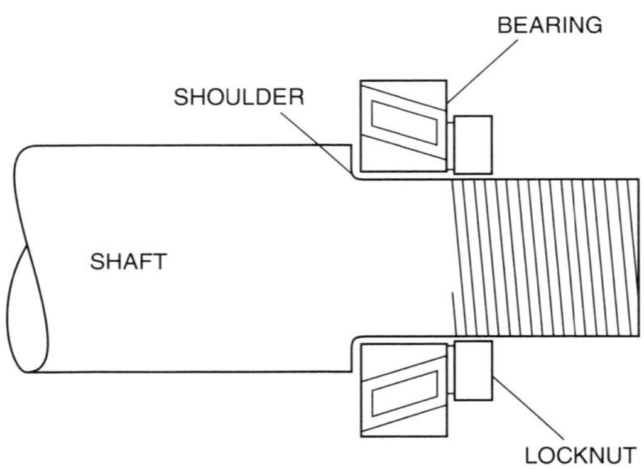


306F13.EPS

*Figure 13* ◆ Bearing locked in position.

*Step 14* Tighten the locknut according to the manufacturer's specifications to load the bearing.

## 4.2.0 Installing Thrust Bearings Using the Press Mounting Method

Thrust bearings support only axial, or thrust, loads. They may be mounted vertically or horizontally. Thrust bearings may be roller, ball, or plain. To mount a thrust bearing, one race is fitted to the shaft, and the other race is fitted to the housing. The shaft race is usually press-fitted, and the housing race is slip-fitted. The thrust bearing is usually pressed on the shaft in an arbor press or a hydraulic press; then the shaft is installed in the equipment, fitting the bearing into the housing.

Follow these steps to install a thrust bearing using the press mounting method.

*Step 1* Clean the shaft thoroughly.

*Step 2* Inspect the shaft, and remove any burrs or nicks.

*Step 3* Polish the shaft lightly, using an emery cloth.

*Step 4* Determine which race will be fitted to the shaft and which race will be fitted to the housing. Refer to the manufacturer's installation procedures and specifications to find any critical measurements and specifications.

*Step 5* Measure the shaft and housing in several places, using a micrometer, to ensure that the diameter of the shaft and other parts are within specifications for the bearing being installed and to ensure that the shaft is not out of round.

*Step 6* Remove the bearing from its protective wrapping.

> **CAUTION**
> The bearing must be slipped onto the shaft and quickly moved to its proper position.

*Step 7* Ensure that the press is clean.

*Step 8* Place the bearing on the press so that the shaft race is well-supported.

*Step 9* Position the shaft in the bearing.

*Step 10* Lower the press ram so that it touches the shaft.



*Instructor's Notes:*

6.12 MILLWRIGHT ◆ LEVEL THREE

*Step 11* Apply pressure to slip the shaft into the bearing.

**CAUTION**

Keep the shaft square with the bearing at all times. If the shaft is cocked during the pressing operation, the bearing will gouge the shaft.

*Step 12* Release pressure once the bearing is in the proper position on the shaft.

*Step 13* Install the shaft in the equipment, pushing the bearing into the housing.

*Step 14* Install the housing covers.

## 4.3.0 Installing Spherical Roller Bearings Using Hydraulic Nut or Locknut

Spherical roller bearings are a type of taper-bored bearing. A taper-bored bearing is either mounted on a tapered shaft or on a tapered sleeve. As the bearing is forced onto the shaft or sleeve, the clearance between the races and the rolling elements is reduced. The bearing clearance must be controlled when forcing the bearing onto the shaft. To control the clearance, measure it before installation and during the tightening process. The amount of initial clearance is reduced according to the tables supplied by the bearing manufacturer. The bearing is tightened until the proper clearance differential is achieved.

Follow these steps to install a spherical roller bearing on a tapered shaft using a hydraulic nut or a locknut.

*Step 1* Clean the shaft thoroughly.

*Step 2* Inspect the shaft, and remove any burrs or nicks.

*Step 3* Measure the shaft in several places, using a micrometer, to ensure that it is the proper size and that it is not out of round.

*Step 4* Oil the shaft lightly.

*Step 5* Remove the bearing from its protective wrapping.

*Step 6* Set the bearing on a clean, level surface.

*Step 7* Measure the bearing clearance using a feeler gauge (*Figure 14*), and record the reading. Clearance should be checked by starting with the thinnest feeler blade and using progressively thicker blades until one will not go. The last blade before the no-go is the measurement of the clearance.

*Step 8* Slip the bearing onto the shaft, and push it as far as possible by hand.

*Step 9* Screw a hydraulic nut or locknut onto the shaft against the bearing inner race (*Figure 15*).

*Step 10* Tighten the nut. If using a hydraulic pump, turn the hydraulic nut and tighten the bearing; if installing a locknut, use a spanner wrench.

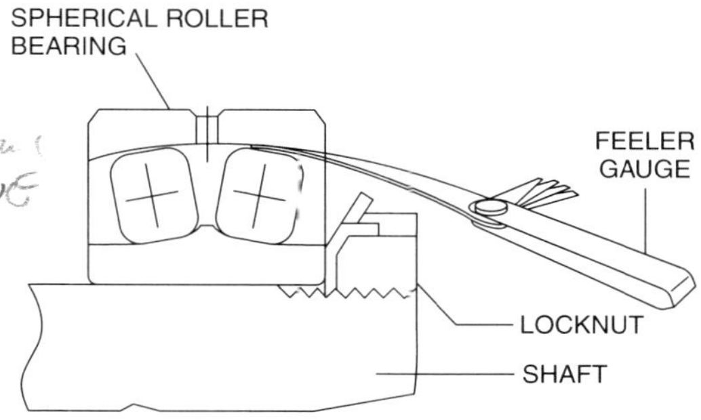


*Figure 14* ◆ Measuring bearing clearance.

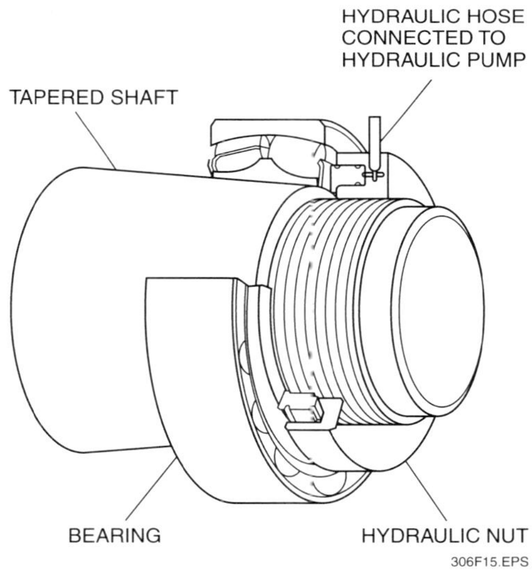


*Figure 15* ◆ Hydraulic nut in place.

**Explain how spherical roller bearings are mounted.**

**Review the steps for installing a spherical roller bearing on a tapered shaft using a hydraulic nut or a locknut.**

**Show Transparency 10 (Figure 14). Explain how to measure the bearing clearance.**

**Show Transparency 11 (Figure 15). Explain how to screw a hydraulic nut or locknut onto the shaft against the bearing inner race.**

**Have trainees review Sections 4.4.0–4.5.0.**

**Ensure that you have everything required to teach this session.**



MODULE 15306-08 ◆ REMOVING AND INSTALLING BEARINGS 6.13

**Explain how pillow block bearings are mounted.**

**Provide a pillow block bearing for trainees to examine.**

**Explain how to install a split housing pillow block bearing.**

**Show Transparency 12 (Figure 17). Review the components of a split housing pillow block bearing.**

*Step 11* Tighten the bearing and measure the clearance alternately until the proper clearance is obtained. Determine the proper clearance from the manufacturer's tables included in the bearing installation instructions.

*Step 12* Release the pressure, and remove the nut.

*Step 13* Install the bearing locknut and washer.

*Step 14* Tighten the bearing locknut to the proper bearing clearance.

> **CAUTION**
> Do not overtighten the locknut because this may change the bearing clearance.

## 4.4.0 Installing Pillow Block Bearings

Pillow blocks bearings (*Figure 16*) are used for independent mounting of antifriction bearings. The bearing is contained in the pillow block housing. The bearing is fitted on the shaft, and the housing is mounted on the equipment. Pillow block bearings can be divided into two groups: split housing and one-piece housing. Pillow blocks are sometimes called plummer blocks.

### *4.4.1 Installing Split Housing Pillow Block Bearings*

The split housing pillow block bearing (*Figure 17*) consists of a base and a cap with a horizontal split. The housing may be plain split or split with gibs or dowels. Housings with gibs or dowels eliminate the possibility of mismatching the cap to the

306F16.EPS

*Figure 16* ◆ Pillow block bearing.

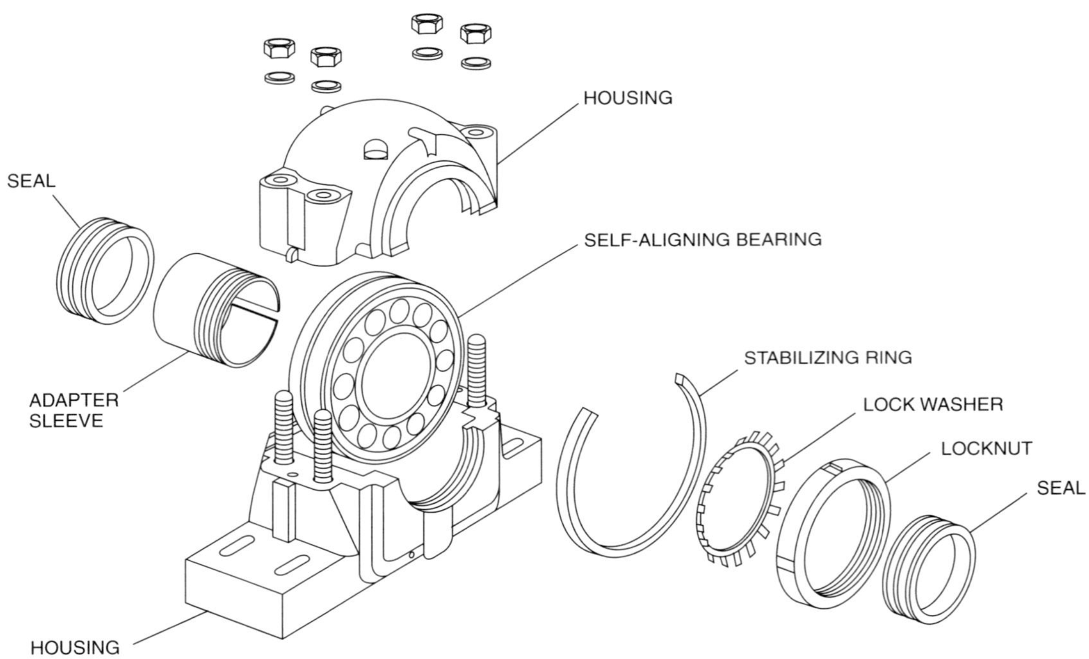


306F17.EPS

*Figure 17* ◆ Split housing pillow block bearing.



Instructor's Notes:

6.14 MILLWRIGHT ◆ LEVEL THREE

base. Plain split housings should be match-marked (*Figure 18*) before disassembly to ensure that the parts are not mismatched when reassembled. When using several pillow blocks of the same size, individually mark each bearing to prevent mixing the parts.

When the pillow block bearing is assembled for installation, it must be determined if the bearing is to be an expansion bearing or a nonexpansion bearing. An expansion bearing is used when the bearing will be subjected to high temperatures. In an expansion bearing, the stabilizing ring is left out and the bearing is installed so that the bearing will be centered in the housing to allow for thermal expansion. In a nonexpansion bearing, the stabilizing ring is installed to hold the bearing in position in the housing and to prevent axial movement.

### *4.4.2 Installing One-Piece Housing Pillow Block Bearings*

Pillow block bearings that have a one-piece housing are completely assembled units. Most such units can be reused. They are slipped over the shaft and bolted in position. They have an extra-long inner race to distribute the load over a wide area on the shaft. Some common designs used in one-piece pillow block bearings are the following:

- Single-groove ball bearings
- Double-taper roller bearings
- Spherical roller bearings

These bearings are self-aligning in the housing. A one-piece pillow block bearing allows for more misalignment than a split housing bearing. The bearings are usually held in place on the shaft with setscrews threaded into a collar and passed through matching holes in the inner ring to engage the shaft or by a self-locking collar that is held in place by a setscrew. The bearing should be installed with the setscrew on the inside, away from the end of the shaft so that any scoring caused by the setscrew contacting the shaft will not hinder the removal of the bearing from the shaft.

## 4.5.0 Installing Angular-Contact Ball Bearings

An angular-contact ball bearing (*Figure 19*) supports thrust load in one direction, sometimes combined with moderate radial load. The bearing has a high thrust-supporting shoulder on the inner ring and a similar high shoulder on the opposite side outer ring. These bearings can be mounted singly or in tandem for constant thrust load in one direction. They can also be mounted in pairs (duplex mounting), either face-to-face or back-to-back, for combined loads. When two or more bearings are mounted together on the shaft, you should be able to turn the outer ring of each bearing individually. Contact angles on angular-contact ball bearings typically range from 30 to 40 degrees.

### *4.5.1 Face-to-Face Mounting*

Face-to-face mounting (*Figure 20*) is used when the bearing takes both thrust and radial loads. This mounting allows for small amounts of misalignment. When mounted face-to-face, both the inner and outer rings are always clamped.

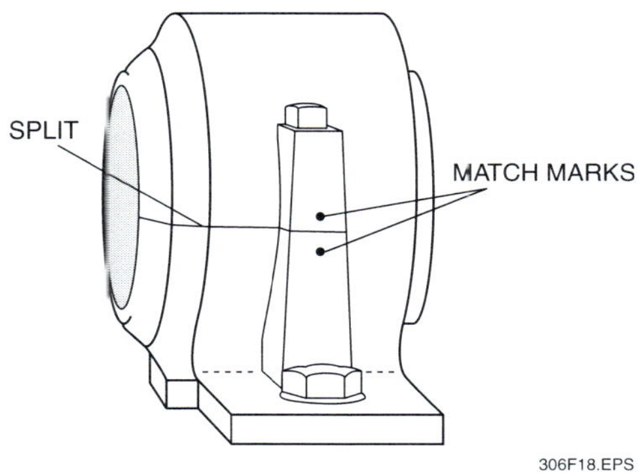


*Figure 18* ◆ Match-marked pillow block bearing.

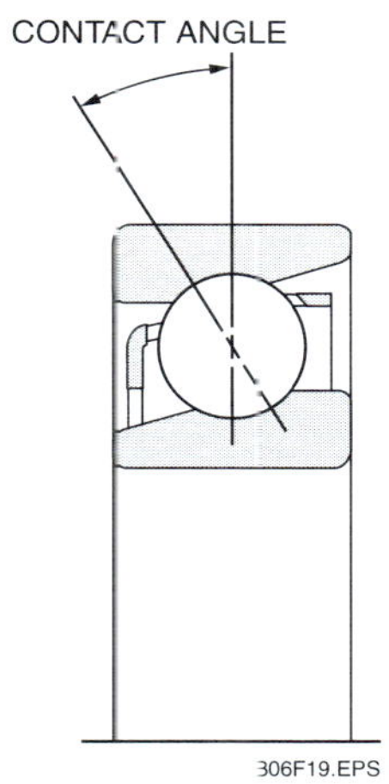


*Figure 19* ◆ Angular-contact ball bearing.

**Explain why plain split housings should be match-marked before disassembly.**

**Show Transparency 13 (Figure 18).**

**Compare expansion bearings to nonexpansion bearings.**

**Explain how to install a one-piece housing pillow block bearing.**

**Explain how angular-contact ball bearings are mounted.**

**Show Transparency 14 (Figure 19). Explain that the contact angle is typically 30°-40°.**

**Show Transparency 15 (Figure 20). Describe face-to-face mounting.**



MODULE 15306-08 ◆ REMOVING AND INSTALLING BEARINGS 6.15

Show Transparency 16 (Figure 21). Discuss back-to-back mounting.

Show Transparency 17 (Figure 22). Discuss tandem mounting.

Ensure that you have everything required to teach this laboratory session.

Demonstrate how to install a new bearing by heating the bearing.

Demonstrate how to install a new bearing by applying pressure.

Have trainees practice installing new bearings by heating or applying pressure. Note the proficiency of each trainee. This laboratory corresponds to Performance Task 2.

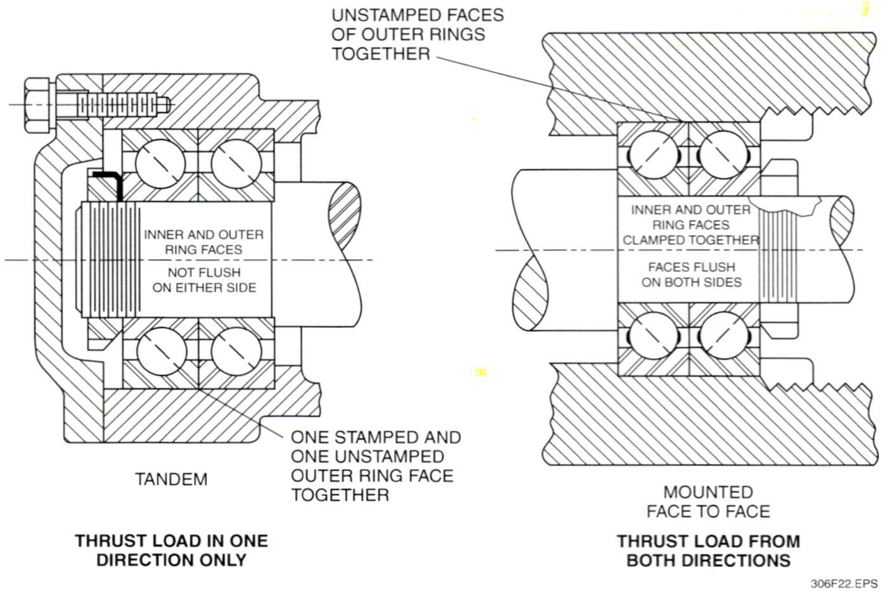


*Figure 22* ◆ Tandem mounting.



Instructor's Notes:

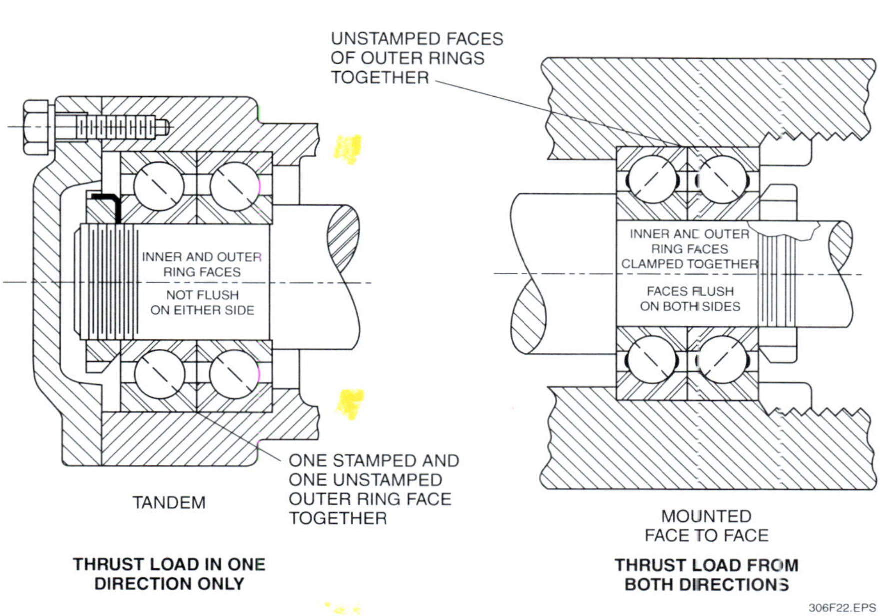


*Figure 22* ◆ Tandem mounting.

**Demonstrate how to install a pillow block bearing.**

**Have trainees practice installing a pillow block bearing. Note the proficiency of each trainee. This laboratory corresponds to Performance Task 3.**



MODULE 15306-08 ◆ REMOVING AND INSTALLING BEARINGS 6.17

Have trainees complete the Review Questions, and go over the answers prior to administering the Module Examination.

## Review Questions

1. The main purpose of a bearing is to _____.
   a. reduce wear
   b. increase efficiency
   c. reduce friction
   d. reduce operating costs

2. It is advisable to shield a bearing when removing it. This is done to _____.
   a. reduce noise
   b. prevent injury
   c. maintain privacy
   d. maintain heat

3. A manual bearing puller is operated using a(n) _____.
   a. wrench
   b. arbor
   c. hydraulic pump
   d. screwdriver

4. An aluminum heating ring should be heated to about _____ before applying it to the bearing to be removed.
   a. 200°F
   b. 300°F
   c. 400°F
   d. 500°F

5. Use a(n) _____ torch to cut a bearing off a shaft.
   a. propane
   b. oxyacetylene
   c. air/acetylene
   d. MAPP gas

6. Flaking observed on a failed bearing is often an indication of _____.
   a. a shock load
   b. vibration
   c. normal fatigue
   d. a broken cam

7. Vibration is a prime cause of _____.
   a. false brinelling
   b. brinelling
   c. misalignment
   d. electric arcing

8. A broken cam is often caused by _____.
   a. an undersized shaft
   b. an oversized shaft
   c. over-lubrication
   d. flexing of the locking screw

9. The most common cause of electric arcing on bearings is _____.
   a. lightning strikes
   b. improper voltage supplied to the machine
   c. electric arc welding
   d. a defective motor capacitor

10. A bearing that is failing due to lack of lubrication is likely to _____.
    a. grind
    b. roar
    c. rattle
    d. whistle

11. Scratches, pits, or scoring on a bearing are evidence of a(n) _____ bearing.
    a. overloaded
    b. misaligned
    c. brinelled
    d. contaminated

12. The two kinds of fits for bearings are _____.
    a. slip fit and tapered fit
    b. slip fit and interference fit
    c. tight fit and loose fit
    d. wedge fit and interference fit

13. When installing bearings, the maximum temperature that the bearing should be subjected to is _____.
    a. 150°F
    b. 212°F
    c. 250°F
    d. 500°F

14. Bearings are sometimes heated for installation by _____.
    a. $NO_2$
    b. LOX
    c. cutting torches
    d. induction heaters



Instructor's Notes:

6.18 MILLWRIGHT ◆ LEVEL THREE

## Review Questions

15. When installing a thrust bearing, the housing race is usually _____.
    a. slip-fitted
    b. interference-fitted
    c. induction-fitted
    d. shrink-fitted

16. The first step in installing a thrust bearing is to _____.
    a. inspect the shaft
    b. clean the shaft
    c. ensure the press is clean
    d. measure the housing covers

17. When installing a spherical roller bearing, the bearing clearance is measured using a(n) _____.
    a. feeler gauge
    b. micrometer
    c. vernier caliper
    d. architect's rule

18. When installing a spherical roller bearing, the bearing is tightened to a clearance based on _____.
    a. the spanner wrench size
    b. a micrometer reading
    c. the bearing manufacturer's literature
    d. the amount of play in the shaft

19. Before installing a pillow block with a plain split housing, always ____ the housings before disassembly.
    a. hand-fit
    b. inspect
    c. clean
    d. match-mark

20. Tandem-mounted angular-contact ball bearings are used in applications that ____.
    a. allow for small amounts of misalignment
    b. support heavy thrust loads
    c. operate at high temperatures
    d. require thrust to be handled in multiple directions



MODULE 15306-08 ◆ REMOVING AND INSTALLING BEARINGS 6.19

Summarize the major concepts presented in the module.

Administer the Module Examination. Record the results on Craft Training Report Form 200, and submit the results to the Training Program Sponsor.

Administer the Performance Test, and fill out Performance Profile Sheets for each trainee. If desired, trainee proficiency noted during laboratory sessions may be used to complete the Performance Test. Record the results on Craft Training Report Form 200, and submit the results to the Training Program Sponsor.

## Summary

Bearings must be removed and installed from time to time, either because they wear out or because they must be removed to disassemble a piece of equipment for repair or maintenance. Bearings are designed to give the longest and best possible service life for a particular application. The service life of a bearing depends on the proper installation and maintenance of the bearing.

When removing a bearing, it is very important to follow the proper procedures to prevent damaging the bearing or the shaft. Sometimes a bearing is removed for maintenance or inspection and is reused on the equipment.

It is important that you keep the bearing being replaced. This helps to determine the cause of failure for predictive and preventive maintenance purposes.

Bearings are precision parts that must be installed as precisely as possible. The method used to install a bearing depends on the type of bearing being installed and on its application.

## Notes



Instructor's Notes:

## Trade Terms Introduced in This Module

***Burr:*** A small, raised, uneven surface.

***Cam:*** A moving piece of machinery used to secure a bearing to a shaft.

***Flaking:*** Cracks in the bearing housing.

***Fluting:*** Long, rounded grooves in the raceway of a bearing caused by electric arcing.

***Spalling:*** The chipping away or breaking of a bearing race.

***Thrust:*** The force applied to the sides of a bearing.



MODULE 15306-08 ◆ REMOVING AND INSTALLING BEARINGS 6.21

# Resources & Acknowledgments

## Additional Resources

This module is intended to present thorough resources for task training. The following reference works are suggested for further study. These are optional materials for continued education rather than for task training.

*Installing and Replacing Bearings,* TPC Training Systems; 310 S. Michigan Avenue, Chicago, IL 60604, (312) 987-4100.

*Care and Maintenance of Bearings, Cat. No 3017/E,* NTN Corporation; 1-3-17 Kyomachibori Nishi-ku, Osaka-shi, Japan.

## Figure Credits

Danaher Tool Group, 306F01

Phoenix Hydraulic Presses, Inc., 306F03 (left)

Dake – Division of JSJ Corporation, 306F03 (right)

NSW Office of Transport Safety Investigations, 306F06

Courtesy of Spectro Inc. and BTS, 306F07

Greenheck Fan Corporation, 306F10

Photos courtesy of SKF USA Inc., 306F11 (top, center)

Easytherm 1, 306F11 (bottom) www.tminduction-us.com

Topaz Publications, Inc., 306F16



Instructor's Notes:

6.22 MILLWRIGHT ◆ LEVEL THREE

# MODULE 15306-08 — TEACHING TIPS

The following are suggested activities or instructional methods to help you teach the material in this module.

## General

When you call on someone to answer a question, the rest of the class relaxes or even tunes out because they expect that the question and answer will take place only between you and the trainee you called on. Instead, use this technique to involve more trainees in answering questions and to keep them on their toes.

1. Ask trainees to define a term or explain a concept.
2. After one trainee has answered, ask a trainee seated nearby if the answer is right. Then ask whether a trainee in the back of the room agrees.
3. Ask trainees to explain why they think an answer is right or wrong.
4. Use the session to clear up incorrect ideas and encourage trainees to learn from their mistakes.

## Section 3.0.0 *Troubleshooting Antifriction Bearings*

This exercise will provide practice in identifying types of bearing failures. Trainees will need pencils and paper. You will need an overhead projector, acetate sheets, and transparency markers. Allow 20 to 30 minutes for this exercise.

1. Divide the class in half. On one sheet write "Team One" on one half, and "Team Two" on the other half.
2. Name and or hold up examples of bearings with various types of failures. Have trainees identify the type of failure that occurred. Each correct answer gets a check on the acetate sheet. An incorrect answer means the other team has the opportunity to answer.
3. Keep score to encourage class participation. To ensure complete participation, each team member can answer no more than two times (adjust according to the number of participants and items to be identified).
4. At the end of the game, answer any questions.

## MODULE 15306-08 — ANSWERS TO REVIEW QUESTIONS

| Answer | | Section |
|---|---|---|
| 1. | c | 1.0.0 |
| 2. | b | 2.0.0 |
| 3. | a | 2.1.0 |
| 4. | d | 2.4.0 |
| 5. | b | 2.5.0 |
| 6. | c | 3.1.0 |
| 7. | a | 3.3.0 |
| 8. | a | 3.6.0 |
| 9. | c | 3.7.0 |
| 10. | d | 3.8.0 |
| 11. | d | 3.9.0 |
| 12. | b | 4.0.0 |
| 13. | c | 4.1.0 |
| 14. | d | 4.1.0 |
| 15. | a | 4.2.0 |
| 16. | b | 4.2.0 |
| 17. | a | 4.3.0 |
| 18. | c | 4.3.0 |
| 19. | d | 4.4.1 |
| 20. | b | 4.5.3 |

## CONTREN® LEARNING SERIES — USER UPDATE

The NCCER makes every effort to keep these textbooks up-to-date and free of technical errors. We appreciate your help in this process. If you have an idea for improving this textbook, or if you find an error, a typographical mistake, or an inaccuracy in NCCER's Contren® textbooks, please write us, using this form or a photocopy. Be sure to include the exact module number, page number, a detailed description, and the correction, if applicable. Your input will be brought to the attention of the Technical Review Committee. Thank you for your assistance.

*Instructors* – If you found that additional materials were necessary in order to teach this module effectively, please let us know so that we may include them in the Equipment/Materials list in the Annotated Instructor's Guide.


**Write:** Product Development and Revision
National Center for Construction Education and Research
3600 NW 43rd St., Bldg. G, Gainesville, FL 32606

**Fax:** 352-334-0932

**E-mail:** curriculum@nccer.org


Craft ______________________ Module Name ______________________

Copyright Date __________ Module Number __________ Page Number(s) __________

Description

______________________________________________

______________________________________________

______________________________________________

______________________________________________

(Optional) Correction

______________________________________________

______________________________________________

______________________________________________

(Optional) Your Name and Address

______________________________________________

______________________________________________

______________________________________________

Module 15307-08

# *Couplings*

## NCCER STANDARDIZED CRAFT TRAINING PROGRAM

The National Center for Construction Education and Research (NCCER) provides a standardized national program of accredited craft training. Key features of the program include instructor certification, competency-based training, and performance testing. The program provides trainees, instructors, and companies with a standard form of recognition through a National Craft Training Registry. The program is described in full in the *Guidelines for Accreditation*, published by NCCER. For more information on standardized craft training, contact the NCCER by writing us at 3600 NW 43rd St., Bldg. G, Gainesville, FL 32606; calling 352-334-0911; or emailing info@nccer.org. More information may be found at our website, www.nccer.org.

## HOW TO USE THIS ANNOTATED INSTRUCTOR'S GUIDE

Each page presents two sections of information. The larger section displays each page exactly as it appears in the Trainee Module. The narrow column ties suggested trainee and instructor actions to each page and provides icons (detailed below) to call your attention to material, safety, audiovisual, or testing requirements. The bottom of each page includes space for your notes.

The **Audiovisual** icon indicates an appropriate time to show a transparency or other audiovisual aid.

The **Classroom** icon prompts you to define a term, stress a point, ask trainees to explain a concept, or give examples.

The **Demonstration** icon directs you to show trainees how to perform tasks.

The **Examination** icon tells you to administer the written module examination.

The **Homework** icon is placed where you may wish to assign reading for the next class, assign a project, or advise trainees to prepare for an examination.

The **Laboratory** icon is used when trainees are to practice performing tasks.

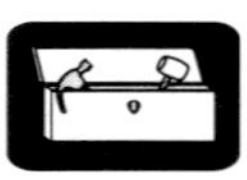
The **Materials** icon is a reminder for you to gather materials needed for classes, labs, and testing.

The **Performance Testing** icon tells you to administer a performance test or a portion thereof.

The **Safety** icon is used to emphasize safety issues. It is often keyed to *Caution* and *Warning!* statements in the Trainee Module.

The **Teaching Tip** icon indicates additional guidance is available, such as how to conduct an exercise, get the most educational value from a field trip, or encourage class participation. Teaching Tips may expand on a feature (*Think About It*, *Did You Know?*) or provide *Quick Quizzes* or similar exercises. You will be referred to the Teaching Tips section at the back of the module if there is additional material.

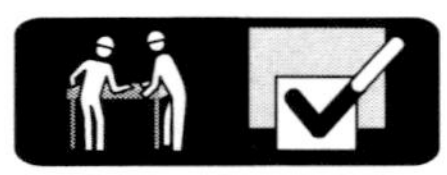
The **Combination** icon indicates that the laboratory listed corresponds with a performance task. If desired, you can note the proficiency of the trainees during the laboratory, and use it to satisfy performance testing requirements.

## PREPARATION

Before teaching this module, you should review the Objectives, Performance Tasks, Materials and Equipment List, and Module Outline. Be sure to allow ample time to prepare your own training or lesson plan and gather all required materials and equipment.

## MODULE OVERVIEW

This module explains how couplings are installed and aligned, and introduces some of the mounting systems used for various couplings.

## PREREQUISITES

Prior to training with this module, it is recommended that the trainee shall have successfully completed *Core Curriculum; Millwright Level One; Millwright Level Two;* and *Millwright Level Three,* Modules 15301-08 through 15306-08.

## OBJECTIVES

Upon completion of this module, the trainee will be able to do the following:

1. Identify and explain coupling types.
2. Install couplings.
3. Remove couplings.

## PERFORMANCE TASKS

Under the supervision of the instructor, the trainee should be able to do the following:

1. Identify and explain coupling types.
2. Install an interference coupling.
3. Install a slip fit coupling.
4. Remove couplings using either the manual or hydraulic method.

## MATERIALS AND EQUIPMENT LIST

Overhead projector and screen
Transparencies
Blank acetate sheets
Transparency pens
Whiteboard/chalkboard
Markers/chalk
Pencils and scratch paper
Appropriate personal protective equipment
Basic trainee tools

Samples of various types of couplings, including:
- Rigid sleeve
- Tubular sleeve
- Gear
- Chain
- Spring
- Tire
- Flexible disc
- Pin and bushing
- Spacer
- Horizontal split and vertical split grid couplings
- Single- and double-joint universal joint couplings

Quick Quiz*
Module Examinations**
Performance Profile Sheets**

* Located in the back of this module
**Located in the Test Booklet

## SAFETY CONSIDERATIONS

Ensure that the trainees are equipped with appropriate personal protective equipment and know how to use it properly. This module requires that the trainees use hand tools. Ensure that trainees are briefed on shop safety procedures.

## ADDITIONAL RESOURCES

This module is intended to present thorough resources for task training. The following reference works are suggested for both instructors and motivated trainees interested in further study. These are optional materials for continued education rather than for task training.

Couplings.ca, http://www.couplings.ca/

Lovejoy, Inc., http://www.lovejoy-inc.com/

## TEACHING TIME FOR THIS MODULE

An outline for use in developing your lesson plan is presented below. Note that each Roman numeral in the outline equates to one session of instruction. Each session has a suggested time period of 2½ hours. This includes 10 minutes at the beginning of each session for administrative tasks and one 10-minute break during the session. Approximately 15 hours are suggested to cover *Couplings*. You will need to adjust the time required for hands-on activity and testing based on your class size and resources. Because laboratories often correspond to Performance Tasks, the proficiency of the trainees may be noted during these exercises for Performance Testing purposes.

| Topic | Planned Time |
|---|---|
| **Session I. Introduction; Rigid Couplings** | |
| A. Introduction | ________ |
| B. Rigid Couplings | ________ |
| 1. Flanged Couplings | ________ |
| 2. Sleeve Couplings | ________ |
| 3. Clamp Couplings | ________ |
| **Session II. Flexible Couplings** | |
| A. Mechanical Flexible Couplings | ________ |
| B. Material Flexible Couplings | ________ |
| **Session III. Soft-Start Couplings** | |
| A. Soft-Start Couplings | ________ |
| B. Fluid Couplings | ________ |
| C. Shot Couplings | ________ |
| D. Clutch-Style Couplings | ________ |
| E. Laboratory<br>Trainees practice identifying and explaining coupling types. This laboratory corresponds to Performance Task 1. | ________ |
| **Session IV. Installing Couplings** | |
| A. Installing Couplings | ________ |
| B. General Coupling Installation Procedures | ________ |
| C. Split Coupling Installation | ________ |
| D. Interference-Fit Installation | ________ |
| E. Laboratory<br>Trainees practice installing an interference coupling. This laboratory corresponds to Performance Task 2. | ________ |
| F. Setting the Coupling Gap | ________ |
| G. Grid Coupling Installation | ________ |
| H. Installing Hydraulic Couplings | ________ |
| I. Laboratory<br>Trainees practice installing a slip fit coupling. This laboratory corresponds to Performance Task 3. | ________ |

**Session V. Removing Couplings**

A. Removing Couplings ____________

B. General Coupling Removal Procedures ____________

C. Mechanical Pullers ____________

D. Hydraulic Removal Method ____________

E. Laboratory ____________
Trainees practice removing couplings using either the manual or hydraulic method. This laboratory corresponds to Performance Task 4.

**Session VI. Review and Testing**

A. Module Review ____________

B. Module Examination ____________

1. Trainees must score 70% or higher to receive recognition from NCCER.
2. Record the testing results on Craft Training Report Form 200, and submit the results to the Training Program Sponsor.

C. Performance Testing ____________

1. Trainees must perform each task to the satisfaction of the instructor to receive recognition from NCCER. If applicable, proficiency noted during laboratory exercises can be used to satisfy the Performance Testing requirements.
2. Record the testing results on Craft Training Report Form 200, and submit the results to the Training Program Sponsor.

## 15307-08

# Couplings

Assign reading of Module 15307-08.

# 15307-08
# *Couplings*

*Topics to be presented in this module include:*



## Overview

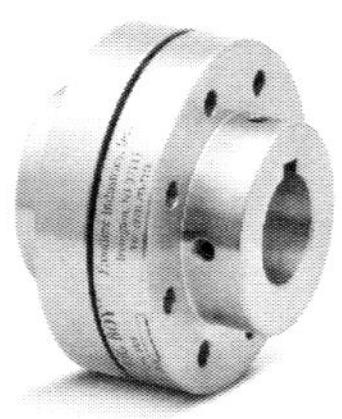

The job of installing and maintaining rotating equipment involves connections between the shafts of the driver and the driven equipment, such as pumps and motors. In this module, you will learn how the couplings are installed and how they are aligned. This module also introduces some of the mounting systems used for various couplings, and teaches you how to mount them.

Instructor's Notes:

## Objectives

When you have completed this module, you will be able to do the following:

1. Identify and explain coupling types.
2. Install couplings.
3. Remove couplings.

## Trade Terms

Angular misalignment
Clutch
Coupling gap
Driven
Driver
Key
Parallel misalignment
Shear pin
Torque
Vortex

## Required Trainee Materials

1. Pencil and paper
2. Appropriate personal protective equipment

## Prerequisites

Before you begin this module, it is recommended that you successfully complete *Core Curriculum; Millwright Level One; Millwright Level Two;* and *Millwright Level Three*, Modules 15301-08 through 15306-08.

This course map shows all of the modules in the third level of the *Millwright* curriculum. The suggested training order begins at the bottom and proceeds up. Skill levels increase as you advance on the course map. The local Training Program Sponsor may adjust the training order.

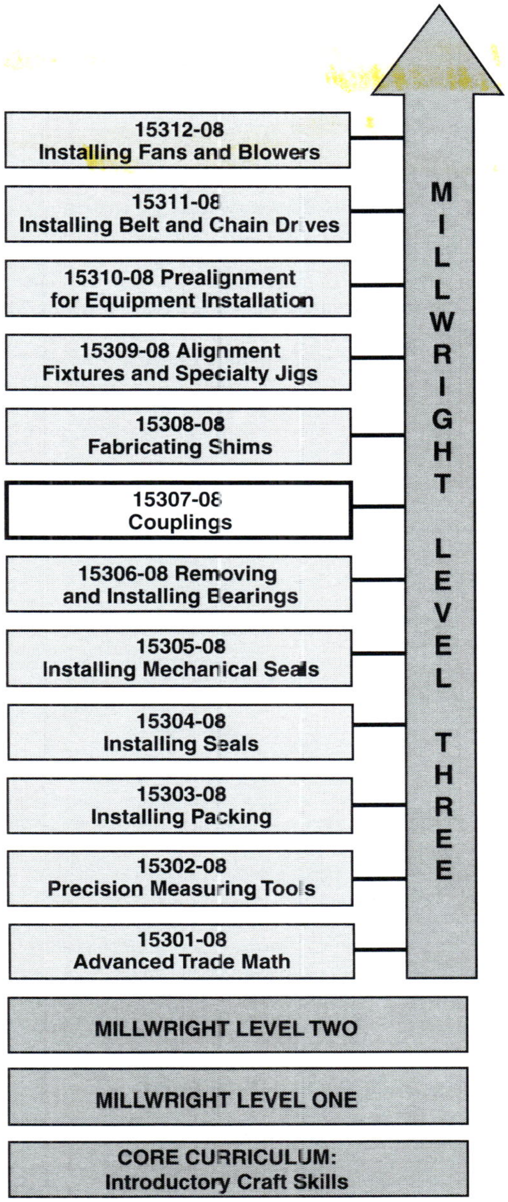


307CMAP.EPS

**Ensure that you have everything required to teach the course. Check the Materials and Equipment list at the front of this module.**

**See the general Teaching Tip at the end of this module.**

**Explain that terms shown in bold are defined in the Glossary at the back of this module.**

**Show Transparency 1, Objectives, and Transparency 2, Performance Tasks. Review the goals of the module, and explain what will be expected of the trainee.**

**Review the modules covered in Level Three and explain how this module fits in.**



MODULE 15307-08 ◆ COUPLINGS 7.1

Explain the general purpose and function of couplings.

Describe how horsepower, size, and speed are required to select the proper coupling. Review the list of additional considerations.

Provide a variety of couplings for trainees to examine.

Describe rigid couplings and explain how they are used.

Discuss flanged couplings and explain how they are used.

Show Transparency 3 (Figure 1).

Describe taper-bored flanged couplings.

Show Transparency 4 (Figure 2).

Discuss clamping hubs and how they operate.

## 1.0.0 ◆ INTRODUCTION

Couplings are used to connect the shaft of a **driver**, such as a motor, to the shaft of a **driven**, such as a pump. Couplings are manufactured in many types and sizes. Some coupling types allow for slight misalignment and end play, or shaft float, between the rotating shafts. Some couplings reduce, or dampen, or absorb vibrations or **torque**. Other couplings insulate the coupling halves from any electrical current transfer, which is common in some motor-generator sets.

In order for equipment to operate properly and efficiently, the proper coupling must be used to connect the driver to the driven. It is also very important that the equipment be properly aligned and the coupling properly installed.

When selecting a coupling for a particular application, a minimum of three items are required to properly select the coupling. These are horsepower, size, and speed. Additional considerations that should also be taken into account include the following:

- Keyways required
- Size of keyways
- Taper on shafts
- Shaft materials
- Torque
- Angular misalignment
- Offset misalignment
- Axial travel
- Distance between shaft ends
- Operating temperature
- Space limitations
- Any other unusual conditions

This module explains some common types of couplings and explains how to install and remove couplings. The three basic types of couplings are rigid, flexible, and soft-start.

## 2.0.0 ◆ RIGID COUPLINGS

As the name implies, rigid couplings rigidly connect the driver and driven shafts. Rigid couplings do not compensate for misalignment and require precise alignment during installation. If a rigid coupling is misaligned and forced together, the drive will be damaged. Even slight misalignment can cause vibration and operating problems. Rigid couplings are manufactured in several styles. The three most common rigid couplings are flanged, sleeve, and clamp.

### 2.1.0 Flanged Couplings

Flanged couplings (*Figure 1*) join the driver and driven shafts, using two mating flanges. One coupling flange fits on the driver shaft, and the other fits on the driven shaft. The connection between the two halves is made by bolting the flanges together. Flanged couplings require **keys** to prevent them from rotating on the shafts and require precise alignment. A **shear pin** can be used in place of a key when connecting equipment that may be subject to binding or an overload condition. In the event of binding, the shear pin, which is made of a much softer material than the connecting shafts, will fracture or break apart, allowing the driver shaft to spin freely, protecting the driver from overload.

Taper-bored flanged couplings, also called compression couplings, are taper-bored and have tapered sleeves that fit on the shafts. The wedge principle is used to tighten the coupling on the shafts. As the two halves of the coupling are pulled together over the tapered sleeve by the flange bolts, the coupling halves are tightened on the shafts (*Figure 2*). These couplings do not require keys and are normally used on small-diameter shafts as they are not suitable for transmitting heavy loads.

Clamping hubs operate on the same principle as the taper-bored couplings. The inner sleeve is split, and is bolted into a tapered housing, so that it is compressed as it is tightened. Flanged couplings, especially the taper-bored couplings, are used for mounting pulleys, for blinding the end of a system, or in any case where a flange must be added or a system must be capable of being disassembled without cutting or welding.

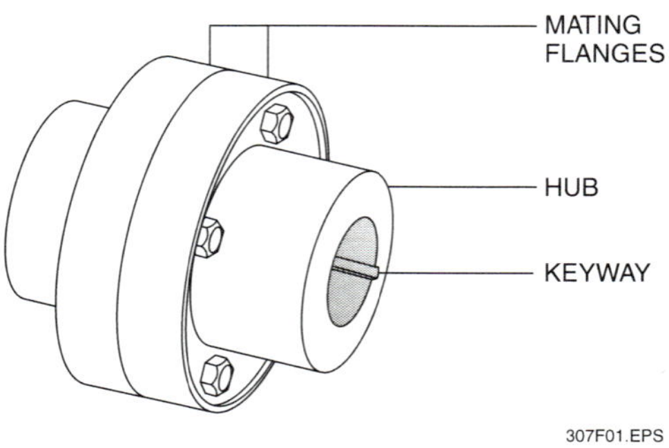


*Figure 1* ◆ Flanged coupling.



Instructor's Notes:

7.2 MILLWRIGHT ◆ LEVEL THREE

## 2.2.0 Sleeve Couplings

There are two types of sleeve couplings that are used to join rotating shafts that are closely aligned. A rigid sleeve coupling with setscrews is available with or without keyways (*Figure 3*). These couplings are solid steel shafts that have two pairs of setscrews typically at ninety-degree angles to secure the coupling to the driver shaft and driven shaft across the X and Y axis of the shafts. A tubular sleeve coupling is also available that consists of a metal tube with slits at each end. This type of sleeve coupling uses clamps on each end to secure the coupling to the shafts. When clamps are tightened on the slitted ends of the coupling, the coupling ends are held securely to the shafts. The clamps are designed with bolts on one side and counterweights on the other side to distribute the load evenly when the shaft is rotating and to reduce vibration. A simpler version uses two halves of a cylinder, split axially. The two halves are bolted together (*Figure 4*).

**Describe rigid sleeve couplings and tubular sleeve couplings. Explain the configurations for each.**

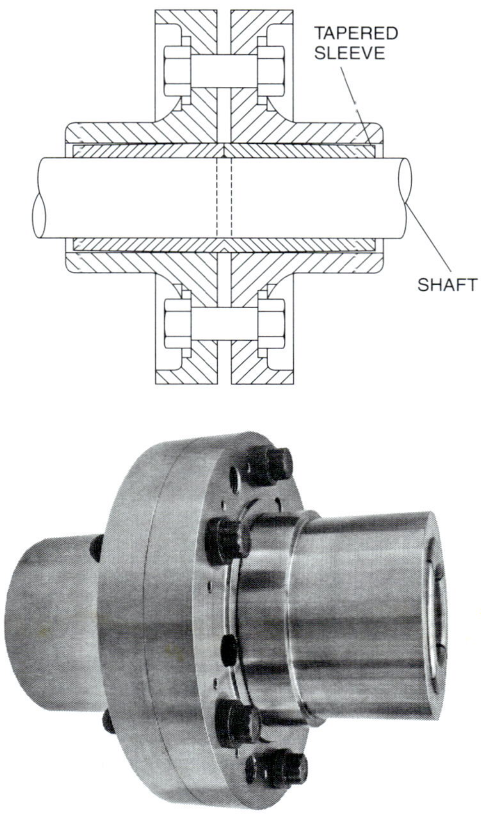


*Figure 2* ◆ Taper-bored flanged coupling.

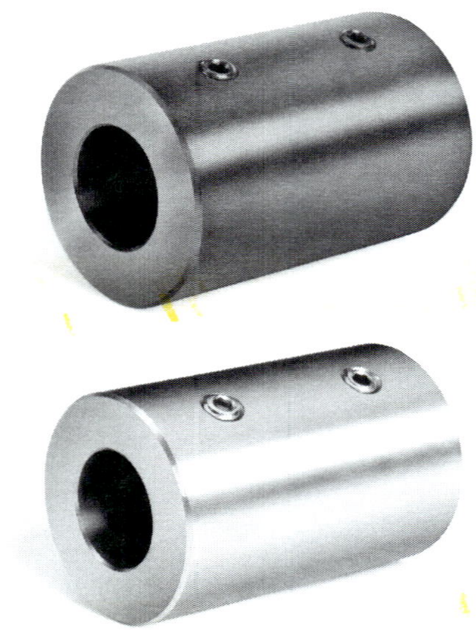


*Figure 3* ◆ Setscrew sleeve couplings.

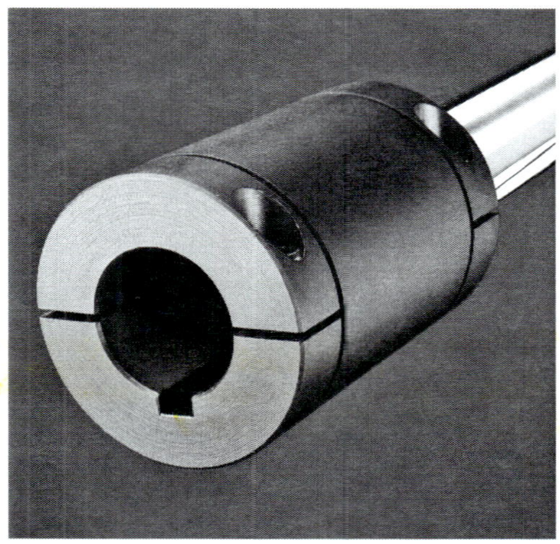


*Figure 4* ◆ Sleeve coupling.

Describe clamp coupling configurations.

Have trainees review Sections 3.0.0–3.2.8.

Ensure that you have everything required to teach this session.

## 2.3.0 Clamp Couplings

Clamp couplings are available in one- or two-piece designs. The two-piece design is used when sleeved couplings are difficult to install. The advantage of a clamp coupling is that it can be installed on shafts that are already in place without moving one of the shafts. Clamp couplings are used when the shafts are the same size and are also used for low-speed drives because of their unbalanced design and weight distribution. They are capable of high axial load and torque capacity, they will not damage the joining shafts, and they are used with or without keys.

Explain why flexible couplings are more commonly used than rigid couplings.

## 3.0.0 ◆ FLEXIBLE COUPLINGS

Flexible couplings are much more commonly used than rigid couplings because they are usually easier to install and maintain, and do not require precise alignment. Some flexible couplings allow for more severe misalignment than others. Although flexible couplings allow for some misalignment, they should be aligned as close as possible using the methods available for maximum coupling life. For instance, a coupling may have a maximum tolerance of .010, but it cannot be used at that amount of misalignment without shortening the life of the coupling.

Flexible couplings should not be used when major **angular misalignment** is known to exist. Deliberate misalignment requires the use of universal joints. Flexible couplings can classified as either mechanical or material types.

Discuss mechanical flexible couplings and list the four major types.

### 3.1.0 Mechanical Flexible Couplings

Mechanical flexible couplings have metal components that may or may not need lubrication. They use the play, or clearance, in a mechanical device, such as chains or gears, to compensate for misalignment. The four major types of mechanical flexible couplings are slider, gear, chain, and grid.

Show Transparency 5 (Figure 5). Discuss slider couplings.

### *3.1.1 Slider Couplings*

Slider couplings allow for angular and **parallel misalignment**. They are designed for low-speed and high-torque applications and are rated up to a maximum speed of 100 rpm. Slider couplings have three pieces: a slider and a two-jawed coupling half for each shaft (*Figure 5*). The slider is driven by one of the two-jawed coupling halves, which in turn drives the other coupling half. Some types have replaceable wear faces on the jaws.

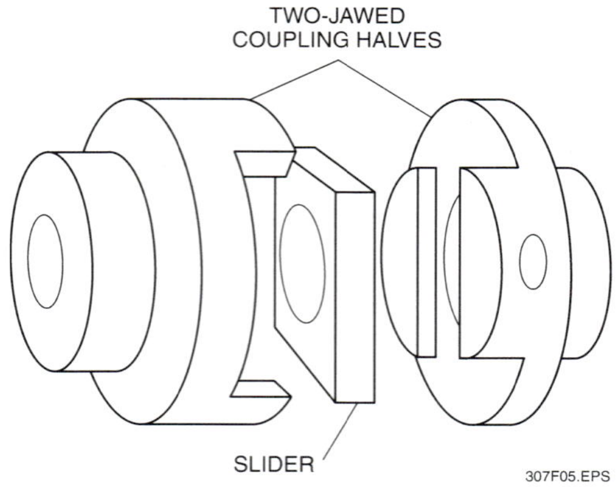


*Figure 5* ◆ Slider coupling.

Describe gear couplings.

### *3.1.2 Gear Couplings*

A gear coupling (*Figure 6*) consists of two coupling halves with external teeth and a mating flange, or sleeve, with internal teeth. The mating flange may be one piece or two, depending on the application. The flange or sleeve may be made of steel or nylon. The advantage of the nylon is that it does not require lubrication.

Gear couplings use keys to prevent them from slipping on the shafts. In perfectly aligned applications, the load is evenly distributed between all the teeth of the coupling. Where misalignment occurs, the load is not evenly distributed. This is why the load rating, or the amount of load that the coupling can carry, drops when misalignment increases.

Gear couplings that are used in high-speed applications are usually balanced and match marked when manufactured. These match marks must be perfectly aligned during assembly. The match-marking process reduces the chance of vibration and alignment problems.

Describe the configuration of chain couplings.

### *3.1.3 Chain Couplings*

Chain couplings consist of three main parts: two hardened sprockets and a chain. The sprockets fit on the driver and driven shafts; the chain is placed around the sprockets, and the ends are fastened together. The clearances between the sprockets, the chain, and in the chain itself compensate for misalignment.

Chain couplings have hardened sprockets for longer wear and are usually enclosed in a sealed cover to keep lubrication on the chain and to keep contaminants out. The chain is usually double-stranded and may be made of case-hardened steel

ALMOST ALWAY DOUBLE



Instructor's Notes:

7.4 MILLWRIGHT ◆ LEVEL THREE

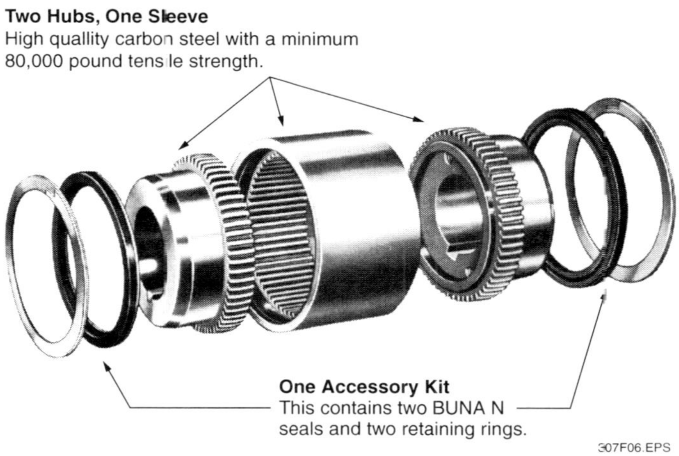


*Figure 6* ◆ Gear coupling.

or nylon. If nylon chain is used, the coupling needs no lubrication.

### *3.1.4 Grid Couplings*

Grid couplings, also called Falk couplings, consist of three basic parts: two slotted hubs and one spring grid. The slotted hubs fit on the driver and driven shafts. The grid spring fits in the slots of the hubs, and the grid encircles the hubs. The flex of the grid and the play between the grid and the slotted hubs compensate for any misalignment. Grid couplings are enclosed in a cover to keep lubrication in and contamination out of the grid and hubs.

There are two basic designs of grid couplings: the horizontal split grid coupling and the vertical split grid coupling (*Figure 7*). The horizontal split

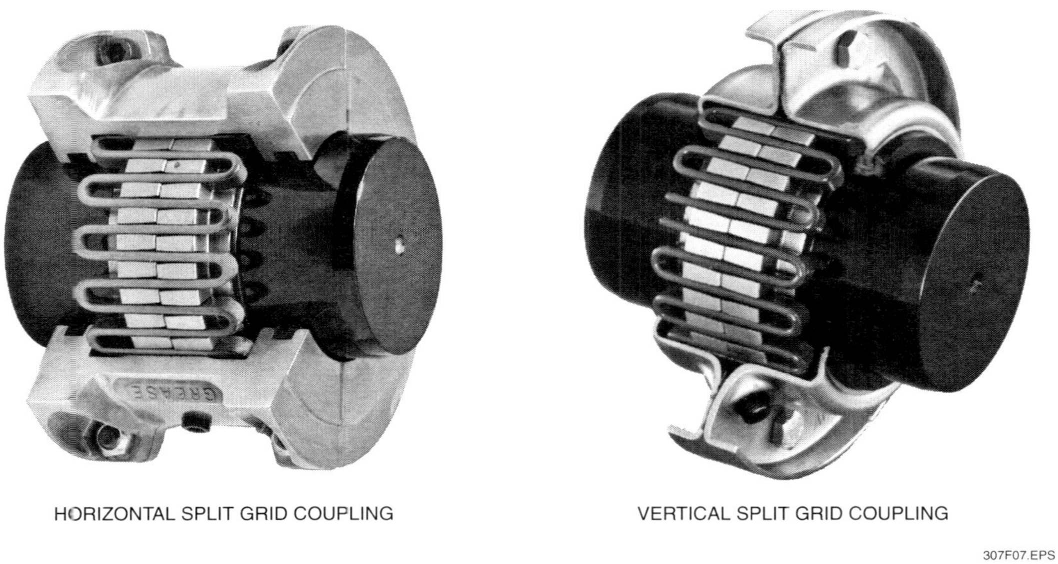


*Figure 7* ◆ Basic grid coupling designs.

**Discuss gear couplings and their parts.**

**Describe horizontal split gear couplings and vertical split gear couplings**

**Describe the flexing elements of material flexible couplings.**

**Introduce common types of material flexible couplings.**

grid coupling is the most versatile design as it allows easy access to the grid member while requiring the least amount of axial space. The vertical split grid coupling is used for higher speed applications, but requires more axial space for grid access and cover assembly.

### 3.2.0 Material Flexible Couplings

Material flexible couplings are designed to allow parts of the couplings to flex to compensate for misalignment. These flexing elements can be made of various materials, such as metal, rubber, or plastic. The life of the coupling depends on the life of the flexible material. As the materials flex, they begin to wear. The more the couplings are misaligned, the more the material is flexed and the faster it wears.

Material flexible couplings with metal flexible members have predictable life expectancies. Those with plastic or other elastomers have poorly defined fatigue limits. Their life expectancy is determined by the amount of misalignment and the operating conditions. There are many types of material flexible couplings. Some of the most common are the following:

- Spider
- Spring
- Tire
- Flexible disc
- Pin and bushing
- Pin and disc
- Spacer
- Universal joint

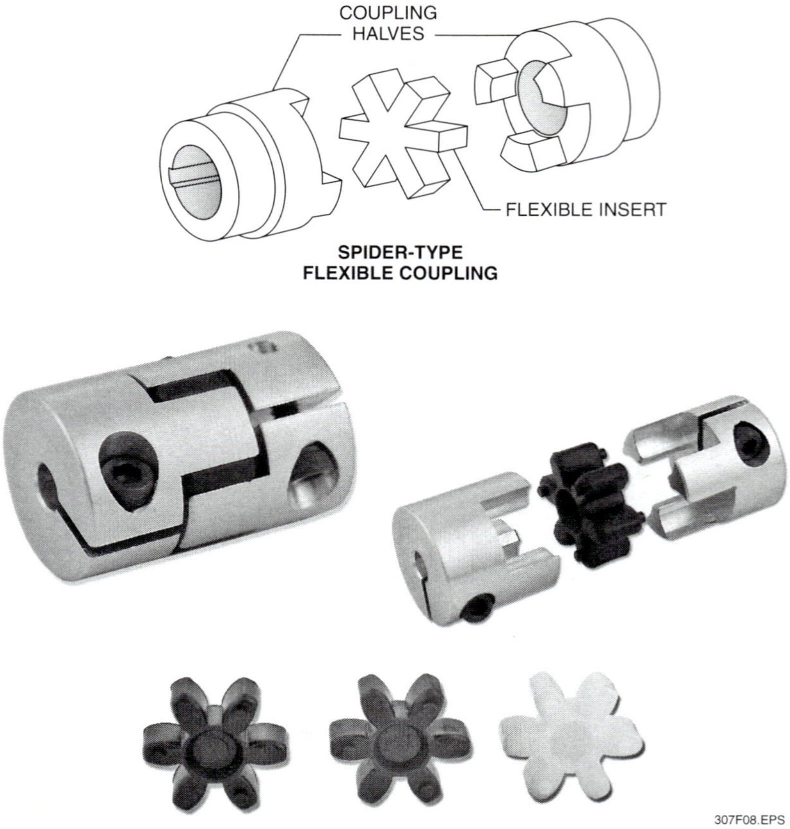


*Figure 8* ◆ Spider coupling.



Instructor's Notes:

7.6 MILLWRIGHT ◆ LEVEL THREE

### 3.2.1 Spider Couplings

Spider couplings, also known as jaw couplings, have two coupling halves and a flexible insert (*Figure 8*). One coupling half drives the insert, which in turn drives the other coupling half. The flexible insert absorbs the misalignment between the two coupling halves. One advantage of the spider coupling is that the halves are interchangeable with other halves and inserts.

### 3.2.2 Spring Couplings

A spring coupling, also known as a Bellows coupling, consists of two coupling hubs and a spring (*Figure 9*). The hubs fit on the driver and the driven shafts, and the spring is fastened between the hubs. As the coupling rotates, the misalignment is compensated for by the flex in the spring. A spring coupling will compensate for a great deal of misalignment, but the greater the misalignment, the more the spring must flex, which reduces coupling life. One advantage of the spring coupling is that the spring can be replaced without moving the hubs on the shafts.

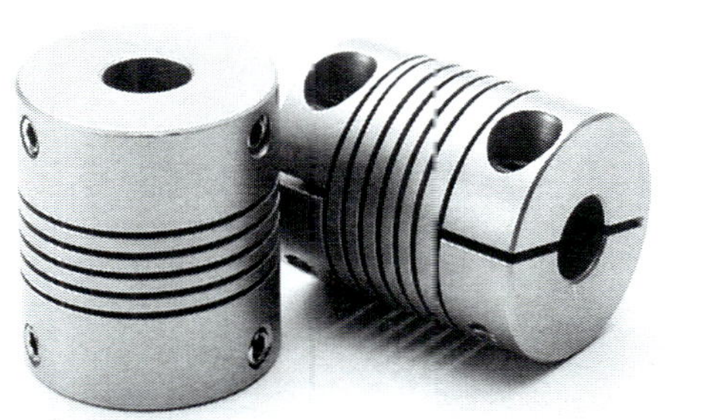

307F09.EPS

*Figure 9* ◆ Spring coupling.

### 3.2.3 Tire Couplings

A tire coupling resembles an automobile tire. It consists of two hubs and a flexible rubber member. The flexible member is attached to both hubs. The tire coupling is a very durable coupling capable of transmitting high horsepower and high speeds. It absorbs misalignment and shock and prevents the passage of electrical current. The flexible member can usually be replaced without moving the coupling on the shafts.

### 3.2.4 Flexible Disc Couplings

A flexible disc coupling (*Figure 10*) has laminated metal discs that absorb misalignment. The discs are bolted to each flange and are connected to each other by flexible spacers supported by a laminated steel center disc. The spring action gives torsional flexibility, and the two side discs compensate for

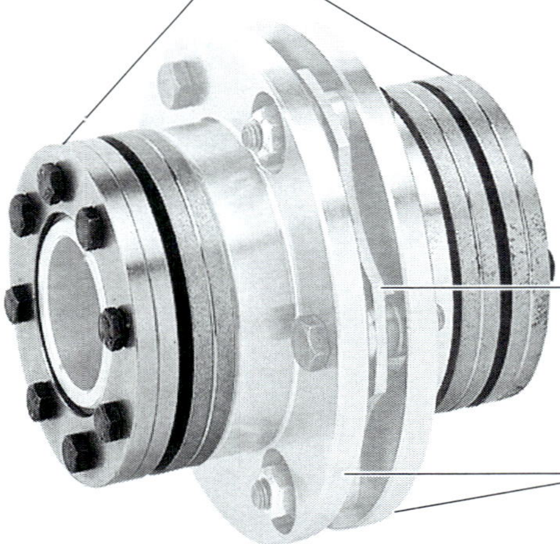


307F10.EPS

*Figure 10* ◆ Flexible disc coupling.



**Show Transparency 6 (Figure 8). Discuss spider couplings and explain how they are used.**

**Describe spring couplings and tire couplings.**

**Explain how flexible disc couplings work.**

MODULE 15307-08 ◆ COUPLINGS 7.7

**Discuss the use of pin and bushing couplings.**

**Describe the configuration of pin and disc couplings.**

**Show Transparency 7 (Figure 12).**

angular and offset misalignment. This type of coupling provides a positive drive in either direction without backlash.

### 3.2.5 Pin and Bushing Couplings

A pin and bushing coupling (*Figure 11*) consists of two coupling halves and a number of steel pins. When the coupling is assembled, the pins are tightly bolted to one coupling half and fit into rubber bushings in the second coupling half. The flex in the pins, along with mating with the rubber bushings, absorbs misalignment.

### 3.2.6 Pin and Disc Couplings

A pin and disc coupling (*Figure 12*) consists of two coupling halves, pins, and a disc. Pins are tightly bolted to both coupling halves. The disc fits between the halves, and the pins fit through the holes in the disc to transmit power. The flexing of the pins and the disc absorbs the misalignment. The disc of the coupling is available in various widths for different applications of horsepower and speed. As a general rule, the wider the disc, the longer the service life, because the force is spread over a larger area.

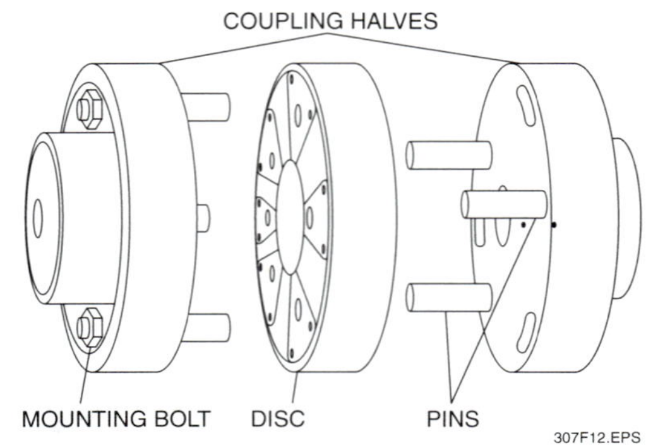


*Figure 12* ◆ Pin and disc coupling.

*Figure 11* ◆ Pin and bushing coupling.



Instructor's Notes:

7.8 MILLWRIGHT ◆ LEVEL THREE

### *3.2.7 Spacer Couplings*

A spacer coupling (*Figure 13*) is a small spacer that is placed between two flexible couplings. Spacer couplings are commonly used on centrifugal pumps where the spacer can be removed so that the flexible couplings can be disassembled without moving the pump or motor on the base. This eliminates the need to realign the couplings after reassembly.

### *3.2.8 Universal Joint Couplings*

A universal joint coupling is used to transmit high torque under conditions of severe misalignment. Universal joint couplings come in two basic styles: single-joint and double-joint. A special splined sleeve, as shown in *Figure 14*, can be used if movement of the driver or the driven is expected to occur. Universal joints are used extensively in vehicles.

## 4.0.0 ◆ SOFT-START COUPLINGS

Soft-start couplings are used in applications where smooth, even starts are needed. Soft-start couplings allow the driving motor to pick up speed before the load is engaged and allow the driven to start slowly and smoothly. They also prevent stalls during overload conditions. The three most common soft-start couplings are the following:

- Fluid
- Shot
- Clutch-style

### 4.1.0 Fluid Couplings

A fluid coupling consists of two members: an impeller or pump (the driving member) and a runner or turbine (the driven member). A fluid coupling transmits power through kinetic energy, or the mass and volume of moving oil. The impeller has fins that extend from the center of the coupling. When in motion, a continuous oil stream is forced outward between the impeller fins and around the circumference of the coupling and is then thrown against the blades of the runner. The centrifugal action of the coupling transmits velocity to the mass of oil.

The oil stream striking against the runner transmits energy to the runner. The oil stream flows through the runner fins, leaves the inner row of fins, and re-enters the inner row of impeller fins. The impeller fins again pick up the oil and restore velocity lost in the runner fins. The process is repeated, continuously providing energy to the runner.

During operation, the driver member and the driven member of the coupling rotate at different speeds, and it is impossible for the fluid to return to the same impeller fins from which it left. The shape and rotation of the impeller and the runner produce a flow path called a **vortex** (*Figure 15*). At startup, when the impeller is rotating and the runner is not, high vortex occurs. The vortex action

307F13.EPS

*Figure 13* ◆ Spacer coupling.

307F14.EPS

*Figure 14* ◆ Double-joint universal joint coupling.

**Explain why spacer couplings are commonly used on centrifugal pumps.**

**Explain that universal joint couplings can be single-joint or double-joint style, and are used extensively in vehicles.**

**Have trainees review Sections 4.0.0–4.3.0.**

**Ensure that you have everything required to teach this session.**

**Discuss applications for soft-start couplings and the three most common types.**

**Explain how fluid couplings transmit power through kinetic energy.**

**Show Transparency 8 (Figure 15). Discuss the vortex action of fluid couplings.**



MODULE 15307-08 ◆ COUPLINGS 7.9

**Describe how shot couplings operate using centrifugal force.**

**Explain the configuration of shot couplings.**

**Discuss how shot couplings provide protection against overloading the drive motor.**

**Show Transparency 9 (Figure 16).**

**Describe the parts of clutch-style couplings and discuss how they operate.**

**Show Transparency 10 (Figure 17).**

**Have trainees practice identifying and explaining coupling types. Note the proficiency of each trainee. This laboratory corresponds to Performance Task 1.**

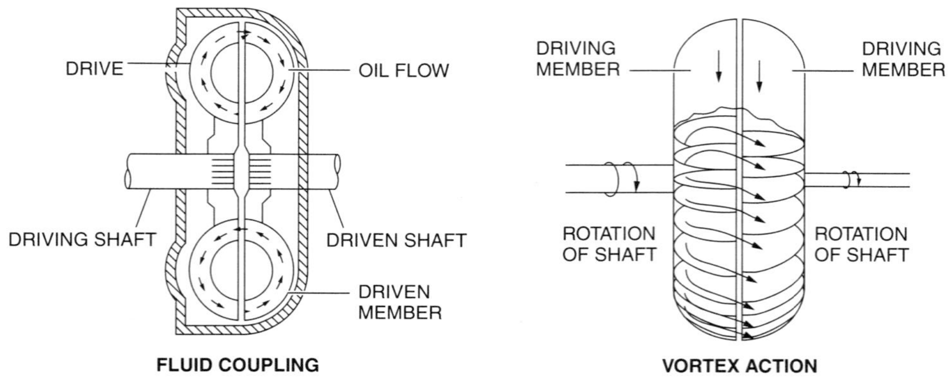


*Figure 15* ◆ Vortex action of fluid coupling.

decreases as the speed of the runner nears the speed of the impeller, and low vortex exists when the impeller and the runner are rotating at nearly equal speeds. The higher the vortex, the greater the driving power as the oil streams from the impeller strike the runner fins at an angle of almost 90 degrees. When the impeller and runner are rotating at the same speeds, there is zero vortex, and the runner is carried in the current of oil.

Fluid couplings do not allow for misalignment. When they are used as a coupling between a driver and a driven machine, a flexible coupling should be used with the fluid coupling to allow for misalignment and shaft float.

## 4.2.0 Shot Couplings

A shot coupling is a centrifugal coupling. It relies on centrifugal force to provide its turning power. A shot coupling (*Figure 16*) consists of a housing, or the driver; a rotor, or the driven; and a quantity of steel shot. The rotor fits inside the housing that contains the steel shot. When the housing starts to turn, centrifugal force throws the steel shot to the perimeter of the housing, where it packs between the housing and the rotor, which transmits power. When speed is established, there is normally no slippage between the housing and the rotor. The amount of power that shot couplings produce can be varied by changing the amount of shot in the coupling.

Shot couplings provide protection against overloading the drive motor, because as the speed drops, the centrifugal force decreases, thereby relieving the load on the motor. Continued overloading and slippage in the coupling generate heat, which shortens coupling life expectancy. Shot couplings do not allow for misalignment and should be used with a flexible coupling.

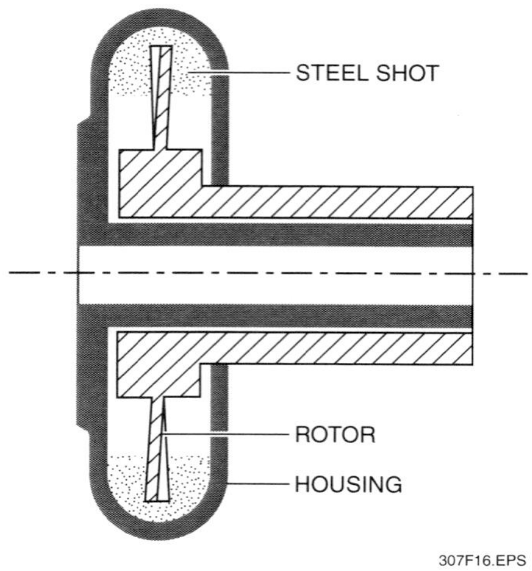


*Figure 16* ◆ Shot coupling.

## 4.3.0 Clutch-Style Couplings

Clutch-style couplings (*Figure 17*) consist of a drum, brake linings, and weights. They allow the driver to come up to partial speed before the load is engaged. When the drive motor is started, centrifugal force causes the weights of the coupling to press against the brake linings, which engage the drum, transmitting force to the driven. The weights may be spring-loaded in some couplings. Spring-loaded weights exert force on the linings only after a certain speed is reached.

Clutch-style couplings allow for minor misalignment, but they will slip when overloaded. Slippage must be controlled, since a slipping coupling generates a great deal of heat. The excessive heat will damage the coupling, so heat-sensing



Instructor's Notes:

7.10 MILLWRIGHT ◆ LEVEL THREE

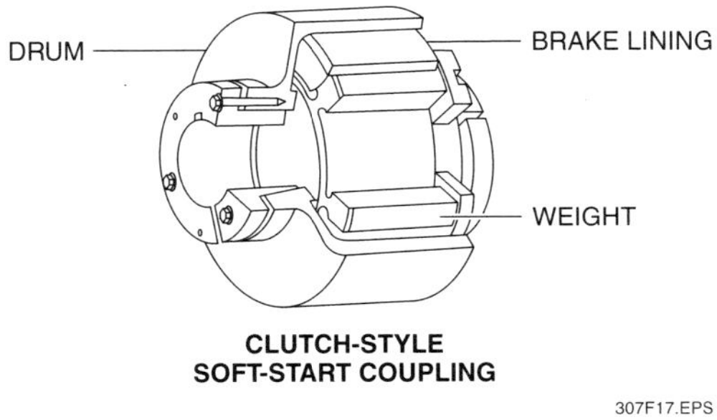


*Figure 17* ◆ Clutch-style coupling.

devices are commonly installed on clutch-style couplings. The heat-sensing device shuts off power if the **clutch** is slipping.

## 5.0.0 ◆ INSTALLING COUPLINGS

In order for a coupling to perform efficiently and give maximum service life, it must be installed properly. Improper installation of couplings can cause equipment damage, excessive vibration, premature wear-out, and coupling failure. Never drive a coupling onto a shaft using a hammer because this will ruin the coupling.

Some couplings are relatively simple to install, while others are more difficult. These sections explain the following methods of coupling installation:

- General coupling installation procedures
- Split coupling installation
- Interference-fit installation
- Press-fit installation
- Setting coupling gap
- Grid coupling installation
- Hydraulic coupling installation

### 5.1.0 General Coupling Installation Procedures

The two types of fits for any equipment mounted on a shaft are the clearance fit and the interference fit. The interference fit, also known as a press fit, forces one object into another whose inside diameter is slightly smaller than the outside diameter of the inner piece. The clearance fit has enough room inside the outer piece for the inner to slide in fairly freely, and then some form of fastening is used to lock the two together. This might include key and setscrew, clamping, or some form of taper-lock.

Regardless of the installation method required for a particular coupling, there are certain general installation procedures that should always be followed. Follow these steps before installing any coupling:

*Step 1* Rotate the shafts of the driver and the driven equipment to ensure that they turn freely and that there are no rough spots or binding. Rough spots or binding indicate that there are problems with the bearings in the driver or driven equipment. These problems must be corrected before installing couplings.

*Step 2* Inspect the shafts to ensure that there are no burrs or high spots. If there are any burrs or high spots, remove them from the shaft, using a honing, grinder, or appropriate tool.

> **CAUTION**
> Hone the shafts only enough to remove the burrs and high spots. Never hone a shaft below its original diameter.

*Step 3* Measure the shafts in several places, using a micrometer or dial indicator, to check for excessive wear and out-of-roundness. If a shaft is excessively worn or out of round, it should be replaced.

*Step 4* Lubricate the shaft lightly, according to specifications.

*Step 5* Check the coupling against the manufacturer's parts list to ensure that you have all of the parts.

*Step 6* Determine what type of coupling installation is being made.

### 5.2.0 Split Coupling Installation

Split couplings are installed after the driver and driven shafts have been aligned and the equipment has been fastened into position. Split couplings are rigid and require precise alignment of the shafts. The coupling is placed around both shafts and bolted in position. Follow these steps to install a split coupling:

*Step 1* Measure the shaft diameter, using a micrometer, to determine the size of the coupling needed.

*Step 2* Select a split coupling of the proper size for the shaft. Split couplings are generally used on applications in which both shafts are the same diameter.



**Have trainees review Sections 5.0.0–5.6.0.**

**Ensure that you have everything required to teach this session.**

**Explain the consequences of improper coupling installation.**

**Introduce the methods listed for installing couplings.**

**Define clearance fit and interference fit.**

**Review the general installation guidelines and procedures.**

**Discuss installation procedures for a split coupling.**

MODULE 15307-08 ◆ COUPLINGS 7.11

**Show Transparency 11 (Figure 18). Explain how the gap allows the coupling to be tightened to full torque value.**

**Show Transparency 12 (Figure 19). Explain that the bolts must be tightened to the proper torque in the right sequence.**

**Describe the interference-fit installation method.**

**Review the steps to install an interference coupling.**

**Demonstrate how to install an interference coupling.**

**Have trainees practice installing an interference coupling. Note the proficiency of each trainee. This laboratory corresponds to Performance Task 2.**

*Step 3* Test-fit the coupling on the shafts to ensure that there is a sufficient gap between the top and bottom halves, approximately $\frac{1}{16}$ to $\frac{1}{8}$ inch on each side. This gap, as shown in *Figure 18*, allows the coupling to be tightened to full torque value in order to achieve full clamping action on the shafts.

*Step 4* Position one half of the coupling on the bottoms of the shafts, and center it where the two shafts meet.

*Step 5* Position the other coupling half on the tops of the shafts.

*Step 6* Insert all of the bolts, and tighten them finger tight.

*Step 7* Tighten the bolts to the proper torque, following the proper tightening sequence, starting at the center and working toward both ends. *Figure 19* shows the proper tightening sequence.

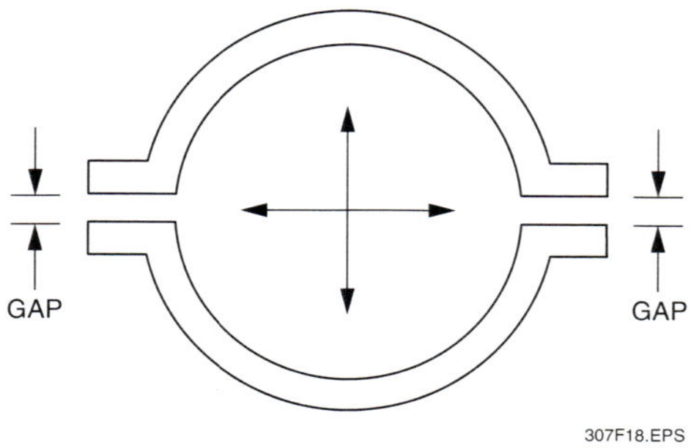


*Figure 18* ◆ Gap.

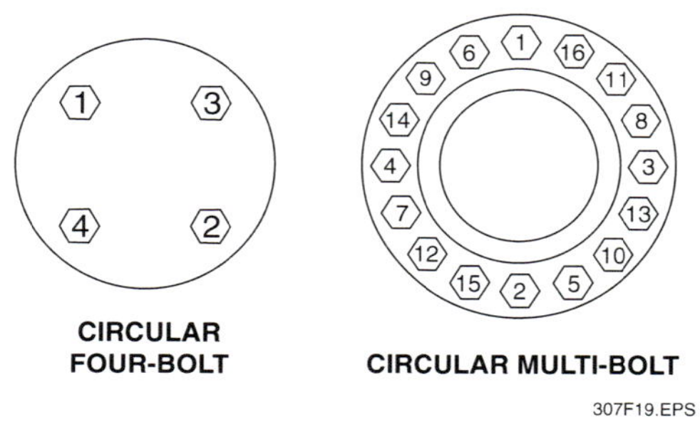


*Figure 19* ◆ Tightening sequence.

> **CAUTION**
> Repeat the tightening sequence two or three times, tightening the bolts a little at a time until the proper torque value is reached. The proper torque value is listed on the manufacturer's specification sheet. Improper tightening sequence or overtightening can break the coupling.

## 5.3.0 Interference-Fit Installation

In the interference-fit method of installation, the coupling has a bore that is slightly smaller than the shaft diameter. The coupling must be either pressed on or preheated to expand the bore before it can be installed on the shaft. As the coupling cools, it shrinks to grip the shaft securely. Follow these steps to install a coupling, using the interference-fit method.

*Step 1* Determine from the manufacturer's installation procedure where the coupling should fit on the shaft.

*Step 2* Clean the shaft thoroughly to remove any foreign material that might interfere with the coupling sliding onto the shaft.

*Step 3* Measure the shaft diameter, using a micrometer, and record the measurement.

*Step 4* Measure the height and width of the key installed on the shaft, using an outside micrometer.

*Step 5* Measure the inside diameter of the coupling, using an inside micrometer, to ensure that the coupling is the proper size to fit the shaft.

> **NOTE**
> The inside diameter of the coupling should be smaller than the outside diameter of the shaft by no more than 0.0015 inch per 1 inch of shaft diameter. For example, a 1-inch shaft requires a coupling that is 0.0015 inch smaller; a 2-inch shaft requires a coupling that is 0.0030 inch smaller.

*Step 6* Install a stop on the shaft that will stop the coupling in its proper position when it is installed. The stop will help ensure that the coupling is positioned properly and quickly. The stop may be a clamp or some other type of stop fastened to the shaft that can be removed after the coupling installation is complete.



Instructor's Notes:

7.12 MILLWRIGHT ◆ LEVEL THREE

*Step 7* Heat the coupling to a temperature of between 200°F and 250°F, using an appropriate method.

**CAUTION**

Be careful not to heat the coupling over 250°F as overheating will damage the coupling.

*Step 8* Measure the inside diameter of the coupling to ensure that it has expanded enough to fit on the shaft. Using a temperature stick can give a more accurate gauge of the heat and expansion.

*Step 9* Remove the coupling from the heating device; ensure the proper alignment and fit of the key and keyway; slip the coupling onto the shaft, and position it against the stop as quickly as possible before it begins to cool and shrink.

**WARNING!**

Wear protective gloves when handling the coupling to prevent burning your hands.

*Step 10* Remove the stop from the shaft.

*Step 11* Provide warning indicators to prevent personnel from touching the hot coupling.

### *5.3.1 Press-Fit Installation*

Some types of interference-fit couplings are mechanically press-fitted onto the shaft. This type of installation requires a hydraulic press to fit the coupling on the shaft. Press-fit installation usually requires that the shaft be removed from the equipment and transported to the press. Follow these steps to install a coupling, using the press-fit method:

*Step 1* Remove the shaft from the equipment, and carry it to the hydraulic press.

*Step 2* Measure the shaft diameter using a micrometer, and record the measurement.

*Step 3* Measure the inside diameter of the coupling, using a micrometer, to ensure that the coupling is the proper size for the shaft.

**NOTE**

For proper fit in a press-fit installation, the inside diameter of the coupling must be 0.0005 per 1-inch diameter of the shaft less than the shaft diameter. For example, a 2.000-inch diameter shaft needs a 1.999-inch bore coupling to press fit:

2.000 (diameter of shaft) – 0.001 (0.0005 × 2)
= 1.999

*Step 4* Mount the shaft in the hydraulic press.

**CAUTION**

The method for mounting the shaft in the hydraulic press depends on the type of press used. The shaft must be mounted in the press so that the shaft is supported when it is under pressure and so that the shaft is not damaged.

*Step 5* Lubricate the shaft and coupling lightly.

*Step 6* Position the coupling on the end of the shaft.

*Step 7* Lower the press ram onto the coupling.

*Step 8* Apply pressure slowly to press the coupling onto the shaft until the coupling is in its proper position.

**CAUTION**

Be careful not to press the coupling past its intended position. Monitor the press pressure gauge, being careful not to apply excessive pressure because this can damage the coupling and the shaft. Do not use a hammer or any other instrument that could damage the coupling.

*Step 9* Release the pressure, and remove the shaft from the press.

*Step 10* Reinstall the shaft in the equipment.

## 5.4.0 Setting the Coupling Gap

The procedure for checking the **coupling gap** is also called the axial alignment of the mating shafts. The importance of setting the proper coupling gap can, in some applications, be critical. For applications where shafts are exposed to thermal expansion or move to a different position under running conditions, the coupling gap is as critical as the alignment in other directions.



**Stress the importance of wearing protective gloves when handling hot couplings.**

**Review the procedure for press-fit installation.**

**Discuss the procedure for setting the coupling gap. Explain why the proper coupling gap can be critical.**

MODULE 15307-08 ◆ COUPLINGS 7.13

**Explain the consequences of an improper coupling gap.**

**Review the steps for checking the coupling gap.**

**Discuss grid coupling installation.**

**Show Transparency 13 (Table 1). Explain how these specifications are used to install grid couplings.**

**Review the steps of the assembly process for a grid coupling.**

An example where the coupling gap is critical is a pump that works with very hot fluids, where the drive shaft expands and contracts a considerable amount. This would also occur with many turbines, compressors, gearboxes, and speed regulators. Other examples include electrical motors with plain bearings that have an excessive amount of endplay.

An improper coupling gap causes excessive axial forces which results in increased bearing load and premature bearing failure. The improper coupling gap can also cause destruction of an electrical motor with a plain bearing. Therefore, it is extremely important to set the correct coupling gap according to the manufacturer's specifications. Follow these steps to check the coupling gap:

*Step 1* Loosen the coupling bolts and check that you have axial clearance between the two coupling halves.

*Step 2* Use a scale, feeler gauge, taper, or an inside micrometer to measure the gap between the coupling halves.

*Step 3* Compare this measurement with the manufacturer's specifications.

*Step 4* To make corrections to the coupling gap, move the driver backward or forward to achieve the proper gap.

## 5.5.0 Grid Coupling Installation

Grid couplings, under a number of brand names, vary in a number of ways. Before beginning the installation process, you must refer to the manufacturer's documentation for specific operating limits (such as rpm limits), installation procedures, coupling gap, specified lubricant types, and required grid or rung spacers. These instructions will vary from manufacturer to manufacturer and must be followed to prevent improper loading of the shaft bearings and premature failures of the driver and/or driven equipment. *Table 1* shows a sample chart that must be used when installing grid couplings.

In this table, installation and operating limits are given for each type of coupling listed. The coupling gap (normal gap), cover fastener torque values, allowable speed, and required lubricant types are also shown.

The following steps are a basic description of the assembly process for a grid coupling. Read and follow the manufacturer's instructions, to prevent damage to or failure of the coupling.

*Step 1* Determine the size and type of all components being used to ensure that everything matches the job specifications. Unpack all of the components, and inspect them to be sure all are correct and undamaged.

*Step 2* Clean all parts, and make sure they are smooth and free of coatings and lubricants.

*Step 3* If the shafts are not interference fits, slide them into the coupling halves equally on both sides, and tighten the setscrews. If an interference fit with the shaft is required, heat the hubs and mount the hubs with the hub teeth flush with the shaft ends. Position the equipment to achieve the appropriate distance between shaft ends.

*Step 4* Most types of grid couplings will be lubricated at this point. However, check your manufacturer's instructions, to be sure of the lubrication requirements.

*Step 5* This is the point when spacers would be attached in most cases

*Step 6* Torque all the fasteners as specified by the manufacturer.

*Step 7* Check the angular and parallel misalignment and make sure that they are within the manufacturer's limits. Remember that if the limits are exceeded, the warranty may be nullified.

*Step 8* Tighten all the setscrews to the torque values specified, and recheck the alignment. If necessary, re-align the coupling.

*Step 9* If specified, apply a lubricant to the grooves before installing the grid. The segments must be turned so that the ends point in the same direction.

*Step 10* Position seals on spacer hubs so that they line up with grooves in cover.

*Step 11* Position gaskets on lower cover half and assemble covers so that match marks are on the same side. If using the coupling in any position other than horizontal, assemble cover halves with the lug and match mark up, or on the high side.

*Step 12* Install the fasteners through the cover halves and torque to the manufacturer's horizontal cover fastener specifications.

*Step 13* Make certain all plugs are inserted and secured before operating the equipment.



Instructor's Notes:

7.14 MILLWRIGHT ◆ LEVEL THREE

Table 1 Grid Coupling Specifications

| Size | Installation Limits | | | Size | Operating Limits | | | Size | Cover Fastener Tightening Torque Values | Allow Speed (rpm) | Lube Wt |
|---|---|---|---|---|---|---|---|---|---|---|---|
| | Parallel Offset-P | Angular (x–y) | Normal Gap ±10% | | Parallel Offset-P | Angular (x–y) | End Float Physical Limit (Min) 2 x F | | In Series Fasteners (lb–in) | | |
| | Max in | Max in | in | | Max in | Max in | in | | | | lb |
| C1020 | 0.006 | 0.003 | 0.125 | C1020 | 0.012 | 0.010 | 0.210 | C1020 | 0.012 | 0.25 | 0.21 |
| C1030 | 0.006 | 0.003 | 0.125 | C1030 | 0.012 | 0.012 | 0.198 | C1030 | 0.012 | 0.30 | 0.198 |
| C1040 | 0.006 | 0.003 | 0.125 | C1040 | 0.012 | 0.013 | 0.211 | C1040 | 0.012 | 0.33 | 0.211 |
| C1050 | 0.008 | 0.004 | 0.125 | C1050 | 0.016 | 0.016 | 0.212 | C1050 | 0.016 | 0.41 | 0.212 |
| C1060 | 0.008 | 0.005 | 0.125 | C1060 | 0.016 | 0.018 | 0.258 | C1060 | 0.016 | 0.46 | 0.258 |
| C1070 | 0.008 | 0.005 | 0.125 | C1070 | 0.016 | 0.020 | 0.259 | C1070 | 0.016 | 0.51 | 0.259 |
| C1080 | 0.008 | 0.006 | 0.125 | C1080 | 0.016 | 0.024 | 0.288 | C1080 | 0.016 | 0.61 | 0.288 |
| C1090 | 0.008 | 0.007 | 0.125 | C1090 | 0.016 | 0.028 | 0.286 | C1090 | 0.016 | 0.71 | 0.286 |
| C1100 | 0.010 | 0.008 | 0.188 | C1100 | 0.020 | 0.033 | 0.429 | C1100 | 0.020 | 0.84 | 0.429 |
| C1110 | 0.010 | 0.009 | 0.188 | C1110 | 0.020 | 0.036 | 0.429 | C1110 | 0.020 | 0.91 | 0.429 |
| C1120 | 0.011 | 0.010 | 0.250 | C1120 | 0.022 | 0.040 | 0.556 | C1120 | 0.022 | 1.02 | 0.556 |
| C1130 | 0.011 | 0.012 | 0.250 | C1130 | 0.022 | 0.047 | 0.551 | C1130 | 0.022 | 1.19 | 0.551 |
| C1140 | 0.011 | 0.013 | 0.250 | C1140 | 0.022 | 0.053 | 0.571 | C1140 | 0.022 | 1.35 | 0.571 |

307T01.EPS



MODULE 15307-08 ◆ COUPLINGS 7.15

Describe the hydraulic installation method.

Demonstrate how to install a slip fit coupling.

Have trainees practice installing a slip fit coupling. Note the proficiency of each trainee. This laboratory corresponds to Performance Task 3.

Have trainees review Sections 6.0.0–6.3.0.

Ensure that you have everything required to teach this session.

Introduce the methods used to remove couplings.

Review the general coupling removal procedures.

Discuss mechanical pullers.

Show Transparency 14 (Figure 20). Explain how manual and hydraulic mechanical pullers are used.

Warn trainees of the dangers of using dry ice to cool a shaft.

> **NOTE**
> Install coupling guard per OSHA or *ASME 815.1* requirements.

### 5.6.0 Installing Hydraulic Couplings

Couplings may be clamped by hydraulic pressure. The clamping action is produced by a hydraulic pump, and the valve on the coupling is shut. The pressure is thus maintained at the coupling. To remove the coupling, the hose is reconnected and the pressure is released. The most common version uses an inner sleeve that has a tapered exterior. The outer sleeve is slid on, and mineral oil is injected into the space. A built-in hydraulic jack forces the outer sleeve onto the taper and creates an interference fit.

The hydraulic coupling pusher is used on a tapered shaft. The coupling is placed on the shaft until it is snug. The pusher is fastened to the threaded end of the shaft, and the pump pushes the coupling onto the shaft until it is locked in place. The pusher is then removed and the threaded collar is attached.

## 6.0.0 ◆ REMOVING COUPLINGS

Couplings must be removed when they wear out or are damaged. They may also need to be removed when the equipment that they are on needs service, repair, or replacement. The following sections explain the following methods of coupling removal:

- General coupling removal procedures
- Mechanical pullers
- Hydraulic removal method

### 6.1.0 General Coupling Removal Procedures

Regardless of the removal method required for a particular coupling, there are certain general removal procedures that should always be followed. Follow these steps before removing any coupling.

*Step 1* Inspect the coupling for wear and damage. If the coupling is worn or damaged, try to determine the reason for the wear or damage so that the situation can be corrected before replacing the coupling again.

*Step 2* Disassemble the coupling.

*Step 3* Inspect the coupling parts for wear and damage. If a part is worn or damaged, try to determine the reason for the wear or damage, such as improper lubrication or misalignment, so that the problem can be corrected.

*Step 4* Clean the coupling hubs thoroughly, using a rag and solvent.

> **NOTE**
> Move the driver or driven equipment if necessary to facilitate coupling removal.

*Step 5* Determine the best method for removing the coupling. The best method may include fabricating a special tool for removal.

### 6.2.0 Mechanical Pullers

Mechanical pullers are the most common method used to remove couplings. Mechanical pullers are available in two types: manual and hydraulic (*Figure 20*). Both are attached to the coupling in the same manner, and both pull the coupling mechanically. The manual type is operated by turning the pressure bolt, using a wrench, to apply pressure to pull the coupling. The hydraulic type uses a hydraulic pump to apply pulling pressure.

If the coupling was installed by the interference-fit method, special procedures may be necessary to loosen it for removal. It is possible to use a heat blanket or a heating coil to expand it before applying pressure to remove it, in which case the coupling may be reusable. If you do not intend to reuse the coupling, it is possible to heat the coupling with a torch over the keyway, if there is one. Specifications and written procedures will usually inform you as to what you may or may not do to a coupling.

The second method is to cool the shaft until it shrinks enough to allow the coupling to be removed. This is done by applying dry ice to the shaft or by using liquid nitrogen.

> **WARNING!**
> Dry ice can cause severe burns if touched with bare skin. Be sure to wear protective gloves and use extreme caution to prevent skin contact with the dry ice.



Instructor's Notes:

7.16 MILLWRIGHT ◆ LEVEL THREE

> **NOTE**
> Be careful not to cool the coupling because this will cause the coupling to shrink also.

Follow these steps to remove a coupling, using a mechanical puller.

*Step 1* Select a puller with the correct jaw length and spread for the coupling being pulled.

*Step 2* Inspect the puller to ensure that the jaws are not bent or excessively worn and that the pressure bolt is not stripped.

*Step 3* Ensure that all setscrews or locking screws have been removed from the coupling.

*Step 4* Position the puller on the shaft, and turn the pressure screw, or pump the hydraulic pump, until the screw presses against the shaft.

*Step 5* Check the puller to ensure that the jaws are properly positioned on the coupling.

> **CAUTION**
> The puller jaws must be positioned so that they will provide a straight, even pull on the coupling and not damage it.

*Step 6* Turn the pressure bolt, or pump the hydraulic pump, to begin applying pressure to the coupling.

> **WARNING!**
> Always wear eye protection when using a puller to avoid particles that may break off and injure your eyes.

*Step 7* Continue to apply pressure until the coupling is removed from the shaft. Tap gently on the back of the coupling, if necessary, to aid in the coupling removal.

## 6.3.0 Hydraulic Removal Method

In the hydraulic removal method (*Figure 21*), fluid is pumped in between the coupling and the shaft to expand the coupling and push it off the shaft. To use this method, the shaft must be drilled and tapped for hydraulic removal before the coupling is installed.

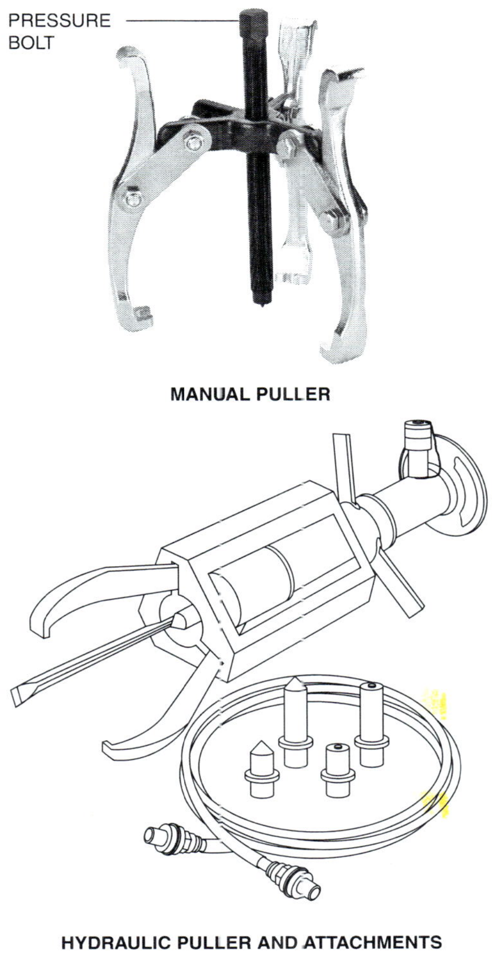


*Figure 20* ◆ Coupling pullers.

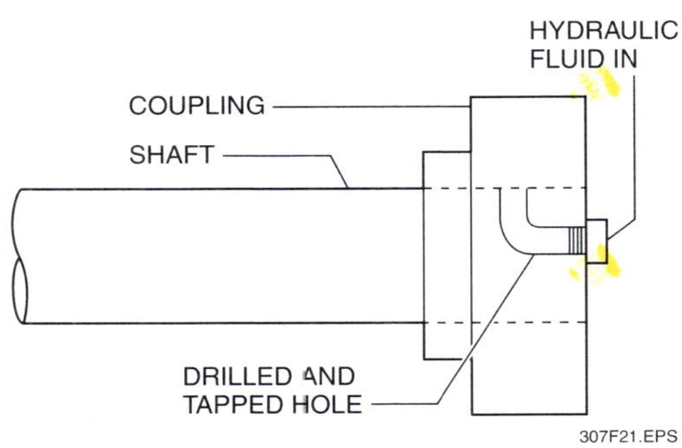


*Figure 21* ◆ Hydraulic removal method.

MODULE 15307-08 ◆ COUPLINGS 7.17

**Review the steps for removing a coupling using a mechanical puller.**

**Stress that eye protection must be worn when using a mechanical puller.**

**Explain the hydraulic removal method.**

**Show Transparency 15 (Figure 21).**

**Review the steps to remove a coupling using the hydraulic removal method.**

**Warn trainees of precautions required when using the hydraulic removal method.**

**Demonstrate how to remove couplings using the manual and hydraulic methods.**

**Have trainees practice removing couplings using either the manual or hydraulic method. Note the proficiency of each trainee. This laboratory corresponds to Performance Task 4.**

**Have trainees complete the Review Questions, and go over the answers prior to administering the Module Examination.**

Follow these steps to remove a coupling, using the hydraulic removal method.

*Step 1* Open the coupling so that the drilled and tapped hole is accessible.

*Step 2* Attach a threaded hydraulic fitting to the shaft.

*Step 3* Attach a hydraulic pump to the fitting.

*Step 4* Stand to one side, and carefully apply hydraulic pressure to the coupling until the coupling comes off the shaft.

> **WARNING!**
> Stand to one side of the coupling while removing it because a coupling under hydraulic pressure can come off the shaft with considerable force and cause personal injury.

## Review Questions

1. An example of a driver is a _____.
   a. pump
   b. compressor
   c. gearbox
   d. motor

2. Shaft end play is also known as _____.
   a. angular misalignment
   b. shaft float
   c. torque
   d. parallel misalignment

3. The advantage of a clamp coupling is that it can be installed on shafts that are already in place without moving one of the shafts.
   a. True
   b. False

4. Flexible couplings are much more commonly used than rigid couplings.
   a. True
   b. False

5. Major angular misalignment of joining shafts requires the use of _____ couplings.
   a. universal joint
   b. rigid
   c. flanged
   d. grid

6. Slider couplings are designed for low-speed and high-torque applications and are rated up to a maximum speed of 100 rpm.
   a. True
   b. False

7. A coupling that has two sprockets is a _____ coupling.
   a. spider
   b. chain
   c. grid
   d. gear

8. Elastomers are used in _____ couplings.
   a. rigid
   b. mechanical
   c. material flexible
   d. flanged

9. A durable coupling that prevents the passage of electrical current is the _____ coupling.
   a. flexible disc
   b. gear
   c. tire
   d. tapered-bore



Instructor's Notes:

7.18 MILLWRIGHT ◆ LEVEL THREE

## Review Questions

10. The coupling that has laminated metal discs that absorb misalignment is the _____ coupling.
    a. spider
    b. pin and disc
    c. spacer
    d. flexible disc

11. The type of coupling that transmits power through kinetic energy, or the mass and volume of moving oil, is the _____ coupling.
    a. fluid
    b. shot
    c. clutch-style
    d. Bellows

12. The type of coupling that has brake linings is the _____ coupling.
    a. spider
    b. spring
    c. clutch-style
    d. spacer

13. Never drive an interference-fit coupling onto a shaft using a hammer because this will ruin the coupling.
    a. True
    b. False

14. Rough spots or binding encountered when rotating the shafts of the driver and driven equipment indicate _____.
    a. bearing problems
    b. horsepower deficiencies
    c. excessive end play
    d. axial misalignment

15. The coupling must be preheated when performing a(n) _____ installation.
    a. split coupling
    b. press fit
    c. interference-fit
    d. grid coupling

16. For proper fit in a press-fit coupling installation, the inside diameter of the coupling must be _____ per 1-inch diameter of the shaft less than the shaft diameter.
    a. 0.0005
    b. 0.005
    c. 0.05
    d. 0.5

17. Excessive axial forces that can result in increased bearing load and premature bearing failure can be caused by improper _____.
    a. bearing load
    b. parallel misalignment
    c. coupling gap
    d. shaft taper

18. When installing couplings, the coupling gap values should be checked against the _____.
    a. manufacturer's specifications
    b. operating policy
    c. MSDS
    d. employee handbook

19. Couplings should be removed using _____.
    a. vise grip pliers
    b. a ball peen hammer
    c. mechanical pullers
    d. a pipe wrench

20. If a coupling was installed using the interference-fit method, it can be removed by either heating the coupling or _____.
    a. heating the shaft
    b. cooling the shaft
    c. rotating the coupling
    d. cooling the coupling



MODULE 15307-08 ◆ COUPLINGS 7.19

Summarize the major concepts presented in the module.

Administer the Module Examination. Record the results on Craft Training Report Form 200, and submit the results to the Training Program Sponsor.

Administer the Performance Test, and fill out Performance Profile Sheets for each trainee. If desired, trainee proficiency noted during laboratory sessions may be used to complete the Performance Test. Record the results on Craft Training Report Form 200, and submit the results to the Training Program Sponsor.

## Summary

Couplings are used to connect the shaft of a driver to the shaft of a driven. Couplings are manufactured in many types and sizes. Some coupling types allow for slight misalignment and end play, or shaft float, between the rotating shafts. Some couplings reduce, or dampen, or absorb vibrations or torque. Other couplings insulate the coupling halves from any electrical current transfer, which is common in some motor-generator sets. In order for equipment to operate properly and efficiently, the proper coupling must be used to connect the driver to the driven. It is also very important that the equipment is properly aligned and that the coupling is properly installed.

## Notes



Instructor's Notes:

______

______

______

7.20 MILLWRIGHT ◆ LEVEL THREE

## Trade Terms Introduced in This Module

***Angular misalignment:*** The condition that occurs when two shafts are at an angle to each other.

***Clutch:*** A device used to engage or disengage a load from a driver.

***Coupling gap:*** The space between the shaft faces within the couplings.

***Driven:*** The device being driven. The driven may be a gear case, pump, or generator.

***Driver:*** The prime mover of a system. The driver is usually a motor.

***Key:*** A device that fits between a coupling and a shaft to prevent slippage.

***Parallel misalignment:*** The condition that occurs when two shafts are misaligned so that their axes never intersect.

***Shear pin:*** A metal key installed between a drive shaft and a coupling or gear. It is designed to break in the event of a mechanical overload, preventing other, more expensive parts of the drive train from being damaged.

***Torque:*** A turning or twisting force measured in foot-pounds (ft-lb), inch pounds (in-lb), or kilogram-meters (kgf-m).

***Vortex:*** Fluid flow involving rotation about an axis.

# Resources & Acknowledgments

## Additional Resources

This module is intended to present thorough resources for task training. The following reference works are suggested for further study. These are optional materials for continued education rather than for task training.

Couplings.ca, http://www.couplings.ca/

Lovejoy, Inc.,. http://www.lovejoy-inc.com/

## Figure Credits

"Anderson Clamp Hub" Patented Coupling Corporation of America, Jacobus, PA, Flexxor Couplings, Ultraflex Couplings, 307F02 (photo)

Climax Metal Products Company, 307F03

Stafford Manufacturing Corp., 307F04

TB Wood's Incorporated, 307F06

Lovejoy Inc., 307F07

Advanced Antivibration Components, 307F09 (photo), www.vibrationmounts.com

Stainless steel coupling by Helical Products Co., Inc., the original manufacturer of beam style couplings, 307F09, www.heli-cal.com

Photo courtesy of ODG, 307F10 (photo)

Frontline Industries, Inc., 307F11, 307F13

Ringfeder Corporation, 307F14

Danaher Tool Group, 307F20 (photo)



Instructor's Notes:

7.22 MILLWRIGHT ◆ LEVEL THREE

# MODULE 15307-08 — TEACHING TIPS

The following are suggested activities or instructional methods to help you teach the material in this module.

## General

When you call on someone to answer a question, the rest of the class relaxes or even tunes out because they expect that the question and answer will take place only between you and the trainee you called on. Instead, use this technique to involve more trainees in answering questions and to keep them on their toes.

1. Ask trainees to define a term or explain a concept.
2. After one trainee has answered, ask a trainee seated nearby if the answer is right. Then ask whether a trainee in the back of the room agrees.
3. Ask trainees to explain why they think an answer is right or wrong.
4. Use the session to clear up incorrect ideas and encourage trainees to learn from their mistakes.

## Sections 1.0.0–6.0.0

### *Trade Terms Quiz*

This Quick Quiz will familiarize trainees with the trade terms commonly used to describe couplings. You will need photocopies of the quiz provided on the following page. Trainees will need pencils. If you allow trainees to use the Trainee Guide, decrease the amount of time you give them to complete the quiz.

1. Make a photocopy of the quiz for each trainee.
2. Give trainees between 5 and 10 minutes to complete the quiz.
3. Go over the answers to the quiz.
4. Ask trainees if they have questions.

### *Answers to Quick Quiz*

1. c
2. f
3. e
4. b
5. h
6. d
7. g
8. a

## Quick Quiz

### *Trade Terms*

For each description listed, identify the term that the text best describes. Write the corresponding letter in the blank provided.

_____ 1. The space between the shaft faces within the couplings is the _____.

_____ 2. The _____ is designed to break in the event of a mechanical overload.

_____ 3. Slider couplings allow for angular and _____.

_____ 4. A(n) _____ is used to engage or disengage a load from a driver.

_____ 5. The shape and rotation of an impeller and a runner produce a flow path called a(n) _____.

_____ 6. A(n) _____ fits between a coupling and a shaft to prevent slippage.

_____ 7. Some couplings reduce, dampen, or absorb _____.

_____ 8. Flexible couplings should *not* be used when major _____ is known to exist.

a. angular misalignment

b. clutch

c. coupling gap

d. key

e. parallel misalignment

f. shear pin

g. torque

h. vortex

## MODULE 15307-08 — ANSWERS TO REVIEW QUESTIONS

| Answer | | Section |
|---|---|---|
| 1. | d | 1.0.0 |
| 2. | b | 1.0.0 |
| 3. | a | 2.3.0 |
| 4. | a | 3.0.0 |
| 5. | a | 3.0.0 |
| 6. | a | 3.1.1 |
| 7. | b | 3.1.3 |
| 8. | c | 3.2.0 |
| 9. | c | 3.2.3 |
| 10. | d | 3.2.4 |
| 11. | a | 4.1.0 |
| 12. | c | 4.3.0 |
| 13. | a | 5.0.0 |
| 14. | a | 5.1.0 |
| 15. | c | 5.3.0 |
| 16. | a | 5.3.1 |
| 17. | c | 5.4.0 |
| 18. | a | 5.4.0 |
| 19. | c | 6.2.0 |
| 20. | b | 6.2.0 |

## CONTREN® LEARNING SERIES — USER UPDATE

The NCCER makes every effort to keep these textbooks up-to-date and free of technical errors. We appreciate your help in this process. If you have an idea for improving this textbook, or if you find an error, a typographical mistake, or an inaccuracy in NCCER's Contren® textbooks, please write us, using this form or a photocopy. Be sure to include the exact module number, page number, a detailed description, and the correction, if applicable. Your input will be brought to the attention of the Technical Review Committee. Thank you for your assistance.

*Instructors* – If you found that additional materials were necessary in order to teach this module effectively, please let us know so that we may include them in the Equipment/Materials list in the Annotated Instructor's Guide.

**Write:** Product Development and Revision
National Center for Construction Education and Research
3600 NW 43rd St., Bldg. G, Gainesville, FL 32606

**Fax:** 352-334-0932

**E-mail:** curriculum@nccer.org

Craft ______________ Module Name ______________

Copyright Date ______________ Module Number ______________ Page Number(s) ______________

Description ______________

______________

______________

______________

(Optional) Correction ______________

______________

______________

(Optional) Your Name and Address ______________

______________

______________

Module 15308-08

# *Fabricating Shims*

## NCCER STANDARDIZED CRAFT TRAINING PROGRAM

The National Center for Construction Education and Research (NCCER) provides a standardized national program of accredited craft training. Key features of the program include instructor certification, competency-based training, and performance testing. The program provides trainees, instructors, and companies with a standard form of recognition through a National Craft Training Registry. The program is described in full in the *Guidelines for Accreditation*, published by NCCER. For more information on standardized craft training, contact the NCCER by writing us at 3600 NW 43rd St., Bldg. G, Gainesville, FL 32606; calling 352-334-0911; or emailing info@nccer.org. More information may be found at our website, www.nccer.org.

## HOW TO USE THIS ANNOTATED INSTRUCTOR'S GUIDE

Each page presents two sections of information. The larger section displays each page exactly as it appears in the Trainee Module. The narrow column ties suggested trainee and instructor actions to each page and provides icons (detailed below) to call your attention to material, safety, audiovisual, or testing requirements. The bottom of each page includes space for your notes.

The **Audiovisual** icon indicates an appropriate time to show a transparency or other audiovisual aid.

The **Classroom** icon prompts you to define a term, stress a point, ask trainees to explain a concept, or give examples.

The **Demonstration** icon directs you to show trainees how to perform tasks.

The **Examination** icon tells you to administer the written module examination.

The **Homework** icon is placed where you may wish to assign reading for the next class, assign a project, or advise trainees to prepare for an examination.

The **Laboratory** icon is used when trainees are to practice performing tasks.

The **Materials** icon is a reminder for you to gather materials needed for classes, labs, and testing.

The **Performance Testing** icon tells you to administer a performance test or a portion thereof.

The **Safety** icon is used to emphasize safety issues. It is often keyed to *Caution* and *Warning!* statements in the Trainee Module.

The **Teaching Tip** icon indicates additional guidance is available, such as how to conduct an exercise, get the most educational value from a field trip, or encourage class participation. Teaching Tips may expand on a feature (*Think About It*, *Did You Know?*) or provide *Quick Quizzes* or similar exercises. You will be referred to the Teaching Tips section at the back of the module if there is additional material.

The **Combination** icon indicates that the laboratory listed corresponds with a performance task. If desired, you can note the proficiency of the trainees during the laboratory, and use it to satisfy performance testing requirements.

## PREPARATION

Before teaching this module, you should review the Objectives, Performance Tasks, Materials and Equipment List, and Module Outline. Be sure to allow ample time to prepare your own training or lesson plan and gather all required materials and equipment.

## MODULE OVERVIEW

This module explains the use of shims, how shims are made, and the types of shims available.

## PREREQUISITES

Prior to training with this module, it is recommended that the trainee shall have successfully completed *Core Curriculum; Millwright Level One; Millwright Level Two;* and *Millwright Level Three,* Modules 15301-08 through 15307-08.

## OBJECTIVES

Upon completion of this module, the trainee will be able to do the following:

1. Identify and explain types of shim stock.
2. Identify and explain shim materials.
3. Fabricate shims.

## PERFORMANCE TASKS

Under the supervision of the instructor, the trainee should be able to do the following:

1. Identify types of shim stock.
2. Identify shim materials.
3. Fabricate shims.

## MATERIALS AND EQUIPMENT LIST

Overhead projector and screen
Transparencies
Blank acetate sheets
Transparency pens
Whiteboard/chalkboard
Markers/chalk
Pencils and scratch paper
Appropriate personal protective equipment
Basic trainee tools

Various samples of shim stock, including:
- Hard
- Roll
- Flat
- Laminated
- Ring

Precut shims
Module Examinations*
Performance Profile Sheets*

* Located in the Test Booklet

## SAFETY CONSIDERATIONS

Ensure that the trainees are equipped with appropriate personal protective equipment and know how to use it properly. This module requires that the trainees use hand tools. Ensure that trainees are briefed on shop safety procedures.

## ADDITIONAL RESOURCES

This module is intended to present thorough resources for task training. The following reference works are suggested for both instructors and motivated trainees interested in further study. These are optional materials for continued education rather than for task training.

Precision Brand Products, Inc., http://www.precisionbrand.com/

SPIROL International Corporation, http://www.spirol.com/

## TEACHING TIME FOR THIS MODULE

An outline for use in developing your lesson plan is presented below. Note that each Roman numeral in the outline equates to one session of instruction. Each session has a suggested time period of 2½ hours. This includes 10 minutes at the beginning of each session for administrative tasks and one 10-minute break during the session. Approximately 5 hours are suggested to cover *Fabricating Shims*. You will need to adjust the time required for hands-on activity and testing based on your class size and resources. Because laboratories often correspond to Performance Tasks, the proficiency of the trainees may be noted during these exercises for Performance Testing purposes.

| **Topic** | **Planned Time** |
|---|---|
| **Session I. Introduction; Shim Stock; Shim Materials; Fabricating Shims** | |
| A. Introduction | ____________ |
| B. Shim Stock | ____________ |
| C. Laboratory<br>Trainees practice identifying types of shim stock. This laboratory corresponds to Performance Task 1. | ____________ |
| D. Shim Materials | ____________ |
| E. Laboratory<br>Trainees practice identifying shim materials. This laboratory corresponds to Performance Task 2. | ____________ |
| F. Fabricating Shims | ____________ |
| G. Laboratory<br>Trainees practice fabricating shims. This laboratory corresponds to Performance Task 3. | ____________ |
| **Session II. Review and Testing** | |
| A. Module Review | ____________ |
| B. Module Examination<br>1. Trainees must score 70% or higher to receive recognition from NCCER.<br>2. Record the testing results on Craft Training Report Form 200, and submit the results to the Training Program Sponsor. | ____________ |
| C. Performance Testing<br>1. Trainees must perform each task to the satisfaction of the instructor to receive recognition from NCCER. If applicable, proficiency noted during laboratory exercises can be used to satisfy the Performance Testing requirements.<br>2. Record the testing results on Craft Training Report Form 200, and submit the results to the Training Program Sponsor. | ____________ |

15308-08

# Fabricating Shims

Assign reading of Module 15308-08.

# 15308-08
# *Fabricating Shims*

*Topics to be presented in this module include:*



## Overview

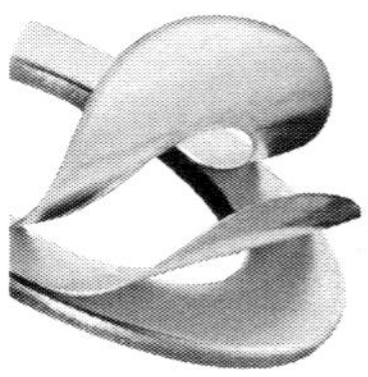

Shims are used to transfer the weight of a machine to its support. In the process of setting and aligning machines, shims are used to support and align the baseplate on the floor and the machine on the ground. In this module, you will learn how shims are made as well as the types of shims that are available.

Instructor's Notes:

## Objectives

When you have completed this module, you will be able to do the following:

1. Identify and explain types of shim stock.
2. Identify and explain shim materials.
3. Fabricate shims.

## Trade Terms

Laminated shims

## Required Trainee Materials

1. Pencil and paper
2. Appropriate personal protective equipment

## Prerequisites

Before you begin this module, it is recommended that you successfully complete *Core Curriculum; Millwright Level One; Millwright Level Two;* and *Millwright Level Three*, Modules 15301-08 through 15307-08.

This course map shows all of the modules in the third level of the *Millwright* curriculum. The suggested training order begins at the bottom and proceeds up. Skill levels increase as you advance on the course map. The local Training Program Sponsor may adjust the training order.

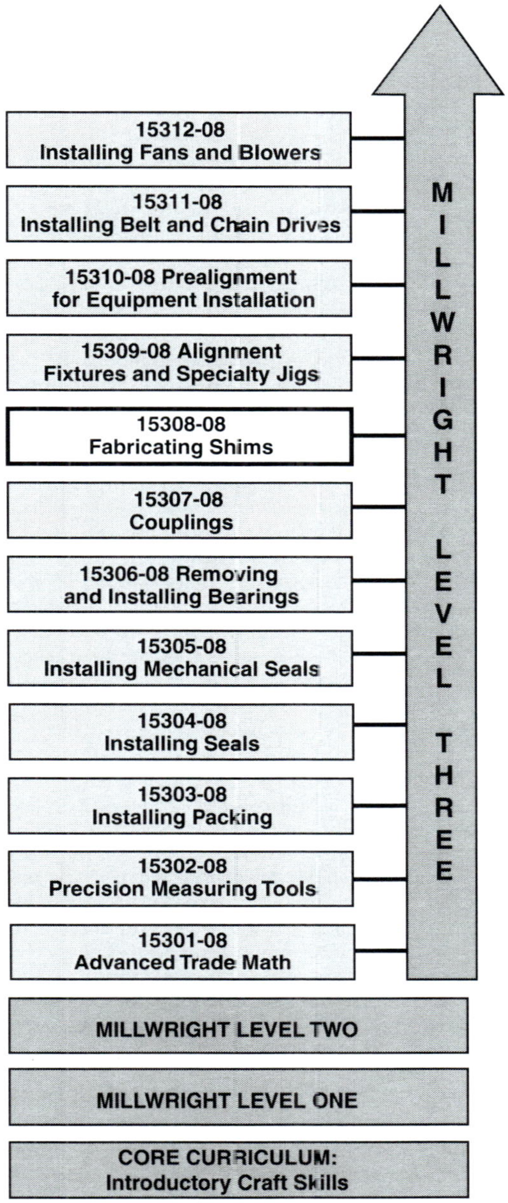


308CMAP.EPS



Ensure that you have everything required to teach the course. Check the Materials and Equipment list at the front of this module.

See the general Teaching Tip at the end of this module.

Explain that terms shown in bold are defined in the Glossary at the back of this module.

Show Transparency 1, Objectives, and Transparency 2, Performance Tasks. Review the goals of the module, and explain what will be expected of the trainee.

Review the modules covered in Level Three and explain how this module fits in.

MODULE 15308-08 ◆ FABRICATING SHIMS 8.1

**Explain how shims are used in millwright work.**

**Discuss materials commonly used to make shims.**

**Introduce the different types of shim stock.**

**Describe hard shim stock and its uses.**

**Discuss roll shim stock and precautions for unrolling it.**

**Discuss the available assortment of flat shim stock.**

**Describe laminated shim stock.**

**Provide samples of hard, roll, flat, and laminated shim stock for trainees to examine.**

## 1.0.0 ◆ INTRODUCTION

Shims are pieces of solid material that are used to fill a space between wood, metal, or synthetic objects. The most common material for shims used in millwright work is metal.

For aligning the bases of machines which are to be bolted to a baseplate, shims are cut to a specific standard shape from standard thickness stock so that the adjustment in height is measurable. The shape is designed to go on both sides of a bolt, because that is where the machine is supported and the point at which that support can be adjusted.

Shims and precision shim stock can be purchased that is accurate to within 0.0001 of an inch in thickness. Shims are available in carbon steel, stainless steel, and brass to allow for use in corrosive environments. Plastic shims are made for low-pressure machinery and to dampen vibration. They are often color-coded by thickness. Shims are frequently made with abrasive waterjetting equipment for accuracy and to prevent burrs.

It is best for millwrights to manufacture shims using accurate stock rather than to buy them. The skills and knowledge gained in the process are well worth the time expended.

## 2.0.0 ◆ SHIM STOCK

Shim materials come in many forms, called shim stock. The type of shim stock that is used on a job is determined by cost, convenience, and special applications. Shim stock is often purchased in bulk to save money when there are a lot of applications for one type of shim material. Precut and specially shaped shims are sometimes used because they save time and are convenient. Special applications where corrosive conditions exist or extremely heavy weight must be supported sometimes require the use of special shims.

Usually, engineering specifications state the type of shim to be used on a job. Always refer to the engineering specifications before selecting shims. Check all stock and precuts for accuracy.

The following sections describe these common types of shim stock:

- Hard
- Roll
- Flat
- Laminated
- Precut
- Ring shims and arbor spacers

### 2.1.0 Hard Shim Stock

Hard shim stock is flat bar or sheet stock that is mostly made from carbon or stainless steel. It is ordered in bulk for use by many crafts on the job site. Hard shims are cut in various thicknesses and sizes from this stock. Hard shims are primarily used to set bed plates, hold machinery for grout plates, and prelevel a soleplate for a piece of equipment.

### 2.2.0 Roll Shim Stock

Metal shim stock is available in packaged rolls of various thicknesses. Rolls are available in sizes from 6 inches by 50 inches to 12 inches by 120 inches. Many job sites maintain an inventory of the most commonly used sizes and thicknesses. The shim stock is taken from the roll, laid out, and cut to the required size and shape. Packaged shim stock is a common method of maintaining shim stock inventory.

> **WARNING!**
> Extra care must be taken when removing the shipping band from roll shim stock, as the corners and edges are sharp and could cause cuts when the roll uncoils.

### 2.3.0 Flat Shim Stock

Shim stock is also available in flat sheets. Sheet shim stock is available in metric and U.S. sizes from 6 inches by 12 inches to 18 inches by 24 inches for smaller quantities of stock. Flat shim stock is available in single sheets as well as in assorted packs. A pack may be an assortment of thicknesses of a single type of shim material or an assortment of thicknesses and materials. Assorted packs with color-coded shims are available in some materials. The color is different for each thickness, and each sheet is usually individually marked with the shim thickness, which eliminates the need for measuring and makes identification and selection much easier. Companies commonly maintain an inventory of flat stock assortments for maintenance situations in which a variety of thicknesses and materials may be needed.

### 2.4.0 Laminated Shim Stock

**Laminated shims** are also available in sheets (*Figure 1*). A laminated sheet consists of numerous layers of shim stock that are from 0.002 to 0.003 of an



Instructor's Notes:

8.2 MILLWRIGHT ◆ LEVEL THREE

inch thick, held together by resin. The shim can be used as is, or the thickness of the shim can be adjusted by peeling off layers using a thin-bladed knife. The type of resin used allows layers to be peeled off without tearing. The shim can be sawed, cut, or bored to produce almost any shape.

## 2.5.0 Precut Shims

Many shim stock manufacturers make precut shims (*Figure 2*). Precut shims come in all common thicknesses, with slots to accommodate most small pump, turbine, compressor, and electric motor applications. Color-coded, precut shims are available for easy thickness identification and selection. Some precut shims are made with tabs for easy installation and removal. Precut shims may be made of any of the shim metals, Teflon®, or plastic. The standard precut sizes are shown in *Table 1*.

## 2.6.0 Ring Shims and Arbor Spacers

Ring shims are thin rings with a plain center hole. They are used for building up gears and bearings and to provide proper clearance between mating parts. Arbor spacers are thin rings with a keyway center hole. They are used for accurate spacing of milling cutters, slitter knives, and gang saws. Ring shims and arbor spacers (*Figure 3*) are usually made of hardened, cold-rolled, low-carbon steel. They are often used in gearshafts to preload bearings.

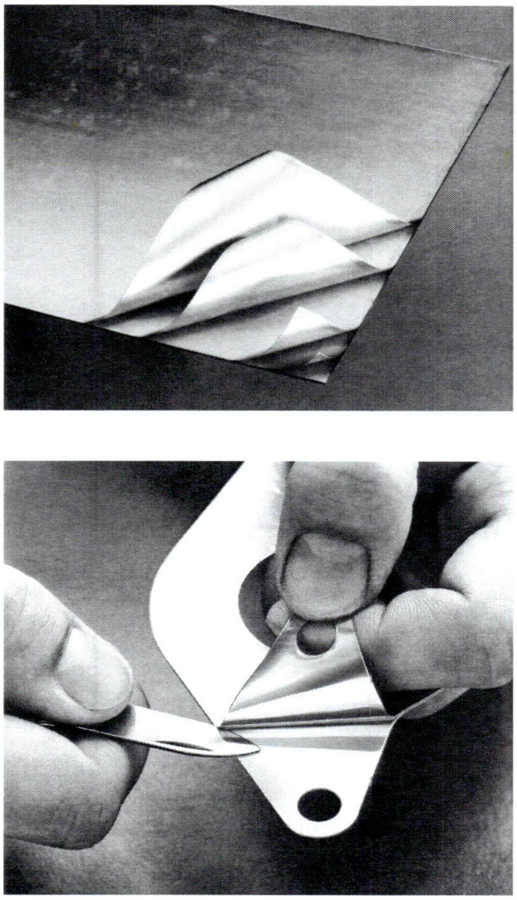

308F01.EPS

*Figure 1* ◆ Laminated shim stock.

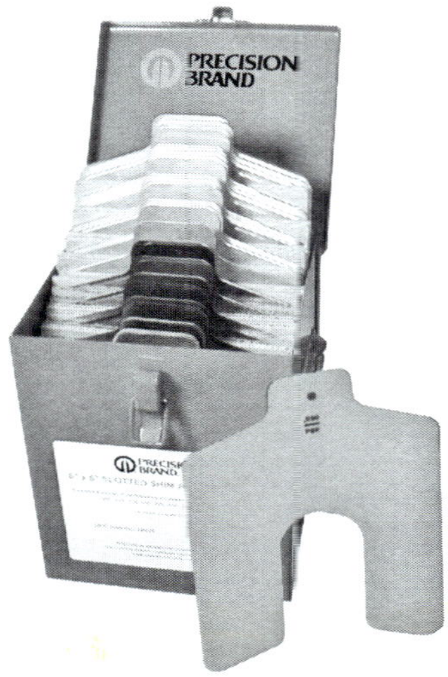


308F02.EPS

*Figure 2* ◆ Precut shims.

**Table 1** Standard Precut Shim Sizes

| Standard Precut Shim Sizes | | | |
|---|---|---|---|
| Shim Size | Max Bolt Size | Slot Size | Shim Dimensions |
| AA | 3/8" | 7/16" | 1 1/2" × 1 1/2" |
| A | 1/2" | 9/16" | 2" × 2" |
| BB | 5/8" | 11/16" | 2 1/2" × 2 1/2" |
| B | 3/4" | 13/16" | 3" × 3" |
| C | 1 1/8" | 1 3/16" | 4" × 4" |
| D | 1 1/2" | 1 9/16" | 5" × 5" |
| E | 1 7/8" | 1 15/16" | 6" × 6" |
| G | 2 1/4" | 2 5/16" | 6" × 6" |

308T01.EPS



**Discuss the variety of precut shims available.**

**Show Transparency 3 (Table 1).**

**Provide samples of precut shims for trainees to examine.**

**Discuss ring shims and arbor spacers and explain how they are used.**

**See the Teaching Tip for Section 2.0.0 at the back of this module.**

**Have trainees practice identifying types of shim stock. Note the proficiency of each trainee. This laboratory corresponds to Performance Task 1.**

Introduce the materials commonly used to make shim stock.

Discuss the use of plastic and Teflon® for shim stock.

Describe the use of brass, steel, stainless steel, and blue tempered steel for shim stock.

Discuss areas where paper shims are used.

Explain how to identify shim materials.

Have trainees practice identifying shim materials. Note the proficiency of each trainee. This laboratory corresponds to Performance Task 2.

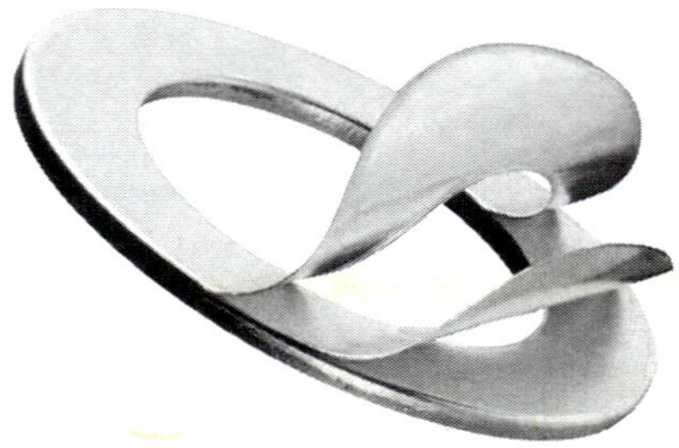

308F03.EPS

*Figure 3* ◆ Ring shim and arbor spacer.

## 3.0.0 ◆ SHIM MATERIALS

A number of materials are used to manufacture shim stock. The type of material used depends on the application and is usually determined by the design engineer. Some of the most common materials used for making shim stock are the following:

- Plastic
- Teflon®
- Brass
- Steel
- Stainless steel
- Blue tempered steel
- Paper

### 3.1.0 Plastic

Plastic shim stock is manufactured in most common thicknesses and in various degrees of hardness. Plastic shim stock is frequently used in chemically corrosive applications and in the food-processing industry, where cleanliness and sanitation are major concerns. Plastic shims are used to support light weights and in situations where the shims required are very thin. The hardness requirements are usually determined by the design engineer and specified to the shim manufacturer. Plastic shims are used for isolation from galvanic current.

### 3.2.0 Teflon®

There are several types of Teflon® materials that are used to manufacture shim stock. Teflon® is commonly used in highly corrosive applications, such as in paper mills. Teflon® also has very good vibration-dampening qualities.

### 3.3.0 Brass

Brass is used in shim applications in explosive atmospheres, such as where gases and other explosive materials are present, because brass does not produce sparks when it rubs against other materials. Brass is also used in highly corrosive conditions. It has excellent forming qualities, and is used extensively in bearings and motors as well as for shimming equipment.

### 3.4.0 Steel

Steel shim stock is usually cold-rolled, low-carbon steel. It is universally used for tool and die alignment, arbor spacers, and shim applications where hardness, flatness, and accuracy are required. It can be punched or sheared with minimal burring and distortion. Steel shims should never be used in explosive atmospheres.

### 3.5.0 Stainless Steel

Stainless steel shim material is usually cold-rolled and is available in many different grades. Stainless steel is the most commonly used shim material. It is widely used in applications where corrosion-resistance is required, in nuclear and conventional power plants, and in oil and gas refineries.

### 3.6.0 Blue Tempered Steel

Blue tempered steel is high-carbon, hardened-spring steel. It is used for various types of coiled and flat springs, such as those used in electrical assemblies and in steel rules. Blue tempered steel is preferred by tool and die makers for templates and for shim applications where toughness and high fatigue strength are required. Because of the hardness of blue tempered steel, it cannot be cut with shears.

### 3.7.0 Paper

Paper shims are used for pumps and gearboxes. They are made from oil-resistant paper, and when the bolts have been fully tightened, the excess paper is removed.

## 4.0.0 ◆ FABRICATING SHIMS

Fabricating shims includes selecting the material to be used, laying out the shape, cutting the shape, and finishing the shim. For accurate leveling and aligning of equipment and machinery, the shims



Instructor's Notes:

8.4 MILLWRIGHT ◆ LEVEL THREE

must be cut and finished properly. A millwright will need to cut shims of many different shapes for different applications. The following procedure explains how to fabricate shims to level a baseplate at the anchor bolts. Follow these steps to fabricate shims:

*Step 1* Determine the type of material the shim is to be made of. For new installations, the shim material is usually specified by the design engineer. In maintenance situations, you should use the same type of material that was originally used, or contact your supervisor to determine what type to use. The type of shim material is often specified on the work order for a particular job.

*Step 2* Determine the size that the shim is required to be. The shim should cover at least 90 percent of the area being shimmed, or according to engineering specifications.

*Step 3* Determine the thickness the shim is required to be. Always use the thickest shim that will do the job. One thick shim will provide much better results than several thinner shims.

*Step 4* Cut a blank a little larger than the intended shim size from the proper type and thickness of shim stock.

> **WARNING!**
> Shim stock is very thin and can have very sharp edges. Wear gloves and use extreme care when handling and cutting shim stock to prevent cutting your hands.

*Step 5* Determine what size slot is required in the shim. The slot should be $\frac{1}{16}$ to $\frac{1}{8}$ of an inch larger than the bolt the shim will fit around.

*Step 6* Lay out the shape of the shim. Leave a tab on the shim. The tab will make it easier to install and remove the shim. Round all corners to reduce the chance of getting cut by the shim when handling it.

*Step 7* Drill or punch a hole in the center of the shim for the slot.

*Step 8* Cut the slot, using snips or scissors. Cut very smoothly and precisely. Do not close the snips or scissors all the way when cutting, because this will make burrs. Cut a little at a time and as neatly as possible. Very thick shims can be cut using a jig saw, a band saw, or a nibbler for sheet metal.

*Step 9* Cut out the shape of the shim. Cut precisely, and round all corners. Be sure to leave the tab.

*Step 10* Place the shim on an anvil, steel table, or other suitable hard surface, and flatten it using a small hammer.

> **CAUTION**
> Use light blows when hammering the shim, and hammer it only enough to flatten it. Excessive hammering can affect the thickness of the shim.

*Step 11* Dress the shim, using flat and rounded files or hones to remove all burrs and very sharp edges.

*Step 12* Thoroughly clean the shim, using solvent and a rag.

*Step 13* Measure the thickness of the shim, using a micrometer.

*Step 14* Mark the shim's thickness onto the shim using a permanent marking pen.

*Figure 4* shows the finished shim.

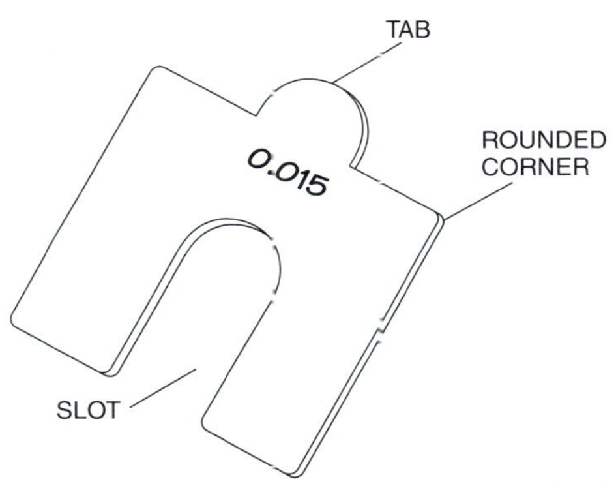


*Figure 4* ◆ Finished shim.

**Explain that fabricating shims includes selecting the material, laying out the shape, cutting the shape, and finishing the shim.**

**Review the example steps to fabricate shims to level a baseplate at the anchor bolts.**

**Explain that gloves must be worn when handling shim stock.**

**Show Transparency 4 (Figure 4). Describe the finished shim.**

**Demonstrate how to fabricate shims.**

**Have trainees practice fabricating shims. Note the proficiency of each trainee. This laboratory corresponds to Performance Task 3.**



MODULE 15308-08 ◆ FABRICATING SHIMS 8.5

**Have trainees complete the Review Questions, and go over the answers prior to administering the Module Examination.**

## Review Questions

1. Hard shims are cut from _____.
   a. granite
   b. tempered steel
   c. iron plate
   d. flat carbon steel bar or stainless steel

2. Metal shim stock is sold in sheets, starting at _____.
   a. 3 inches by 6 inches
   b. 6 inches by 12 inches
   c. 24 inches by 48 inches
   d. 4 feet by 8 feet

3. A laminated shim sheet is composed of several layers of shim stock, held together with _____.
   a. epoxy
   b. magnetism
   c. resin
   d. surface tension

4. Laminated shim stock thickness can be adjusted by _____.
   a. grinding off layers
   b. peeling off layers
   c. milling
   d. bandsawing

5. Ring shims are usually used for _____.
   a. setting machine bases
   b. building up gears and bearings
   c. accurate spacing of milling cutters
   d. setting soleplates

6. Arbor spacers are thin rings with a _____.
   a. set of bolt holes around the ring
   b. set of notches around the edge
   c. keyway in the center hole
   d. built-up edge

7. Plastic shims are commonly used in _____.
   a. setting soleplates
   b. corrosive applications
   c. very heavy equipment
   d. high temperature environments

8. Brass is used in applications in _____.
   a. food service
   b. wet areas
   c. explosive atmospheres
   d. nuclear plants

9. The most commonly used shim material is _____.
   a. brass
   b. Teflon®
   c. tempered steel
   d. stainless steel

10. The slot for a bolt should be cut _____ than the bolt.
   a. smaller
   b. the same size as
   c. $\frac{1}{16}$ to $\frac{1}{8}$ inch larger
   d. $\frac{1}{2}$ inch larger



Instructor's Notes:

8.6 MILLWRIGHT ◆ LEVEL THREE

# Summary

Shims are used to give precise spacing to machine parts, as well as to provide full-level support for machine bases. Shim stock is made from brass and both carbon and stainless steel, as well as Teflon® and plastics.

Shims are available from commercial manufacturers in several configurations for different purposes. Laminated shims and shim stock allow the millwright to peel off thin layers until the precise thickness is achieved. Ring shims allow machine parts to be spaced for clearance. Precut machine base shims are made in standard kits to allow a base to be set quickly and easily.

The ability to fabricate shims from stock allows the millwright to produce the shims as needed for a given application. Depending on the material, shims can be cut with bandsaws, wet abrasive cutters, or by hand with snips.

# Notes



**Summarize the major concepts presented in the module.**

**Administer the Module Examination. Record the results on Craft Training Report Form 200, and submit the results to the Training Program Sponsor.**

**Administer the Performance Test, and fill out Performance Profile Sheets for each trainee. If desired, trainee proficiency noted during laboratory sessions may be used to complete the Performance Test. Record the results on Craft Training Report Form 200, and submit the results to the Training Program Sponsor.**

## Trade Terms Introduced in This Module

***Laminated shims:*** Layers of shims that are bonded together.



Instructor's Notes:

# Resources & Acknowledgments

## Additional Resources

This module is intended to present thorough resources for task training. The following reference works are suggested for further study. These are optional materials for continued education rather than for task training.

Precision Brand Products, Inc., http://www.precisionbrand.com/

SPIROL International Corporation, http://www.spirol.com/

## Figure Credits

Precision Brand Products, Inc., 308F01 (top), 308F02, 308F03

SPIROL International Corporation, 308F01 (bottom)



MODULE 15308-08 ◆ FABRICATING SHIMS 8.9

# MODULE 15308-08 — TEACHING TIPS

The following are suggested activities or instructional methods to help you teach the material in this module.

## General

When you call on someone to answer a question, the rest of the class relaxes or even tunes out because they expect that the question and answer will take place only between you and the trainee you called on. Instead, use this technique to involve more trainees in answering questions and to keep them on their toes.

1. Ask trainees to define a term or explain a concept.
2. After one trainee has answered, ask a trainee seated nearby if the answer is right. Then ask whether a trainee in the back of the room agrees.
3. Ask trainees to explain why they think an answer is right or wrong.
4. Use the session to clear up incorrect ideas and encourage trainees to learn from their mistakes.

## Section 2.0.0 *Shim Stock*

This exercise will help familiarize trainees with shim stock. Trainees will need pencils and paper. You will need to arrange workstations with a different type of shim stock at each station. Allow 20 to 30 minutes for this exercise. This exercise corresponds to Performance Task 1.

1. Discuss the various types of shim stock with trainees.
2. Have trainees rotate between the various workstations and identify the type of shim stock at each station.
3. After all trainees have visited each station, have one trainee identify the type of shim stock at their station. Answer any questions trainees may have.

## MODULE 15308-08 — ANSWERS TO REVIEW QUESTIONS

| Answer | Section |
|---|---|
| 1. d | 2.1.0 |
| 2. b | 2.3.0 |
| 3. c | 2.4.0 |
| 4. b | 2.4.0 |
| 5. b | 2.6.0 |
| 6. c | 2.6.0 |
| 7. b | 3.1.0 |
| 8. c | 3.3.0 |
| 9. d | 3.5.0 |
| 10. c | 4.0.0 |

## CONTREN® LEARNING SERIES — USER UPDATE

The NCCER makes every effort to keep these textbooks up-to-date and free of technical errors. We appreciate your help in this process. If you have an idea for improving this textbook, or if you find an error, a typographical mistake, or an inaccuracy in NCCER's Contren® textbooks, please write us, using this form or a photocopy. Be sure to include the exact module number, page number, a detailed description, and the correction, if applicable. Your input will be brought to the attention of the Technical Review Committee. Thank you for your assistance.

*Instructors* – If you found that additional materials were necessary in order to teach this module effectively, please let us know so that we may include them in the Equipment/Materials list in the Annotated Instructor's Guide.


**Write:** Product Development and Revision
National Center for Construction Education and Research
3600 NW 43rd St., Bldg. G, Gainesville, FL 32606

**Fax:** 352-334-0932

**E-mail:** curriculum@nccer.org


Craft ______ Module Name ______

Copyright Date ______ Module Number ______ Page Number(s) ______

Description ______

______

______

______

(Optional) Correction ______

______

______

(Optional) Your Name and Address ______

______

______

Module 15309-08

# *Alignment Fixtures and Specialty Jigs*

## NCCER STANDARDIZED CRAFT TRAINING PROGRAM

The National Center for Construction Education and Research (NCCER) provides a standardized national program of accredited craft training. Key features of the program include instructor certification, competency-based training, and performance testing. The program provides trainees, instructors, and companies with a standard form of recognition through a National Craft Training Registry. The program is described in full in the *Guidelines for Accreditation*, published by NCCER. For more information on standardized craft training, contact the NCCER by writing us at 3600 NW 43rd St., Bldg. G, Gainesville, FL 32606; calling 352-334-0911; or emailing info@nccer.org. More information may be found at our website, www.nccer.org.

## HOW TO USE THIS ANNOTATED INSTRUCTOR'S GUIDE

Each page presents two sections of information. The larger section displays each page exactly as it appears in the Trainee Module. The narrow column ties suggested trainee and instructor actions to each page and provides icons (detailed below) to call your attention to material, safety, audiovisual, or testing requirements. The bottom of each page includes space for your notes.

The **Audiovisual** icon indicates an appropriate time to show a transparency or other audiovisual aid.

The **Classroom** icon prompts you to define a term, stress a point, ask trainees to explain a concept, or give examples.

The **Demonstration** icon directs you to show trainees how to perform tasks.

The **Examination** icon tells you to administer the written module examination.

The **Homework** icon is placed where you may wish to assign reading for the next class, assign a project, or advise trainees to prepare for an examination.

The **Laboratory** icon is used when trainees are to practice performing tasks.

The **Materials** icon is a reminder for you to gather materials needed for classes, labs, and testing.

The **Performance Testing** icon tells you to administer a performance test or a portion thereof.

The **Safety** icon is used to emphasize safety issues. It is often keyed to *Caution* and *Warning!* statements in the Trainee Module.

The **Teaching Tip** icon indicates additional guidance is available, such as how to conduct an exercise, get the most educational value from a field trip, or encourage class participation. Teaching Tips may expand on a feature (*Think About It*, *Did You Know?*) or provide *Quick Quizzes* or similar exercises. You will be referred to the Teaching Tips section at the back of the module if there is additional material.

The **Combination** icon indicates that the laboratory listed corresponds with a performance task. If desired, you can note the proficiency of the trainees during the laboratory, and use it to satisfy performance testing requirements.

## PREPARATION

Before teaching this module, you should review the Objectives, Performance Tasks, Materials and Equipment List, and Module Outline. Be sure to allow ample time to prepare your own training or lesson plan and gather all required materials and equipment.

## MODULE OVERVIEW

This module explains how to make mounting fixtures for alignment fixtures and how to use and set up fixtures for reverse alignment.

## PREREQUISITES

Prior to training with this module, it is recommended that the trainee shall have successfully completed *Core Curriculum; Millwright Level One; Millwright Level Two;* and *Millwright Level Three,* Modules 15301-08 through 15308-08.

## OBJECTIVES

Upon completion of this module, the trainee will be able to do the following:

1. Identify and explain types of jigs.
2. Fabricate jigs.
3. Set up jigs.

## PERFORMANCE TASKS

Under the supervision of the instructor, the trainee should be able to do the following:

1. Identify types of jigs.
2. Fabricate one or more of the following:
   - Angle iron jig
   - Chain jig
   - Christmas tree jig
   - Piano wire jig
3. Set up three or more of the following:
   - Angle iron jig
   - Chain jig
   - Complex reverse-indicator jig
   - Christmas tree jig
   - Piano wire jig

## MATERIALS AND EQUIPMENT LIST

Overhead projector and screen
Transparencies
Blank acetate sheets
Transparency pens
Whiteboard/chalkboard
Markers/chalk
Pencils and scratch paper
Appropriate personal protective equipment
Basic trainee tools

Sample jigs, including:
  Angle iron
  Chain
  Christmas tree
  Piano wire
  Complex reverse-indicator
Module Examinations*
Performance Profile Sheets*

* Located in the Test Booklet

## SAFETY CONSIDERATIONS

Ensure that the trainees are equipped with appropriate personal protective equipment and know how to use it properly. This module requires trainees to work with hand tools. Ensure that trainees are briefed on shop safety procedures.

## ADDITIONAL RESOURCES

This module is intended to present thorough resources for task training. The following reference work is suggested for both instructors and motivated trainees interested in further study. This is optional material for continued education rather than for task training.

L.S. Starrett Company, http://www.starrett.com

## TEACHING TIME FOR THIS MODULE

An outline for use in developing your lesson plan is presented below. Note that each Roman numeral in the outline equates to one session of instruction. Each session has a suggested time period of 2½ hours. This includes 10 minutes at the beginning of each session for administrative tasks and one 10-minute break during the session. Approximately 10 hours are suggested to cover *Alignment Fixtures and Specialty Jigs*. You will need to adjust the time required for hands-on activity and testing based on your class size and resources. Because laboratories often correspond to Performance Tasks, the proficiency of the trainees may be noted during these exercises for Performance Testing purposes.

| Topic | Planned Time |
|---|---|
| **Session I. Introduction; Types of Jigs** | |
| A. Introduction | ____________ |
| B. Types of Jigs | ____________ |
| C. Laboratory<br>Trainees practice identifying types of jigs. This laboratory corresponds to Performance Task 1. | ____________ |
| **Session II. Fabricating Jigs** | |
| A. Fabricating Jigs | ____________ |
| B. Laboratory<br>Trainees practice fabricating one or more of the following: angle iron jig, chain jig, Christmas tree jig, and/or piano wire jig. This laboratory corresponds to Performance Task 2. | ____________ |
| **Session III. Setting Up Jigs** | |
| A. Setting Up Jigs | ____________ |
| B. Laboratory<br>Trainees practice setting up three or more of the following: angle iron jig, chain jig, complex reverse-indicator jig, Christmas tree jig, and/or piano wire jig. This laboratory corresponds to Performance Task 3. | ____________ |
| **Session IV. Review and Testing** | |
| A. Module Review | ____________ |
| B. Module Examination<br>1. Trainees must score 70% or higher to receive recognition from NCCER.<br>2. Record the testing results on Craft Training Report Form 200, and submit the results to the Training Program Sponsor. | ____________ |
| C. Performance Testing<br>1. Trainees must perform each task to the satisfaction of the instructor to receive recognition from NCCER. If applicable, proficiency noted during laboratory exercises can be used to satisfy the Performance Testing requirements.<br>2. Record the testing results on Craft Training Report Form 200, and submit the results to the Training Program Sponsor. | ____________ |

15309-08

# Alignment Fixtures and Specialty Jigs

Assign reading of Module 15309-08.

# 15309-08
# *Alignment Fixtures and Specialty Jigs*

*Topics to be presented in this module include:*



## Overview

In this module, you will learn to make mounting fixtures for alignment fixtures, especially for dial indicators. Some of these fixtures can also be used to mount laser alignment equipment. You will also learn to use and set up these fixtures for reverse alignment.

Instructor's Notes:

## Objectives

When you have completed this module, you will be able to do the following:

1. Identify and explain types of jigs.
2. Fabricate jigs.
3. Set up jigs.

## Trade Terms

Indicator sag
Piano wire
Rigid

## Required Trainee Materials

1. Pencil and paper
2. Appropriate personal protective equipment

## Prerequisites

Before you begin this module, it is recommended that you successfully complete *Core Curriculum; Millwright Level One; Millwright Level Two;* and *Millwright Level Three*, Modules 15301-08 through 15308-08.

This course map shows all of the modules in the third level of the *Millwright* curriculum. The suggested training order begins at the bottom and proceeds up. Skill levels increase as you advance on the course map. The local Training Program Sponsor may adjust the training order.

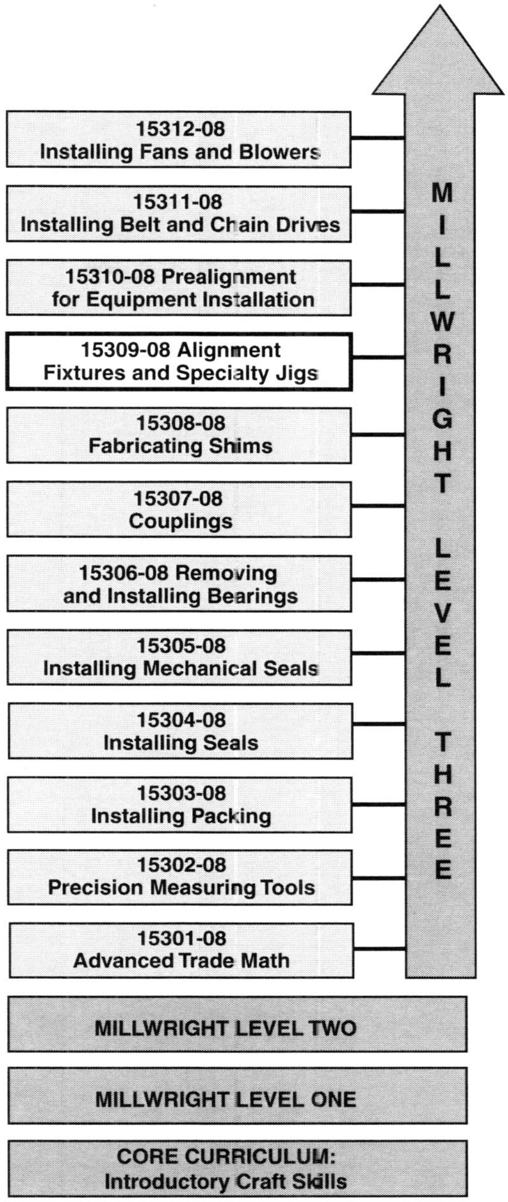


309CMAP.EPS



Ensure that you have everything required to teach the course. Check the Materials and Equipment list at the front of this module.

See the general Teaching Tip at the end of this module.

Explain that terms shown in bold are defined in the Glossary at the back of this module.

Show Transparency 1, Objectives, and Transparency 2, Performance Tasks. Review the goals of the module, and explain what will be expected of the trainee.

Review the modules covered in Level Three and explain how this module fits in.

MODULE 15309-08 ◆ ALIGNMENT FIXTURES AND SPECIALTY JIGS 9.1

**Explain the function of alignment fixtures and specialty jigs.**

**Discuss the role of dial indictors in alignment.**

**Introduce common types of jigs used for alignment.**

**Describe the configuration of angle iron jigs.**

**Show Transparency 3 (Figure 2).**

## 1.0.0 ◆ INTRODUCTION

Alignment fixtures and specialty jigs are used to mount dial indicators on the equipment being aligned. There are many types of jigs. Some jigs are used to set up **piano wire** for alignment purposes. Although there are a few jigs available on the market, most jigs are fabricated by the millwright for specific needs.

Dial indicators are used extensively in conventional alignment, also called rim and face alignment, and in reverse-indicator alignment. The dial indicators are mounted, or set up, in numerous ways to produce the readings needed to perform the alignment procedures. This module explains common types of jigs, how they are made, and how they are mounted, or set up. When it is possible, magnetic bases are commercially available. These sit on a flat, horizontal surface (*Figure 1*).

## 2.0.0 ◆ TYPES OF JIGS

The type of jig needed for an alignment job depends on the physical setup of the equipment and on the degree of accuracy required in the alignment procedure. A jig may be as simple as a piece of angle iron that is drilled and tapped to mount the dial indicator for rough alignment, or it may be a specially made tool for precise alignment. The following are common types of jigs used in alignment:

- Angle iron
- Chain
- Complex reverse-indicator
- Christmas tree
- Piano wire

### 2.1.0 Angle Iron Jigs

An angle iron jig consists of two correctly sized pieces of angle iron and straps to mount the iron on the shafts (*Figure 2*). The two pieces of angle iron have rods on which to mount the dial indicators. These rods are either welded on or tapped into the angle near the end. Angle iron jigs, among other uses, are used for reverse dial-indicating.

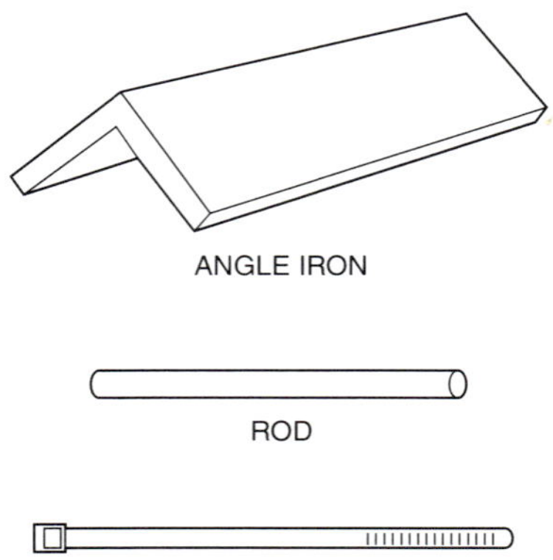


*Figure 2* ◆ Angle iron jig components.

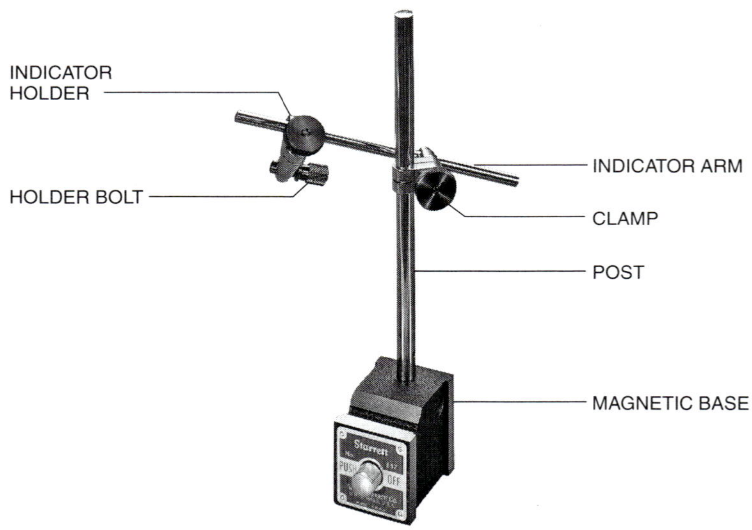


*Figure 1* ◆ Dial indicator magnetic base.



Instructor's Notes:

9.2 MILLWRIGHT ◆ LEVEL THREE

## 2.2.0 Chain Jigs

A chain jig consists of a block, a chain, an adjusting bolt, and a wing nut. The block is drilled and tapped to facilitate mounting of the indicators and is notched to fit the shaft. The chain is attached to the adjusting bolt, which fits in a hole through the block. There is a hook or pin on the opposite side of the block on which to hook the chain. Chain jigs are used for conventional alignment and laser alignment. *Figure 3* shows the components of two types of chain jigs.

## 2.3.0 Complex Reverse-Indicator Jigs

The complex reverse-indicator jig (*Figure 4*) is more complicated than the other types and is very accurate and easy to use. Two of these jigs are used at the same time to perform reverse-indicator alignment. A complex reverse-indicator jig consists of a clamping device with bolts, a mounting pin, and an indicator support device. This type of jig is much more accurate over long distances because it is more **rigid** and has less **indicator sag** than other jigs. The components of this jig are usually made on a lathe or milling machine.

## 2.4.0 Christmas Tree Jigs

A Christmas tree jig (*Figure 5*), which has a shape similar to that of a Christmas tree, is a specialty jig that is used to indicate a long distance between the shafts being aligned. It is used for conventional alignment. The Christmas tree jig must be checked both for runout and sag as the long extension will magnify the effect of both.

## 2.5.0 Piano Wire Jigs

Piano wire is used extensively to align pieces of equipment on a common line and to set pieces of equipment at a common elevation. If the alignment is to be accurate, the piano wire must be precisely set and very taut to prevent sag. Piano wire jigs (*Figure 6*) are used to pull the wire taut and to make minute adjustments to the alignment of the wire. Two jigs are needed to set and adjust a wire. A piano wire jig consists of a steel plate with three upright members. The two outside members are welded to the plate and support a threaded rod. The center member is free-floating on the threaded rod. Adjusting nuts are used to position the center member right or left on the rod. The piano wire sits in a notch in the center member. The jig has a turnbuckle that is used to anchor the wire to an eyebolt and to apply tension to the wire.

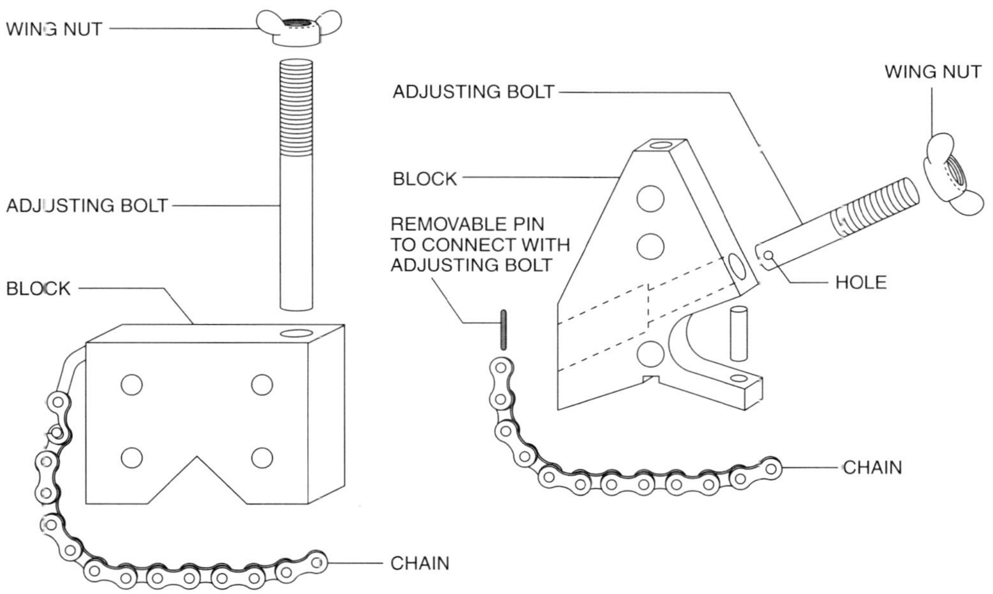


*Figure 3* ◆ Chain jig components.



**Discuss the components of chain jigs.**

**Show Transparency 4 (Figure 3).**

**Explain that an advantage of the complex reverse-indicator jig is its accuracy.**

**Show Transparency 5 (Figure 4).**

**Describe the Christmas tree jig and explain how it is used.**

**Show Transparency 6 (Figure 5).**

**Discuss how piano wire is used to align pieces of equipment.**

**Describe the piano wire jig and explain how it is used.**

**Show Transparency 7 (Figure 6).**

**Provide various types of jigs for trainees to examine.**

MODULE 15309-08 ◆ ALIGNMENT FIXTURES AND SPECIALTY JIGS 9.3

See the Teaching Tip for Section 2.0.0 at the end of this module.

Have trainees practice identifying types of jigs. Note the proficiency of each trainee. This laboratory corresponds to Performance Task 1.

Have trainees review Sections 3.0.0–3.5.0.

Ensure that you have everything required to teach this session.

Explain that most alignment fixtures and specialty jigs are fabricated by millwrights for specific purposes.

Walk through the procedure for fabricating angle iron jigs.

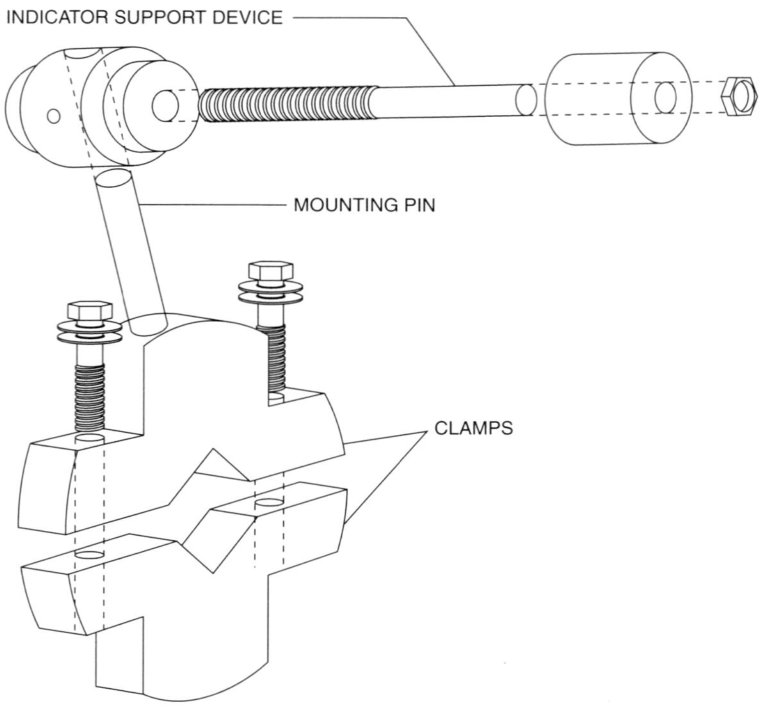


*Figure 4* ◆ Complex reverse-indicator jig components.

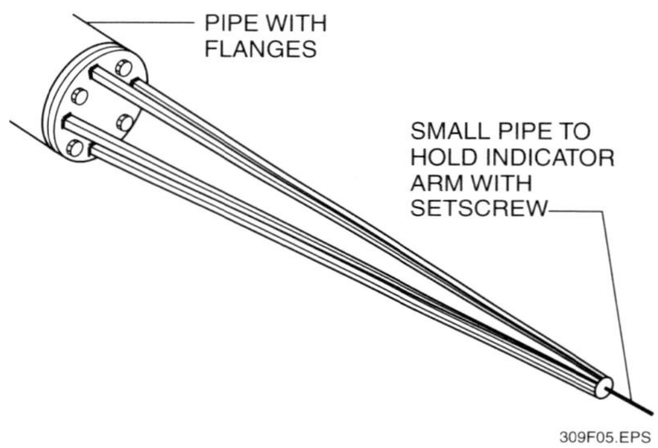


*Figure 5* ◆ Christmas tree jig.

## 3.0.0 ◆ FABRICATING JIGS

Most alignment fixtures and specialty jigs are fabricated by the millwright for specific purposes. The type and size jig needed depends on the alignment procedure to be performed and the physical space available to set up the jig. Equipment with very large shafts is aligned by laser or optical alignment methods. Equipment with very small shafts is uncommon. The majority of alignment jobs that you will perform using jigs will have shafts that range from 1 to 6 inches in diameter. Therefore, you can fabricate a few common jigs that can be used for the majority of the alignment jobs that you will perform. When fabricating a jig, try to minimize the indicator sag. The following sections explain how to fabricate the following jigs:

- Angle iron
- Chain
- Complex reverse-indicator
- Christmas tree
- Piano wire

### 3.1.0 Fabricating Angle Iron Jigs

Angle iron jigs should be made of steel angle iron. The angle must be thick enough so that it is very rigid. A $2 \times 2 \times \frac{1}{4}$-inch angle is a common size for shafts ranging from 3 to 6 inches in diameter.



Instructor's Notes:

9.4 MILLWRIGHT ◆ LEVEL THREE

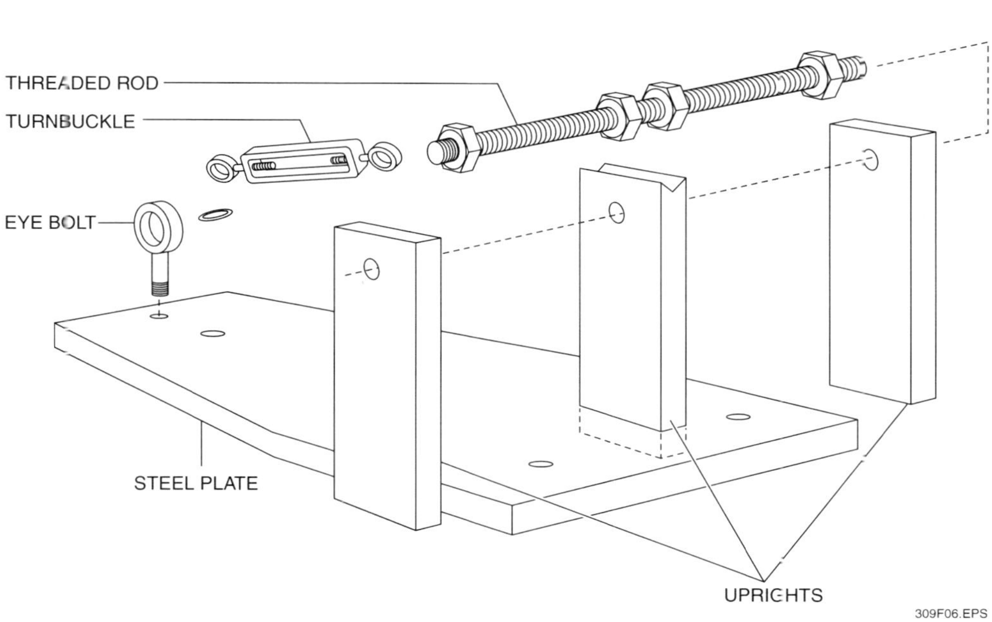


*Figure 6* ◆ Piano wire jig.

Smaller angle iron should be used for shafts smaller than 3 inches. Angle iron jigs are commonly used to align shafts before the couplings are installed. Follow these steps to fabricate an angle iron jig:

*Step 1* Cut a 4-inch piece of angle iron.

> **NOTE**
> Four inches is a standard length for the angle iron jig. Certain applications may require that the angle iron be shorter.

*Step 2* Cut a 3-inch piece of 5/16-inch steel rod.

> **NOTE**
> The standard size of a Starrett dial indicator mounting post is 5/16 inch. The mounting hardware of most dial indicators will fit a 5/16-inch rod.

*Step 3* Position the 5/16-inch rod approximately ½ inch back from the end and on the corner of the angle iron (*Figure 7*).

*Step 4* Hold the rod square to the angle, and weld it into position.

*Step 5* Clean the angle and rod to remove any nicks, burrs, and weld slag.

*Step 6* Select a mounting strap that will fit around the jig and the shaft being aligned. The mounting strap may be a metal tie wrap, hose clamp, or any other relatively strong, adjustable strap.

### 3.2.0 Fabricating Chain Jigs

The chain jig design may be simple or complex. The most important considerations when choosing a chain jig design are the versatility of the jig and its ability to firmly fit on the shaft or coupling. The following sections contain drawings showing how to fabricate two types of chain jigs. These jigs are for shafts or couplings of 3 to 6 inches in diameter. Smaller jigs can be made for smaller shafts and couplings. Jig A is a simple block jig with holes tapped for the dial indicator hardware (*Figure 8*). Jig B is more complicated to make, but it is more versatile (*Figure 9*).

**Show Transparency 8 (Figure 7). Describe the rod position.**

**Discuss the considerations for choosing a chain jig design.**

**Describe the fabrication of chain jig A.**

**Describe the fabrication of chain jig B.**

**Show Transparencies 9 and 10 (Figures 8 and 9). Answer any questions trainees may have about fabricating chain jigs.**



MODULE 15309-08 ◆ ALIGNMENT FIXTURES AND SPECIALTY JIGS 9.5

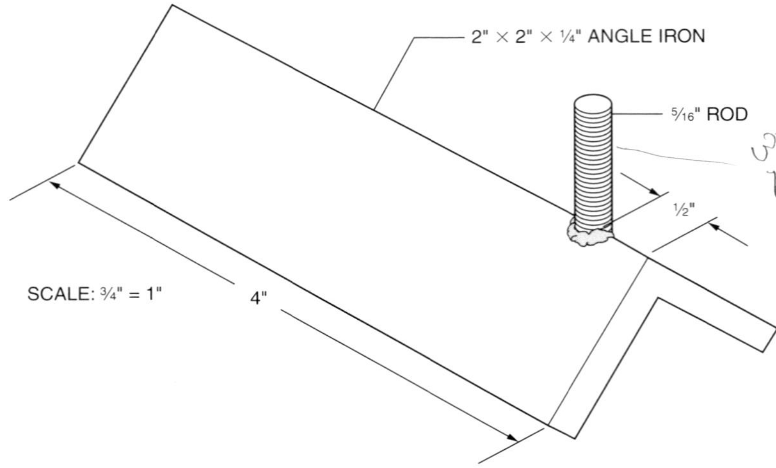


309F07.EPS

*Figure 7* ◆ Rod position on angle iron.

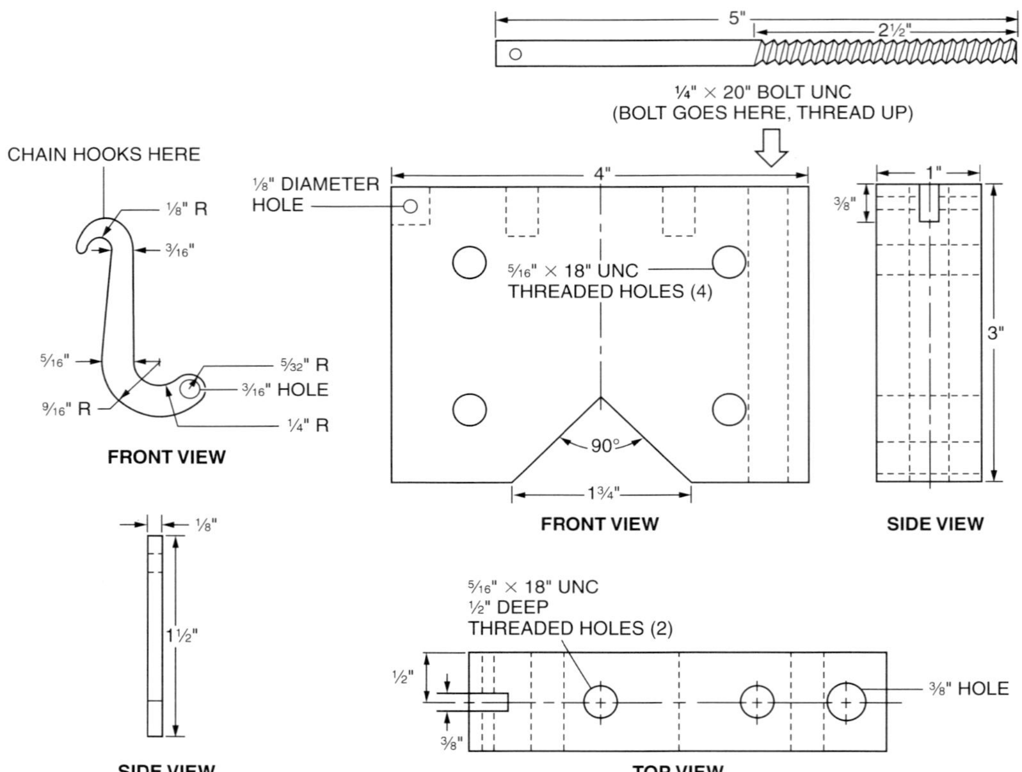


309F08.EPS

*Figure 8* ◆ Fabricating chain jig A.



Instructor's Notes:

9.6 MILLWRIGHT ◆ LEVEL THREE

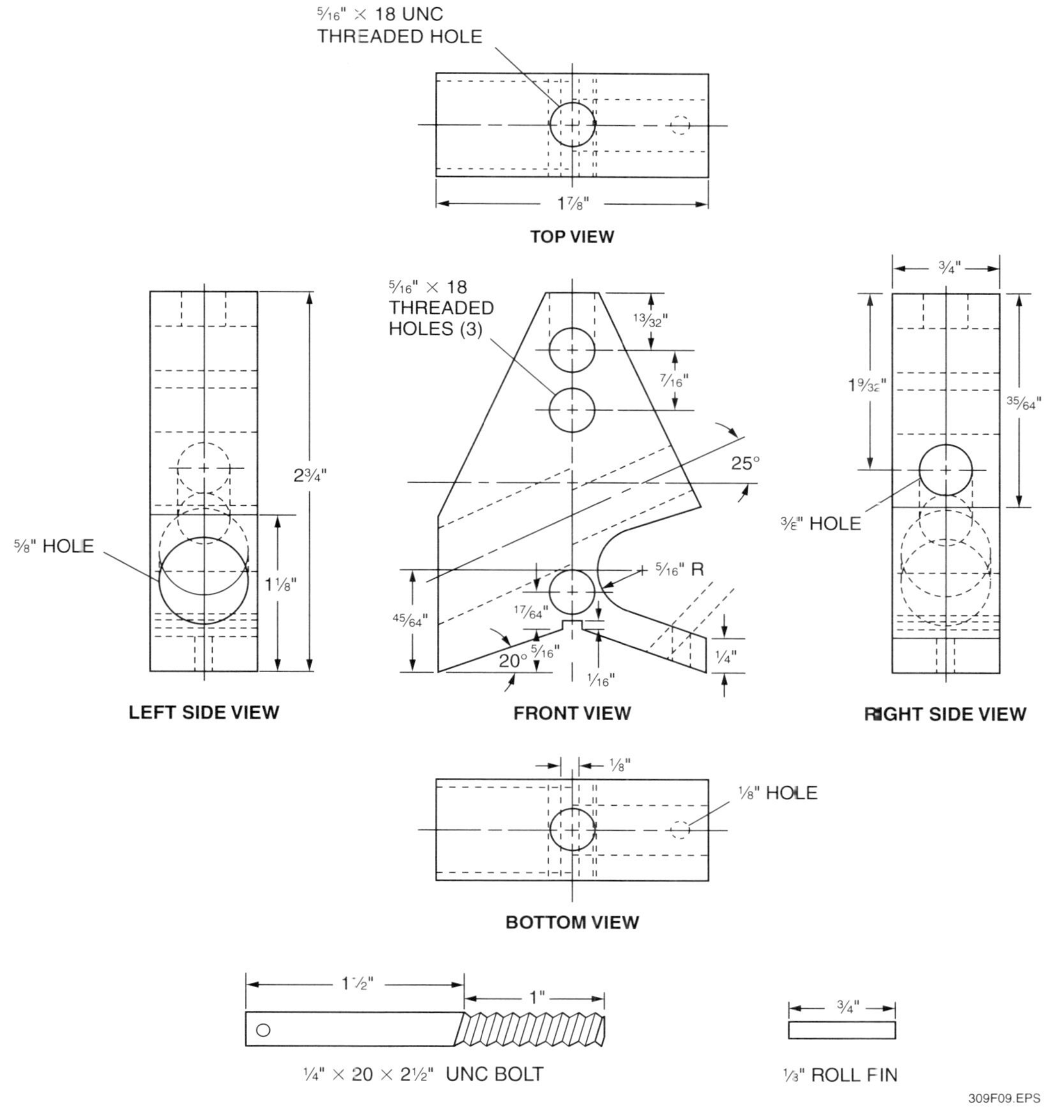


*Figure 9* ◆ Fabricating chain jig B.

## 3.3.0 Fabricating Complex Reverse-Indicator Jigs

The complex reverse-indicator jig (*Figure 10*) has a complicated design, but it is very sturdy and has very little indicator sag over long distances. These jigs are used in pairs for reverse-indicating. The length of the indicator support device can be adjusted to facilitate indicating over various distances. This jig is usually made on a lathe and a bandsaw or mill.

## 3.4.0 Fabricating Christmas Tree Jigs

The Christmas tree jig must be accurately made and must be very rigid to help reduce indicator sag at the end of the jig. This type of jig is limited in its applications since there are only approxi-



**Discuss the fabrication of complex reverse-indicator jigs.**

**Show Transparency 11 (Figure 10). Answer any questions trainees may have.**

**Discuss the fabrication of Christmas tree jigs.**

**Show Transparency 12 (Figure 11). Answer any questions trainees may have.**

MODULE 15309-08 ◆ ALIGNMENT FIXTURES AND SPECIALTY JIGS 9.7

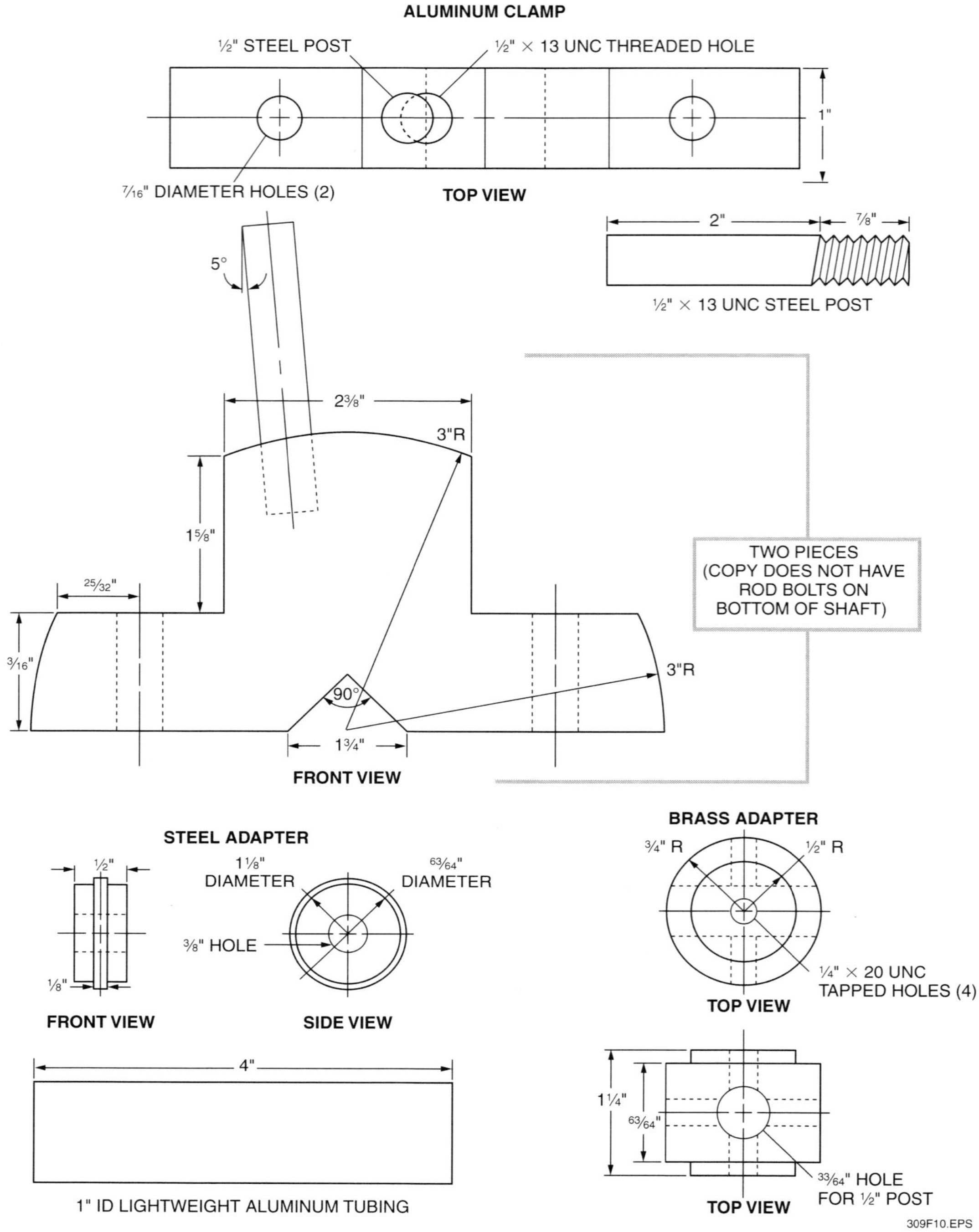


*Figure 10* ◆ Fabricating a complex reverse-indicator jig.



Instructor's Notes:

9.8 MILLWRIGHT ◆ LEVEL THREE

mately 6 inches of adjustment in the distance that it can accommodate. Therefore, it is necessary to fabricate a jig for each application that varies more than 6 inches in distance (*Figure 11*).

### 3.5.0 Fabricating Piano Wire Jigs

The piano wire jig has a simple design (*Figure 12*). It must be strong to provide a firm base for the piano wire. The jig must provide the tension necessary to pull the wire taut, and it must be adjustable for precise alignment of the wire.

## 4.0.0 ◆ SETTING UP JIGS

Properly setting up jigs is crucial to the success of any alignment job. If the jigs are not properly set up and adjusted, the indicators may give inconsistent readings, and the dial indicator may become damaged. The following sections illustrate how to set up the following jigs:

- Angle iron
- Chain
- Complex reverse-indicator
- Christmas tree
- Piano wire

### 4.1.0 Setting Up Angle Iron Jigs

Angle iron jigs are relatively simple to set up (*Figure 13*). Angle iron jigs are used for reverse dial-indicating, so two jigs are used. The angle is strapped to the shafts in positions that facilitate proper placement of the dial indicators. When properly set up, one indicator is exactly on top of the machine to be moved (MTBM) shaft and the

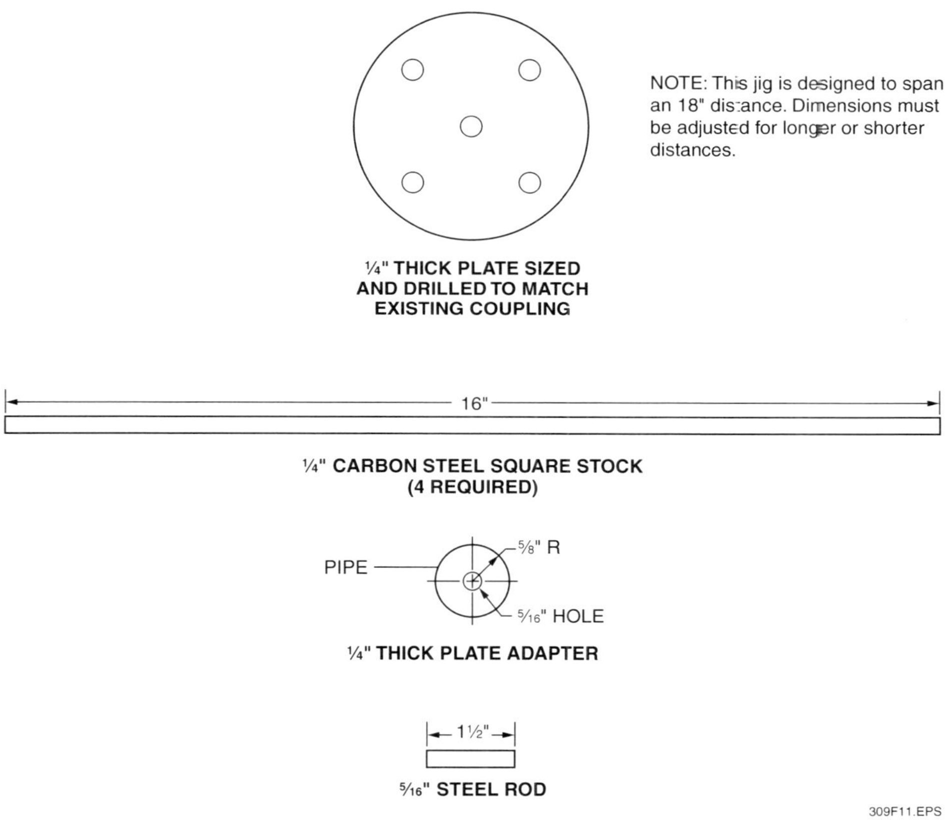


*Figure 11* ◆ Fabricating a Christmas tree jig.



**Discuss the fabrication of piano wire jigs.**

**Show Transparency 13 (Figure 12).**

**Demonstrate how to fabricate one or more of the following jigs: angle iron, chain, Christmas tree, and/or piano wire.**

**Have trainees practice fabricating one or more of the following jigs: angle iron, chain, Christmas tree, and/or piano wire. Note the proficiency of each trainee. This laboratory corresponds to Performance Task 2.**

**Have trainees review Sections 4.0.0–4.5.0.**

**Ensure that you have everything required to teach this session.**

**Discuss the consequences of improper jig setup.**

**Explain how to set up angle iron jigs.**

**Show Transparency 14 (Figure 13).**

MODULE 15309-08 ◆ ALIGNMENT FIXTURES AND SPECIALTY JIGS 9.9

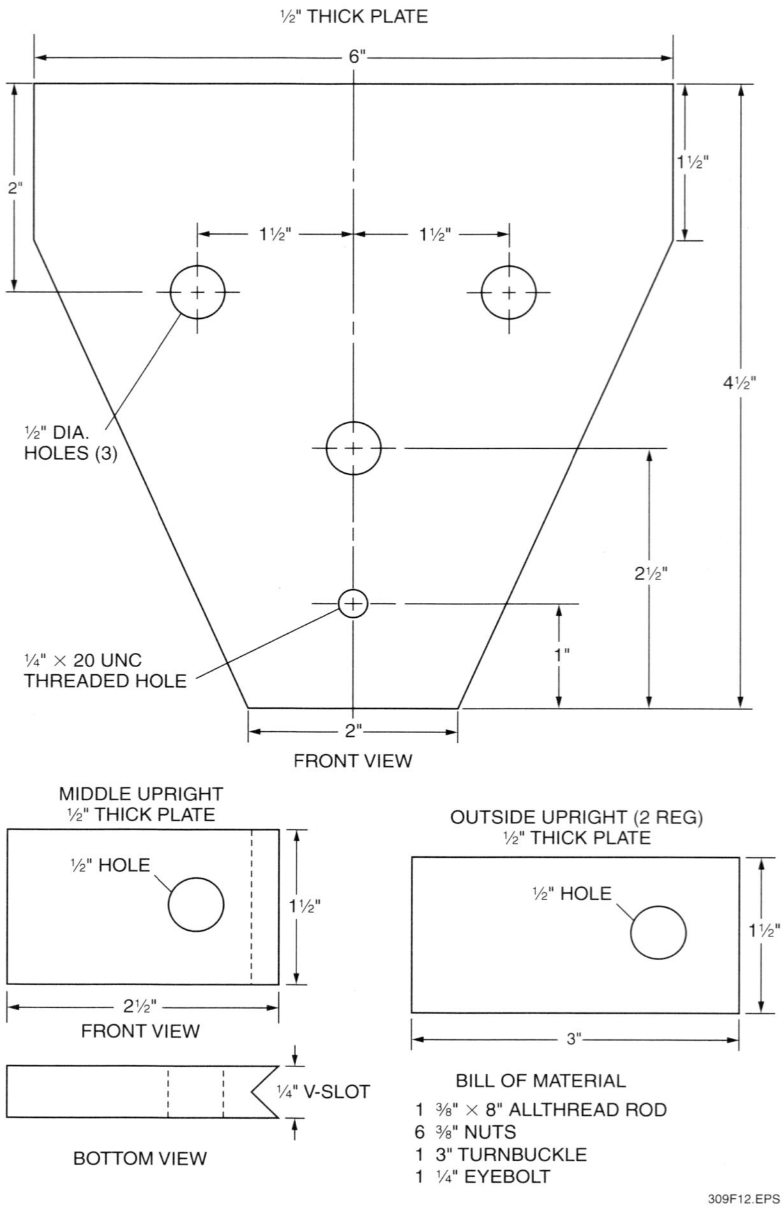


*Figure 12* ◆ Fabricating a piano wire jig.



Instructor's Notes:

9.10 MILLWRIGHT ◆ LEVEL THREE

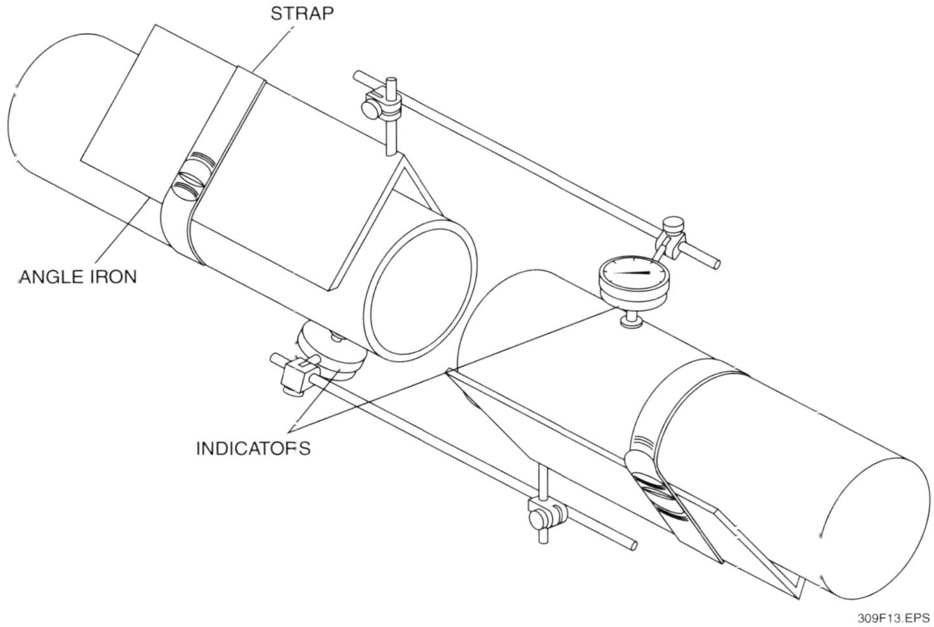


*Figure 13* ◆ Setting up an angle iron jig.

other is exactly on the bottom of the driven shaft (diametrically opposed). The jig that is clamped to the driving shaft supports the indicator that is on the driven shaft and vice versa. The indicators should be positioned vertically so that the feet are in full contact with the shaft surface. Contact point pressure should be enough so that the stem is at the midpoint of its travel range. Positioning the indicators at an angle may result in inconsistent readings.

Dial indicators are very fragile and are easily damaged. Do not bump the foot against the shaft. This may jam the stem and damage the indicator.

## 4.2.0 Setting Up Chain Jigs

Setting up a chain jig is a simple matter of clamping the jig on the coupling and attaching the indicator hardware and indicators (*Figure 14*). The jig is positioned on the coupling, and the chain is wrapped around the coupling and hooked to the jig. Then the adjusting nut is tightened until the chain is tight around the coupling and the jig is rigidly set. This setup is for conventional alignment, so two indicators are used on the same jig. The indicator hardware and indicators are attached and set. One indicator is set to contact the face of the coupling half opposite the jig. The other is set to contact the rim of the same coupling half. The indicators should be positioned at the top of the coupling half.

## 4.3.0 Setting Up Complex Reverse-Indicator Jigs

Setting up the complex reverse-indicator jig is more complicated than setting up the chain jig. This setup is for reverse-dial indicator alignment, so two jigs and two indicators are used. One jig is clamped around the driving coupling half, and the other is clamped around the driven coupling half. The two jigs are positioned so that the mounting pin of one jig is in line with the flat on top of the other jig (*Figure 15*).

The indicator hardware and indicators are mounted next. One indicator and its hardware are mounted on the driving coupling jig so that the indicator contacts the flat on top of the driven coupling jig. The other indicator and its hardware are mounted on the driven coupling jig so that the indicator contacts the flat on top of the driving coupling jig (*Figure 16*).

**Explain how to set up a chain jig.**

**Show Transparency 15 (Figure 14).**

**Explain how to set up a complex reverse-indicator jig.**

**Show Transparencies 16 and 17 (Figures 15 and 16).**



MODULE 15309-08 ◆ ALIGNMENT FIXTURES AND SPECIALTY JIGS 9.11

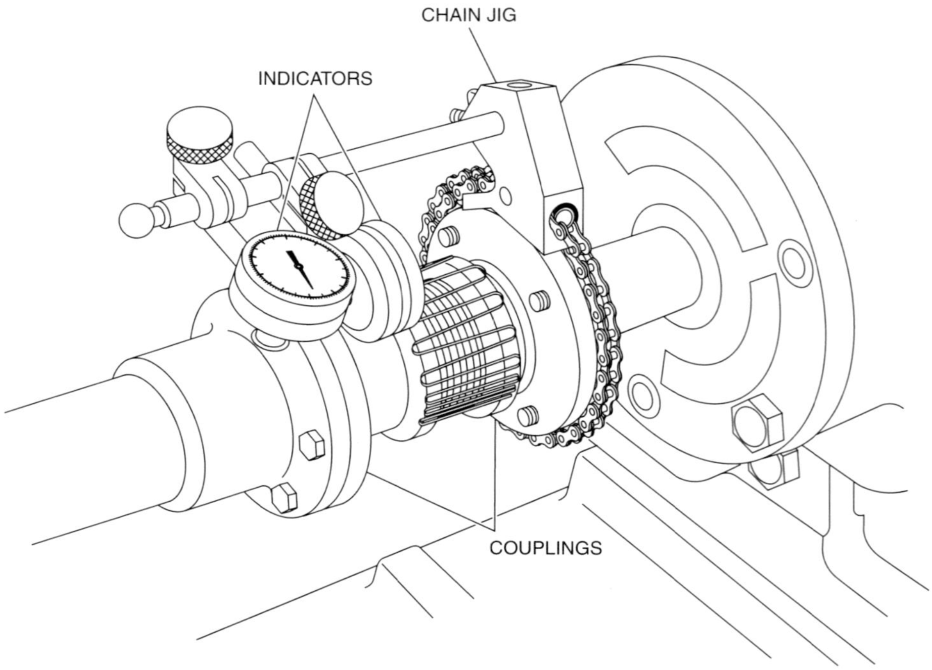


*Figure 14* ◆ Setting up chain jig.

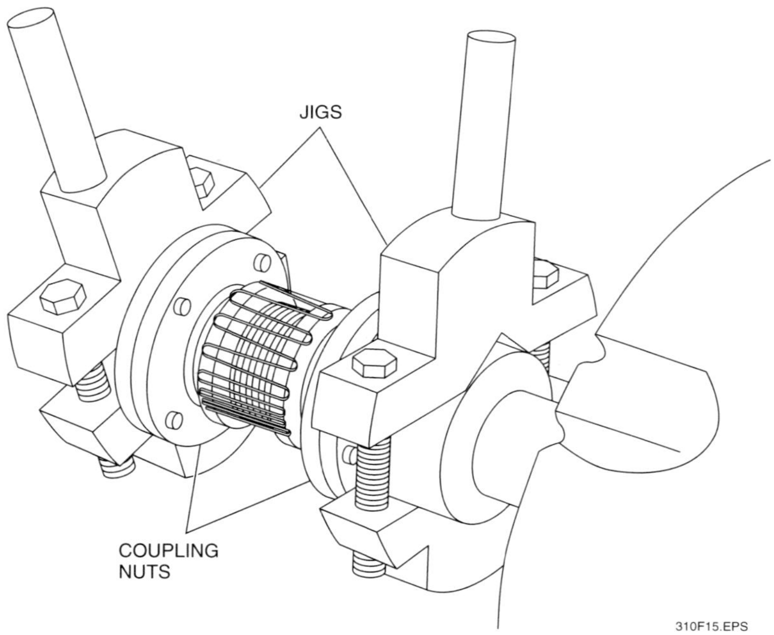


*Figure 15* ◆ Mounting complex reverse-indicator jigs.

9.12 MILLWRIGHT ◆ LEVEL THREE

Instructor's Notes:

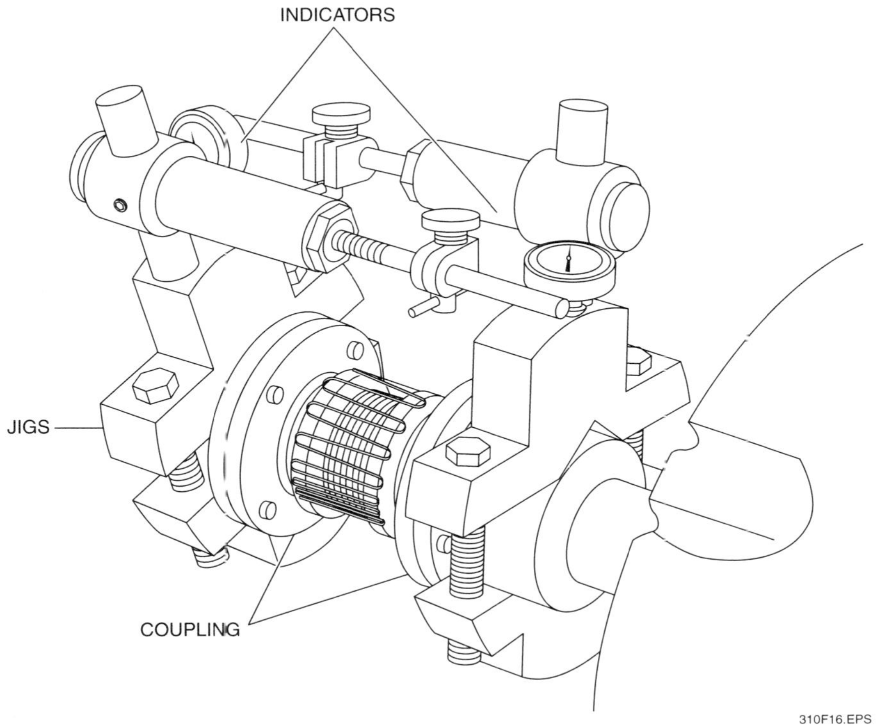


*Figure 16* ◆ Mounting indicators.

## 4.4.0 Setting Up Christmas Tree Jigs

The Christmas tree jig is used only for conventional alignment and is easy to set up (*Figure 17*). The jig is bolted to the driven coupling half so that it extends to within a few inches of the driving shaft or coupling half. The indicator hardware and indicator are mounted on the end of the Christmas tree jig so that the indicator contacts the outside of the shaft or coupling half. The Christmas tree jig can be checked for runout and sag on a lathe.

## 4.5.0 Setting Up Piano Wire Jigs

The piano wire jig can be set up in two ways. One setup is used to align machinery and equipment to a common line on the floor. In this setup, the jigs are set as closely as possible to established benchmarks on the floor and are then anchored to the floor. The piano wire is then strung between the jigs and attached to the turnbuckle of each jig. Then the center upright member of each jig is adjusted to precisely align the wire with the benchmarks on the floor. The turnbuckles are tightened to pull the wire taut.

Piano wire will sag a certain amount over a distance. That sag is determined by the wire diameter, the tension on the wire, and the span. The sag varies directly as the span, and inversely as the tension and the diameter. A tension in the range of 60 foot pounds will minimize sag, although it cannot be completely eliminated.

The second setup is used to align machinery and equipment to a common elevation. In this setup, the jigs are bolted or clamped to two columns or other vertical members as closely as possible to established elevation marks. The piano wire is then strung between the jigs and attached to the turnbuckle of each jig. Then the center upright member of each jig is free-moving and is adjusted to precisely align the wire with the elevation marks. The turnbuckles are tightened to pull the wire taut. *Figure 18* shows how to set up a piano wire jig.



**Describe how to set up a Christmas tree jig.**

**Show Transparency 18 (Figure 17).**

**Discuss the two ways a piano wire jig can be set up.**

**Show Transparency 19 (Figure 18).**

**Demonstrate how to set up three or more of the following jigs: angle iron, chain, complex reverse-indicator, Christmas tree, and/or piano wire.**

**Have trainees practice setting up three or more of the following jigs: angle iron, chain, complex reverse-indicator, Christmas tree, and/or piano wire. Note the proficiency of each trainee. This laboratory corresponds to Performance Task 3.**

MODULE 15309-08 ◆ ALIGNMENT FIXTURES AND SPECIALTY JIGS 9.13

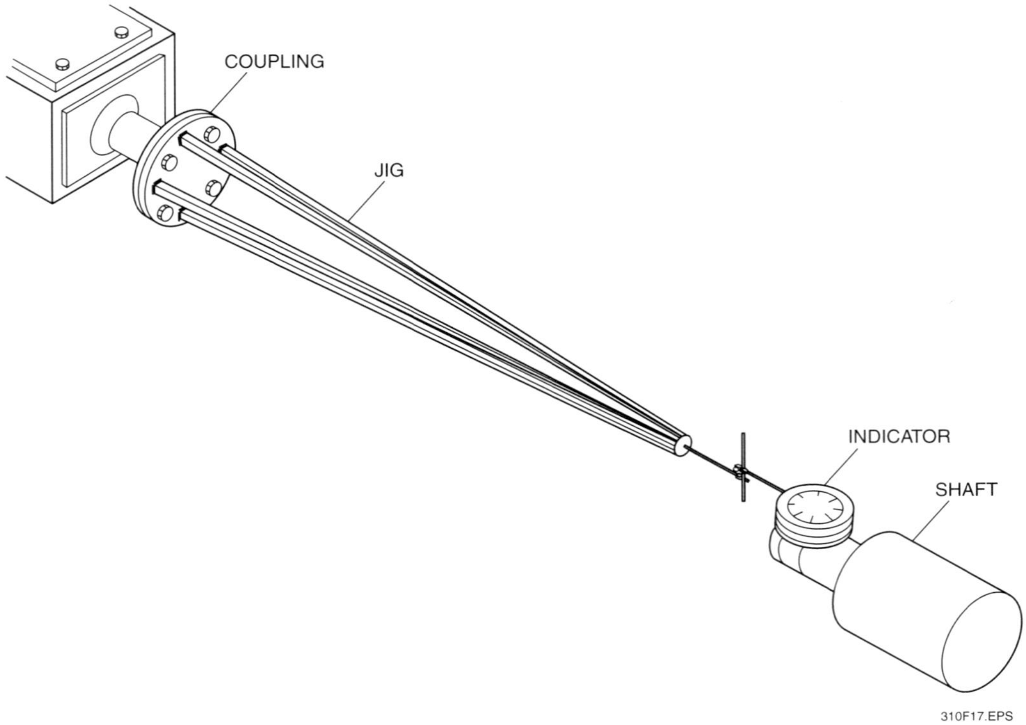


*Figure 17* ◆ Setting up a Christmas tree jig.

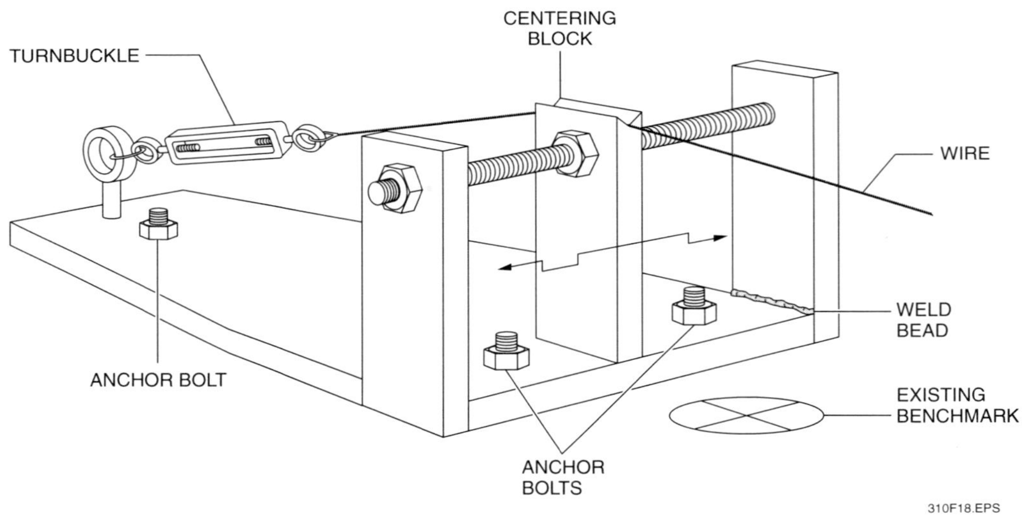


*Figure 18* ◆ Setting up a piano wire jig.



Instructor's Notes:

9.14 MILLWRIGHT ◆ LEVEL THREE

# Review Questions

1. Dial indicators are used for _____ alignment.
   a. laser and theodolite
   b. machine baseplate
   c. conventional and reverse-indicator
   d. optical

2. The chain jig is used for _____ alignment.
   a. positional
   b. positive
   c. reverse-indicator
   d. conventional

3. Complex reverse-indicator jigs are _____ than other jigs.
   a. harder to use
   b. more rigid
   c. less accurate
   d. less often available

4. On a piano wire jig, the wire must be _____.
   a. very taut
   b. loose
   c. at right angle to the plate
   d. welded to the plate

5. The type and size of jig needed depends in part on the _____.
   a. number of jigs needed
   b. physical space available
   c. accuracy required
   d. direction of the shaft

6. Except for turbine installation, equipment with very large shafts is usually aligned with _____.
   a. chain jigs
   b. Christmas tree jigs
   c. piano wire jigs
   d. laser or optical equipment

7. The mounting hardware on most dial indicators will fit a _____.
   a. ½-inch bolt
   b. ¾-inch rod
   c. $\frac{5}{16}$-inch rod
   d. 1-inch bolt

8. One of the most important considerations in choosing a chain jig design is _____.
   a. price
   b. ability to fit firmly on the shaft
   c. length of the chain
   d. the size of the bolt

9. When the angle iron jigs are clamped for reverse indication, the indicator on one jig is _____.
   a. not touching either shaft
   b. in contact with the other shaft
   c. in contact with the shaft it is clamped on
   d. at an angle to the shaft

10. Positioning the indicators at an angle may result in _____.
    a. bent indicator contact
    b. indicator confusion
    c. inconsistent readings
    d. correct readings



Have trainees complete the Review Questions, and go over the answers prior to administering the Module Examination.

MODULE 15309-08 ◆ ALIGNMENT FIXTURES AND SPECIALTY JIGS 9.15

**Summarize the major concepts presented in the module.**

**Administer the Module Examination. Record the results on Craft Training Report Form 200, and submit the results to the Training Program Sponsor.**

**Administer the Performance Test, and fill out Performance Profile Sheets for each trainee. If desired, trainee proficiency noted during laboratory sessions may be used to complete the Performance Test. Record the results on Craft Training Report Form 200, and submit the results to the Training Program Sponsor.**

## Summary

The fabrication of different special jigs for dial indicators, lasers, and piano wire requires knowledge of the way the alignment methods are to be used. The types of jigs that are used for dial indicators can be used for laser alignment as well. Standard manufactured equipment, such as magnetic-based stands, are also available, but are no more accurate than the shop-made ones.

It is critical to determine indicator sag if you are using a dial indicator. It is equally important to determine wire sag with piano wire jigs.

## Notes



Instructor's Notes:

## Trade Terms Introduced in This Module

***Indicator sag:*** The displacement reading on the dial indicator created by the weight of the dial indicator and the jig acting against the stiffness of the material from which the jig is made.

***Piano wire:*** A high-strength wire that can be drawn very tight to make a straight line.

***Rigid:*** Stiff; not able to bend or flex.

# Resources & Acknowledgments

## Additional Resources

This module is intended to present thorough resources for task training. The following reference work is suggested for further study. This is optional material for continued education rather than for task training.

L.S. Starrett Company, http://www.starrett.com

## Figure Credits

L.S. Starrett Company, 309F01



Instructor's Notes:

# MODULE 15309-08 — TEACHING TIPS

The following are suggested activities or instructional methods to help you teach the material in this module.

## General

When you call on someone to answer a question, the rest of the class relaxes or even tunes out because they expect that the question and answer will take place only between you and the trainee you called on. Instead, use this technique to involve more trainees in answering questions and to keep them on their toes.

1. Ask trainees to define a term or explain a concept.
2. After one trainee has answered, ask a trainee seated nearby if the answer is right. Then ask whether a trainee in the back of the room agrees.
3. Ask trainees to explain why they think an answer is right or wrong.
4. Use the session to clear up incorrect ideas and encourage trainees to learn from their mistakes.

## Section 2.0.0 *Types of Jigs*

This exercise will help familiarize trainees with various types of jigs. Trainees will need pencils and paper. You will need to arrange workstations with a different type of jig at each station. Allow 20 to 30 minutes for this exercise. This exercise corresponds to Performance Task 1.

1. Discuss the various types of jigs with trainees.
2. Have trainees rotate between the various workstations and identify the type of jig at each station.
3. After all trainees have visited each station, have one trainee identify the type of jig at their station. Answer any questions trainees may have.

## MODULE 15309-08 — ANSWERS TO REVIEW QUESTIONS

| Answer | Section |
|---|---|
| 1. b | 1.0.0 |
| 2. d | 2.2.0 |
| 3. a | 2.3.0 |
| 4. b | 2.4.0 |
| 5. c | 2.5.0 |
| 6. b | 3.0.0 |
| 7. c | 4.3.0 |
| 8. b | 3.3.0 |
| 9. d | 4.1.0 |
| 10. a | 4.1.0 |
| 11. b | 2.3.0 |
| 12. c | 2.2.0 |
| 13. b | 3.4.0 |
| 14. a | 4.5.0 |
| 15. d | 3.0.0 |

## CONTREN® LEARNING SERIES — USER UPDATE

The NCCER makes every effort to keep these textbooks up-to-date and free of technical errors. We appreciate your help in this process. If you have an idea for improving this textbook, or if you find an error, a typographical mistake, or an inaccuracy in NCCER's Contren® textbooks, please write us, using this form or a photocopy. Be sure to include the exact module number, page number, a detailed description, and the correction, if applicable. Your input will be brought to the attention of the Technical Review Committee. Thank you for your assistance.

*Instructors* – If you found that additional materials were necessary in order to teach this module effectively, please let us know so that we may include them in the Equipment/Materials list in the Annotated Instructor's Guide.


**Write:** Product Development and Revision
National Center for Construction Education and Research
3600 NW 43rd St., Bldg. G, Gainesville, FL 32606

**Fax:** 352-334-0932

**E-mail:** curriculum@nccer.org


Craft ______________________ Module Name ______________________

Copyright Date __________ Module Number __________ Page Number(s) __________

Description

______________________________________________

______________________________________________

______________________________________________

______________________________________________

(Optional) Correction

______________________________________________

______________________________________________

______________________________________________

(Optional) Your Name and Address

______________________________________________

______________________________________________

______________________________________________

Module 15310-08

# *Prealignment for Equipment Installation*

## NCCER STANDARDIZED CRAFT TRAINING PROGRAM

The National Center for Construction Education and Research (NCCER) provides a standardized national program of accredited craft training. Key features of the program include instructor certification, competency-based training, and performance testing. The program provides trainees, instructors, and companies with a standard form of recognition through a National Craft Training Registry. The program is described in full in the *Guidelines for Accreditation*, published by NCCER. For more information on standardized craft training, contact the NCCER by writing us at 3600 NW 43rd St., Bldg. G, Gainesville, FL 32606; calling 352-334-0911; or emailing info@nccer.org. More information may be found at our website, www.nccer.org.

## HOW TO USE THIS ANNOTATED INSTRUCTOR'S GUIDE

Each page presents two sections of information. The larger section displays each page exactly as it appears in the Trainee Module. The narrow column ties suggested trainee and instructor actions to each page and provides icons (detailed below) to call your attention to material, safety, audiovisual, or testing requirements. The bottom of each page includes space for your notes.

The **Audiovisual** icon indicates an appropriate time to show a transparency or other audiovisual aid.

The **Classroom** icon prompts you to define a term, stress a point, ask trainees to explain a concept, or give examples.

The **Demonstration** icon directs you to show trainees how to perform tasks.

The **Examination** icon tells you to administer the written module examination.

The **Homework** icon is placed where you may wish to assign reading for the next class, assign a project, or advise trainees to prepare for an examination.

The **Laboratory** icon is used when trainees are to practice performing tasks.

The **Materials** icon is a reminder for you to gather materials needed for classes, labs, and testing.

The **Performance Testing** icon tells you to administer a performance test or a portion thereof.

The **Safety** icon is used to emphasize safety issues. It is often keyed to *Caution* and *Warning!* statements in the Trainee Module.

The **Teaching Tip** icon indicates additional guidance is available, such as how to conduct an exercise, get the most educational value from a field trip, or encourage class participation. Teaching Tips may expand on a feature (*Think About It*, *Did You Know?*) or provide *Quick Quizzes* or similar exercises. You will be referred to the Teaching Tips section at the back of the module if there is additional material.

The **Combination** icon indicates that the laboratory listed corresponds with a performance task. If desired, you can note the proficiency of the trainees during the laboratory, and use it to satisfy performance testing requirements.

## PREPARATION

Before teaching this module, you should review the Objectives, Performance Tasks, Materials and Equipment List, and Module Outline. Be sure to allow ample time to prepare your own training or lesson plan and gather all required materials and equipment.

Prealignment for Equipment Installation
Annotated Instructor's Guide

**Module 15310-08**

## MODULE OVERVIEW

This module explains how to perform preliminary alignment procedures for driver and driven units, including inspecting, setting up, and leveling the equipment. It also covers coupling installation.

## PREREQUISITES

Prior to training with this module, it is recommended that the trainee shall have successfully completed *Core Curriculum; Millwright Level One; Millwright Level Two;* and *Millwright Level Three,* Modules 15301-08 through 15309-08.

## OBJECTIVES

Upon completion of this module, the trainee will be able to do the following:

1. Inspect equipment.
2. Install couplings.
3. Set STAT equipment.
4. Set MTBM equipment.

## PERFORMANCE TASKS

Under the supervision of the instructor, the trainee should be able to do the following:

1. Inspect equipment.
2. Install couplings, using either the key and setscrew or the tapered shaft and locking nut method.
3. Set STAT equipment.
4. Set MTBM equipment.

## MATERIALS AND EQUIPMENT LIST

Overhead projector and screen
Transparencies
Blank acetate sheets
Transparency pens
Whiteboard/chalkboard
Markers/chalk
Pencils and scratch paper
Appropriate personal protective equipment
Basic trainee tools
Samples of split clamping couplings
Samples of tapered clamp hubs
Quick Quiz*
Module Examinations**
Performance Profile Sheets**

* Located in the back of this module
**Located in the Test Booklet

## SAFETY CONSIDERATIONS

Ensure that the trainees are equipped with appropriate personal protective equipment and know how to use it properly. This module requires that the trainees work with hand tools. Ensure that trainees are briefed on shop safety procedures.

## ADDITIONAL RESOURCES

This module is intended to present thorough resources for task training. The following reference works are suggested for both instructors and motivated trainees interested in further study. These are optional materials for continued education rather than for task training.

R+W America L.P., www.rw-america.com

Coupling Corporation of America, www.couplingcorp.com

## TEACHING TIME FOR THIS MODULE

An outline for use in developing your lesson plan is presented below. Note that each Roman numeral in the outline equates to one session of instruction. Each session has a suggested time period of 2½ hours. This includes 10 minutes at the beginning of each session for administrative tasks and one 10-minute break during the session. Approximately 15 hours are suggested to cover *Prealignment for Equipment Installation*. You will need to adjust the time required for hands-on activity and testing based on your class size and resources. Because laboratories often correspond to Performance Tasks, the proficiency of the trainees may be noted during these exercises for Performance Testing purposes.

| Topic | Planned Time |
|---|---|
| **Session I. Introduction; Inspecting Equipment; Setting Up and Leveling Baseplates** | |
| A. Introduction | ________ |
| B. Inspecting Equipment | ________ |
| C. Laboratory<br>Trainees practice inspecting equipment. This laboratory corresponds to Performance Task 1. | ________ |
| D. Setting Up and Leveling Baseplates | ________ |
| **Session II. Installing Couplings** | |
| A. Installing Couplings | ________ |
| B. Laboratory<br>Trainees practice installing couplings, using either the key and setscrew or the tapered shaft and locking nut method. This laboratory corresponds to Performance Task 2. | ________ |
| **Session III. Setting the Driven Piece of Equipment** | |
| A. Setting the Driven Piece of Equipment | ________ |
| B. Laboratory<br>Trainees practice setting STAT equipment. This laboratory corresponds to Performance Task 3. | ________ |
| **Session IV. Setting the Driver** | |
| A. Setting the Driver | ________ |
| B. Laboratory<br>Trainees practice setting MTBM equipment. This laboratory corresponds to Performance Task 4. | ________ |
| **Session V. Review** | |
| A. Module Review | ________ |

**Session VI. Testing**

A. Module Examination ____________

1. Trainees must score 70% or higher to receive recognition from NCCER.
2. Record the testing results on Craft Training Report Form 200, and submit the results to the Training Program Sponsor.

B. Performance Testing ____________

1. Trainees must perform each task to the satisfaction of the instructor to receive recognition from NCCER. If applicable, proficiency noted during laboratory exercises can be used to satisfy the Performance Testing requirements.
2. Record the testing results on Craft Training Report Form 200, and submit the results to the Training Program Sponsor.

15310-08

# Prealignment for Equipment Installation

Assign reading of Module 15310-08.

# 15310-08
# *Prealignment for Equipment Installation*

*Topics to be presented in this module include:*



## Overview

Prealignment is the process used to set machinery close enough to the correct position to allow the actual alignment to take place. This module will teach you how to mount the driver and driven equipment on the baseplate.

Instructor's Notes:

## Objectives

When you have completed this module, you will be able to do the following:

1. Inspect equipment.
2. Install couplings.
3. Set STAT equipment.
4. Set MTBM equipment.

## Trade Terms

Axial movement
Skew
Torsional movement

## Required Trainee Materials

1. Pencil and paper
2. Appropriate personal protective equipment

## Prerequisites

Before you begin this module, it is recommended that you successfully complete *Core Curriculum; Millwright Level One; Millwright Level Two;* and *Millwright Level Three*, Modules 15301-08 through 15309-08.

This course map shows all of the modules in the third level of the *Millwright* curriculum. The suggested training order begins at the bottom and proceeds up. Skill levels increase as you advance on the course map. The local Training Program Sponsor may adjust the training order.

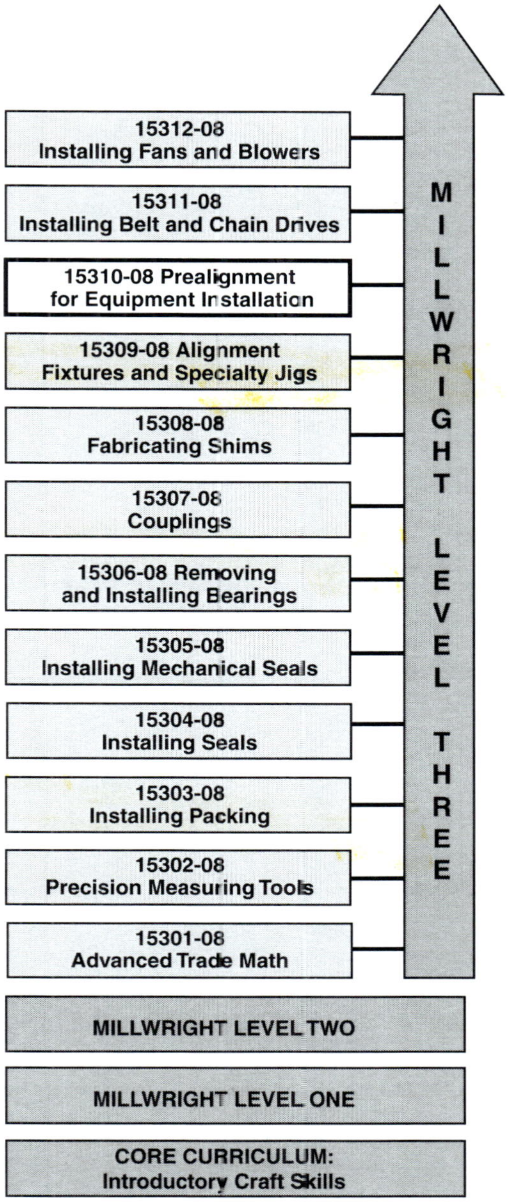


310CMAP.EPS

**Ensure that you have everything required to teach the course. Check the Materials and Equipment list at the front of this module.**

**See the general Teaching Tip at the end of this module.**

**Explain that terms shown in bold are defined in the Glossary at the back of this module.**

**Show Transparency 1, Objectives, and Transparency 2, Performance Tasks. Review the goals of the module, and explain what will be expected of the trainee.**

**Review the modules covered in Level Three and explain how this module fits in.**



MODULE 15310-08 ◆ PREALIGNMENT FOR EQUIPMENT INSTALLATION 10.1

**Define prealignment.**

**Explain the terms driven and driver, STAT and MTBM.**

**Present the steps for inspecting equipment prior to installation and prealignment.**

**Demonstrate how to inspect equipment.**

**Have trainees practice inspecting equipment. Note the proficiency of each trainee. This laboratory corresponds to Performance Task 1.**

**Discuss setting up and leveling the equipment baseplates.**

**Have trainees review Sections 4.0.0–4.3.0.**

**Ensure that you have everything required to teach this session.**

**Describe the clearance method of coupling installation.**

**Review the steps for the key and setscrew method of installation.**

**Show Transparency 3 (Figure 1).**

## 1.0.0 ◆ INTRODUCTION

Prealignment is the preliminary, rough alignment of equipment and machinery that is performed when the equipment is initially set in place. Prealignment enables other craftworkers, such as pipefitters, electricians, and instrument fitters, to perform their work. This module explains how to prealign a driven unit and a driver, also called a static machine (STAT) and a machine to be moved (MTBM).

## 2.0.0 ◆ INSPECTING EQUIPMENT

Before installing and prealigning a piece of equipment, the equipment must be inspected to ensure that it is suitable for service. Follow these steps to inspect a STAT and MTBM before installation:

*Step 1* Inspect the STAT and MTBM for physical damage.

*Step 2* Rotate the STAT and MTBM by hand to ensure that they are not locked up or binding, that there are no flat places in the bearings, and that the shafts are turning true.

> **NOTE**
> The shafts of the machines should turn smoothly. If there are flat places in the bearings, you can feel a roughness as you rotate the shaft. Rough places in the bearing can be caused by many things, including improperly attached welding grounds, or improperly installed metal bearings.

*Step 3* Inspect the equipment baseplate to ensure that it is not warped or damaged.

## 3.0.0 ◆ SETTING UP AND LEVELING BASEPLATES

The first step in prealignment is to set up and level the equipment baseplate. If the baseplate is not correctly oriented, it is very difficult to align the equipment. The purpose of setting up the baseplate is to bring the equipment to the proper height and align it correctly.

A piece of equipment may be set at an assigned angle other than level. However, the angle must be precise to prevent unnecessary stress on the couplings and bearings.

## 4.0.0 ◆ INSTALLING COUPLINGS

The first step in setting two machines is to install couplings on the mating shafts. There are two basic types of couplings: rigid and flexible. As its name indicates, a rigid coupling connects the shafts together rigidly and does not allow for any misalignment or movement. Flexible couplings, on the other hand, allow for a small amount of misalignment and some **torsional movement** and **axial movement**. Two basic methods of coupling installation are clearance installation and interference installation. This section explains clearance installation.

Clearance installation uses a setscrew, clamps, or nut to lock the coupling onto the shaft. The most common methods for clearance installation are the following:

- Key and setscrew method
- Tapered shaft and locking nut method
- Clamping hub or coupling

### 4.1.0 Key and Setscrew Method

With the key and setscrew method, the coupling has a key that fits into the shaft (*Figure 1*). A setscrew then screws through the coupling and presses on the key to lock the coupling onto the shaft.

Follow these steps to install a key and locking screw coupling:

*Step 1* As with any coupling, clean the shaft thoroughly, using solvent and a rag.

*Step 2* Sand the shaft lightly, if specifications permit, using an emery cloth to smooth it.

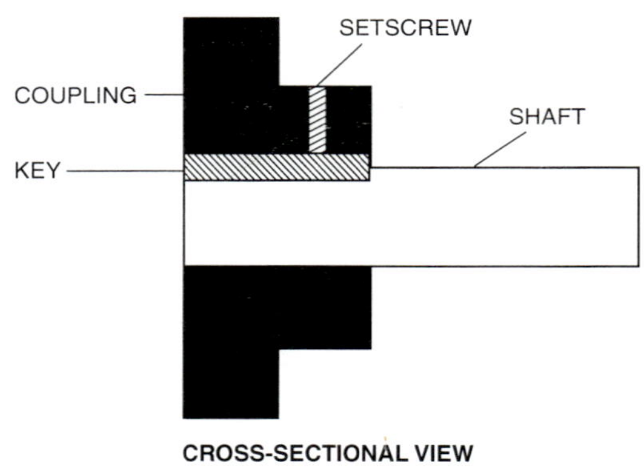


*Figure 1* ◆ Key and setscrew coupling.



Instructor's Notes:

10.2 MILLWRIGHT ◆ LEVEL THREE

*Step 3* Inspect the shaft to ensure that there are no burrs that would interfere with the coupling fitting onto the shaft.

*Step 4* Coat the shaft lightly with special lubricant, if permitted, to make the installation of the coupling easier.

*Step 5* Position the key in the shaft keyway. Check the key for fit or roughness. The keys should be placed so that one half of the keyway remains empty. On couplings, the keys are 180 degrees from each other, unless the couplings are match marked.

*Step 6* Slip the coupling onto the shaft over the key. The coupling should be placed on the shaft by hand or by lightly tapping it with a soft-blow hammer.

**CAUTION**

Do not drive the coupling onto the shaft, as this can damage the shaft and the coupling. If the coupling does not go on easily, the shaft or the coupling must be dressed down until the coupling does go on easily. When installing a coupling on a new shaft, dress the coupling to fit the shaft. When installing a coupling on an old shaft, the shaft can be dressed down to remove burrs or repair damage. Never dress the shaft below its original diameter. Dress down the coupling instead.

*Step 7* Align the end of the coupling flush with the end of the shaft.

*Step 8* Tighten the setscrew to hold the coupling in place.

## 4.2.0 Tapered Shaft and Locking Nut Method

The tapered coupling and locking nut (*Figure 2*) use a wedging principle to hold the coupling onto the shaft. The coupling is slipped onto the shaft and the nut is tightened, causing a wedging action that generates tremendous holding force. This type of coupling also has a key. The combined wedging force and key prevent the coupling from rotating on the shaft.

Follow these steps to install a tapered coupling:

*Step 1* Clean the shaft thoroughly.

*Step 2* Inspect the shaft threads to ensure that they are clean and not damaged.

*Step 3* Install the key in the shaft keyway, and check the key for fit and placement.

*Step 4* Slip the coupling onto the shaft over the key.

*Step 5* Screw the nut onto the end of the shaft.

*Step 6* Tighten the nut to lock the coupling onto the shaft.

**CAUTION**

Do not overtighten the nut because this may damage the nut and the shaft. Use a torque wrench to tighten the nut to the proper torque.

**Describe the tapered shaft and locking nut method of clearance installation.**

**Review the steps for installing a tapered coupling.**

**Show Transparency 4 (Figure 2).**

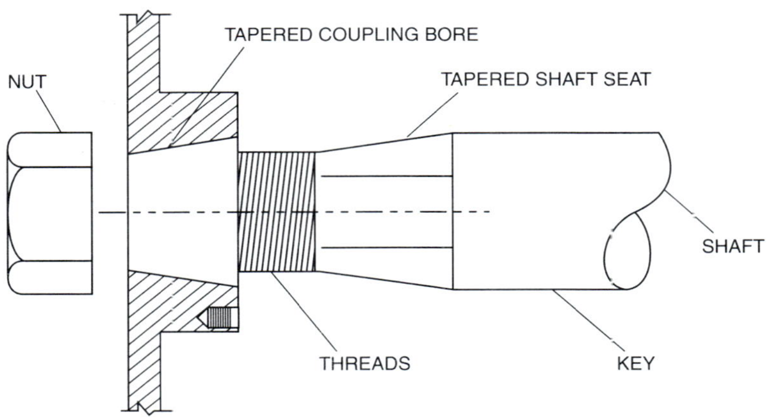


*Figure 2* ◆ Tapered coupling and shaft.



MODULE 15310-08 ◆ PREALIGNMENT FOR EQUIPMENT INSTALLATION 10.3

**Discuss how couplings are clamped onto shafts.**

**Explain how hubs are used for clamping couplings.**

**Provide samples of split clamping couplings and tapered clamp hubs for trainees to examine.**

**Review the steps for clamping couplings.**

**Demonstrate how to install couplings using the key and setscrew and the tapered shaft and locking nut methods.**

**Have trainees practice installing couplings, using either the key and setscrew or the tapered shaft and locking nut method. Note the proficiency of each trainee. This laboratory corresponds to Performance Task 2.**

**Have trainees review Section 5.0.0.**

**Ensure that you have everything required to teach this session.**

### 4.3.0 Clamping Couplings

The other category of couplings is clamped onto the shafts. One such arrangement uses bolts holding split couplings together. Tightening the bolts pulls the couplings tight onto the shaft (*Figure 3*). Another technology uses an asymmetric thread axially to tighten a hub into clamping on the shaft (*Figure 4*). Standard couplings then mount on the hub. The procedure is very similar to any other clearance fit:

*Step 1* Clean the shaft carefully, being careful to remove any lubricants or dirt. Make certain that there are no flat places or burrs, which may skew the clamp mounting.

*Step 2* Clean the inside of the hub and coupling.

*Step 3* For the clamping sleeve, tighten the cross bolts evenly until the sleeve is secure, according to manufacturer's specifications. For the tapered clamp, tighten the load screws by hand until the hub is secure, and torque the screws according to the manufacturer's specification.

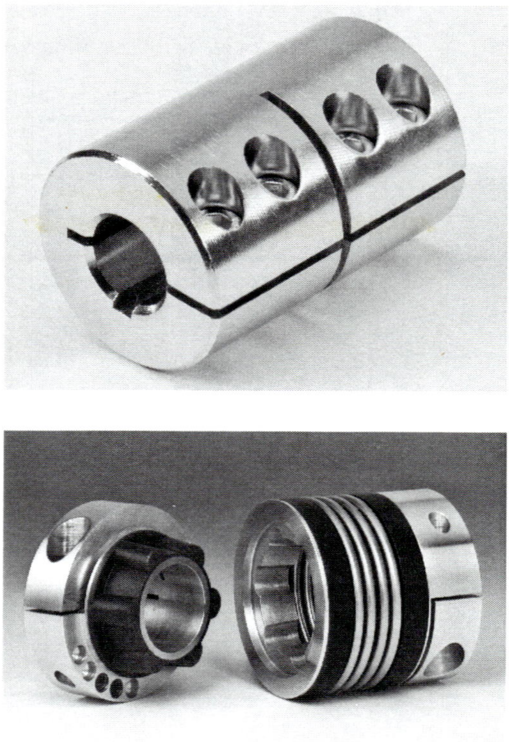

310F03.EPS

*Figure 3* ◆ Split clamping coupling.

Clamping and tapered couplings are generally more accurately centered mountings than the setscrew method, because the setscrew can introduce some slight off-center stress on the coupling. The accuracy of the machining of the shaft and coupling, and the condition and accuracy of the bearings, are the other issues in determining accuracy.

310F04.EPS

*Figure 4* ◆ Tapered clamp hub.



Instructor's Notes:

10.4 MILLWRIGHT ◆ LEVEL THREE

## 5.0.0 ◆ SETTING THE DRIVEN PIECE OF EQUIPMENT

Once the baseplate is set and the couplings are installed, the STAT is ready to be placed in position on the baseplate. This is only a rough alignment for the purpose of drilling and tapping the bolt holes for the equipment. Final, precise alignment is done at a later time. The STAT must be set on the equipment center line and placed to allow room for the driver. Follow these steps to set a driven unit:

*Step 1* Ensure that the baseplate is clean and free of foreign matter that might interfere with the equipment being set level.

> **NOTE**
> The driven unit must be centered on the lengthwise center line and the coupling face parallel to the crosswise center line, leaving ample room for the driver.

*Step 2* Place the machine on the baseplate, aligning it with the equipment center lines (*Figure 5*).

*Step 3* Place shims to allow for future alignment under the STAT base to level the shaft. In the case of a pump, plumb the pump suction. While shimming the STAT base, be careful not to move the STAT out of alignment with the center line. If the STAT has to be moved after the bolts are placed, it may become bolt-bound. The STAT must be shimmed to allow for adjustment.

*Step 4* Punch mark the bolt hole locations, using a transfer punch, through the STAT bolt holes. If not possible, blue the area through the hole.

*Step 5* Remove the STAT from the base.

*Step 6* Drill and tap the bolt holes in the baseplate.

*Step 7* Place the STAT back onto the baseplate and replace the shims.

*Step 8* Install the holddown bolts.

## 6.0.0 ◆ SETTING THE DRIVER

Once the STAT is set and bolted, the MTBM is ready to be set. The MTBM must be set and roughly aligned with the STAT before drilling and tapping the motor bolt holes into the baseplate. Follow these steps to set the MTBM:

*Step 1* Place the MTBM on the baseplate, aligning it with the lengthwise center line.

*Step 2* Move the MTBM toward the STAT until the couplings are close enough together to set the gap between the couplings.

*Step 3* Set the gap between the couplings, using a thickness gauge.

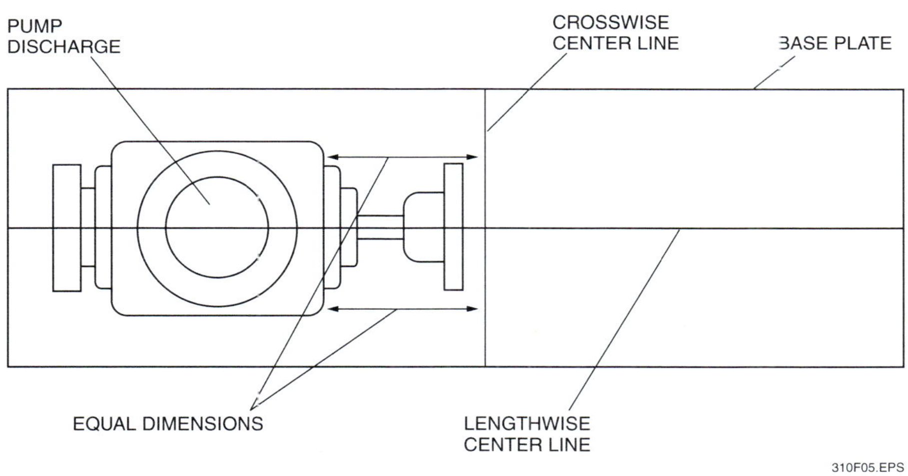


*Figure 5* ◆ Aligning driven piece with center lines.



**Explain how the STAT is placed on the baseplate for prealignment.**

**Review the steps for setting a driven unit.**

**Show Transparency 5 (Figure 5). Explain how to align the driven to equipment center lines.**

**Demonstrate how to set STAT equipment.**

**Have trainees practice setting STAT equipment. Note the proficiency of each trainee. This laboratory corresponds to Performance Task 3.**

**Have trainees review Section 6.0.0.**

**Ensure that you have everything required to teach this session.**

**Explain how the driver is set.**

**Review the steps for setting the MTBM.**

**Show Transparency 6 (Figure 6). Explain how to measure horizontal offset.**

**Explain how to take readings using a dial indicator.**

**Show Transparency 7 (Figure 7).**

*Step 4* Align the couplings as close as possible by eyesight. The gap setting may vary, depending on the type of coupling used. Refer to the coupling manufacturer's specifications for the proper gap setting.

*Step 5* Place a straightedge on the side of the couplings.

*Step 6* Measure the offset misalignment using a thickness gauge (*Figure 6*).

*Step 7* Move the MTBM to correct the offset misalignment.

> **NOTE**
> The offset misalignment of the couplings must be minimal before the MTBM holddown bolt holes are drilled and tapped, because once the bolt holes are drilled, there is very little side-to-side movement of the MTBM. After the offset has been corrected as much as possible with a straightedge and thickness gauge, use a dial indicator to further align the coupling from side to side.

*Step 8* Attach a dial indicator to the MTBM coupling.

*Step 9* Set the dial indicator at the 3-o'clock position on the MTBM half of the coupling.

*Step 10* Set the dial indicator to read zero.

*Step 11* Rotate the couplings until the dial indicator is at the 9-o'clock position on the MTBM half of the coupling.

*Step 12* Read the dial indicator.

*Step 13* Adjust the MTBM and take readings at the 3-o'clock and the 9-o'clock positions until zero readings on both sides of the coupling are achieved (*Figure 7*).

*Step 14* Scribe the outline of the holes carefully.

*Step 15* Remove the MTBM.

*Step 16* Find the hole center.

*Step 17* Punch mark the hole centers.

*Step 18* Remove the MTBM from the baseplate.

> **CAUTION**
> Ensure that the baseplate holes are clearly marked before moving the MTBM, or the entire alignment procedure will have to be repeated.

*Step 19* Drill and tap the baseplate holes.

*Step 20* Place the MTBM back onto the baseplate.

*Step 21* Install the MTBM mounting bolts. The STAT and MTBM do not need further adjustment because they will be properly aligned later.

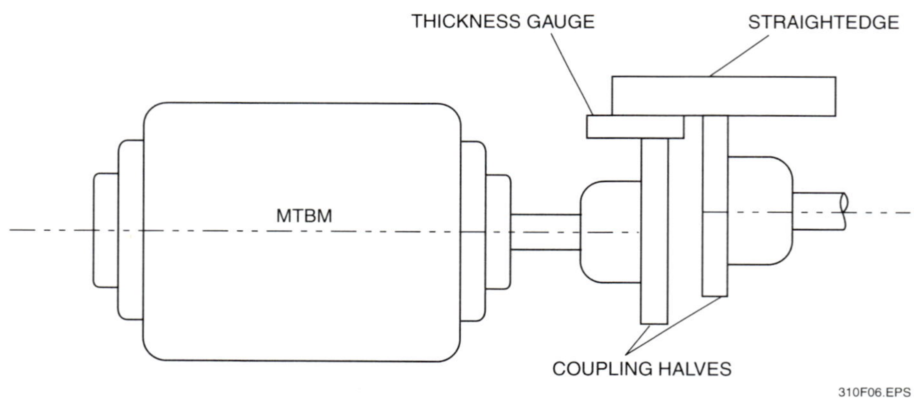


*Figure 6* ◆ Measuring horizontal offset.



*Instructor's Notes:*

10.6 MILLWRIGHT ◆ LEVEL THREE

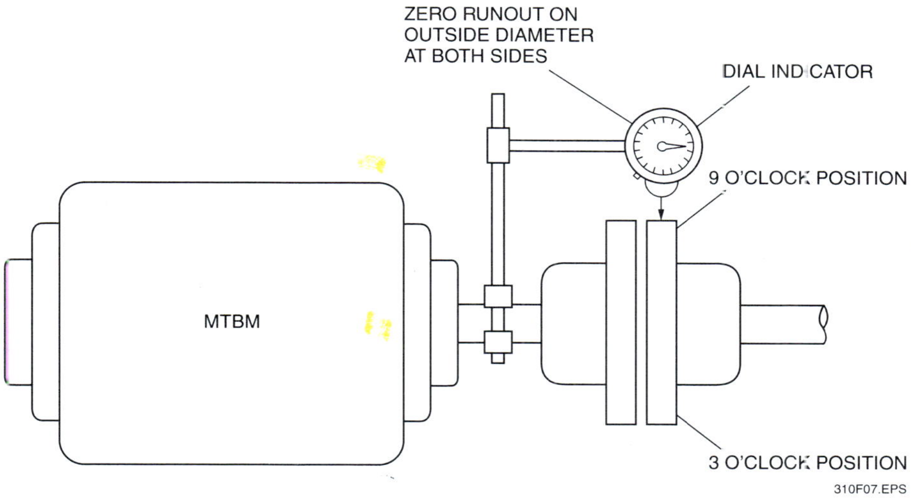


*Figure 7* ◆ Taking readings, using a dial indicator.

Have trainees complete the Review Questions, and go over the answers prior to administering the Module Examination.

## Review Questions

1. Before prealigning a piece of equipment, the equipment must be _____.
   a. plugged in
   b. inspected
   c. connected to the other pieces
   d. roughly aligned

2. MTBMs are checked to ensure they are not binding by _____.
   a. plugging them in
   b. looking at the shaft
   c. turning them by hand
   d. turning them with a wrench

3. Rough places in a MTBM bearing may be caused by _____.
   a. turning them by hand
   b. too much lubricant
   c. improper welding grounds
   d. interference-fit

4. The first step in setting a STAT and MTBM is to _____.
   a. install the MTBM on the baseplate
   b. connect the STAT to the piping
   c. connect the MTBM to the driver
   d. install couplings on the mating shafts

5. The two basic types of couplings are _____.
   a. clamp and set
   b. rigid and flexible
   c. base and tip
   d. MTBM and STAT

6. Flexible couplings allow for some _____.
   a. loose setscrews
   b. soft foot
   c. misalignment
   d. pipe stress

7. Two basic methods of coupling installation are _____.
   a. clearance and interference
   b. welding and clamping
   c. lathe and mill
   d. threading and welding

8. The key should only fill _____.
   a. the keyway completely
   b. half of the keyway
   c. the coupling
   d. the cam

TERRIBLY WORDED

## Review Questions

9. On a clearance fit coupling, the coupling should be placed on the shaft by hand or with a _____.
   a. sledgehammer
   b. hydraulic jack
   c. soft-blow hammer
   d. clamp

10. A tapered coupling has a(n) _____.
    a. setscrew pressing on the key
    b. eccentric thread
    c. nut on the end
    d. clamp on the shaft

11. To tighten the nut on a tapered coupling, use a _____.
    a. box wrench
    b. torque wrench
    c. key
    d. setscrew

12. A clamping or tapered coupling is _____ than the setscrew and key coupling.
    a. less accurate
    b. harder to mount
    c. easier to mount
    d. more accurately centered

13. The STAT must be centered on the _____.
    a. crosswise center line
    b. edge of the baseplate
    c. lengthwise center line
    d. coupling

14. Shims are placed under the STAT base to _____.
    a. move the bolts
    b. achieve elevation
    c. align the center line
    d. align the MTBM

15. If the STAT has to be moved after the bolts are placed, it may become bolt-bound.
    a. True
    b. False

16. Punch mark the bolt hole locations through the _____.
    a. baseplate holes
    b. STAT bolt holes
    c. flange bolt holes
    d. grout

17. Once the STAT has been set, the MTBM must be _____.
    a. parallel
    b. bolted down
    c. coupled
    d. aligned

18. The MTBM must be aligned with the crosswise center line.
    a. True
    b. False

19. Move the MTBM close enough to the STAT to set the _____.
    a. coupling gap
    b. offset
    c. boltholes
    d. shims

20. Set the coupling gap, using a _____.
    a. HI-LO gauge
    b. micrometer
    c. thickness gauge
    d. taper gauge



Instructor's Notes:

10.8 MILLWRIGHT ◆ LEVEL THREE

## Summary

Prealignment is the preliminary, rough alignment of equipment and machinery and is done to allow other craftworkers to perform their work. Equipment should be inspected to ensure that it is in good operating condition before it is installed.

Prealignment includes setting and leveling the baseplate, setting the equipment, installing couplings, performing rough alignment, and drilling and tapping the mounting holes for the equipment.

**Summarize the major concepts presented in the module.**

**Administer the Module Examination. Record the results on Craft Training Report Form 200, and submit the results to the Training Program Sponsor.**

**Administer the Performance Test, and fill out Performance Profile Sheets for each trainee. If desired, trainee proficiency noted during laboratory sessions may be used to complete the Performance Test. Record the results on Craft Training Report Form 200, and submit the results to the Training Program Sponsor.**

## Notes



MODULE 15310-08 ◆ PREALIGNMENT FOR EQUIPMENT INSTALLATION 10.9

## Trade Terms Introduced in This Module

***Axial movement:*** Movement in the direction of a shaft axis.

***Skew:*** A nonparallel and noncoaxial condition; at an angle.

***Torsional movement:*** The rotating movement of a shaft.



Instructor's Notes:

10.10 MILLWRIGHT ◆ LEVEL THREE

# Resources & Acknowledgments

## Additional Resources

This module is intended to present thorough resources for task training. The following reference works are suggested for further study. These are optional materials for continued education rather than for task training.

R+W America L.P., www.rw-america.com

Coupling Corporation of America, www.couplingcorp.com

## Figure Credits

Dayton Superior Products, 310F03 (top)

R+W America L.P., www.rw-america.com, 310F03 (bottom)

"Anderson Clamp Hub" Patented Coupling Corporation of America, Jacobus, PA, Flexxor Couplings, Ultraflex Couplings, 310F04



MODULE 15310-08 ◆ PREALIGNMENT FOR EQUIPMENT INSTALLATION 10.11

## MODULE 15310-08 — TEACHING TIPS

The following are suggested activities or instructional methods to help you teach the material in this module.

### General

When you call on someone to answer a question, the rest of the class relaxes or even tunes out because they expect that the question and answer will take place only between you and the trainee you called on. Instead, use this technique to involve more trainees in answering questions and to keep them on their toes.

1. Ask trainees to define a term or explain a concept.
2. After one trainee has answered, ask a trainee seated nearby if the answer is right. Then ask whether a trainee in the back of the room agrees.
3. Ask trainees to explain why they think an answer is right or wrong.
4. Use the session to clear up incorrect ideas and encourage trainees to learn from their mistakes.

## MODULE 15310-08 — ANSWERS TO REVIEW QUESTIONS

| Answer | | Section |
|---|---|---|
| 1. | b | 2.0.0 |
| 2. | c | 2.0.0 |
| 3. | c | 2.0.0 |
| 4. | d | 4.0.0 |
| 5. | b | 4.0.0 |
| 6. | c | 4.0.0 |
| 7. | a | 4.0.0 |
| 8. | b | 4.1.0 |
| 9. | c | 4.1.0 |
| 10. | c | 4.2.0 |
| 11. | b | 4.2.0 |
| 12. | d | 4.3.0 |
| 13. | c | 5.0.0 |
| 14. | b | 5.0.0 |
| 15. | a | 5.0.0 |
| 16. | b | 5.0.0 |
| 17. | d | 6.0.0 |
| 18 | b | 6.0.0 |
| 19. | a | 6.0.0 |
| 20. | c | 6.0.0 |

## CONTREN® LEARNING SERIES — USER UPDATE

The NCCER makes every effort to keep these textbooks up-to-date and free of technical errors. We appreciate your help in this process. If you have an idea for improving this textbook, or if you find an error, a typographical mistake, or an inaccuracy in NCCER's Contren® textbooks, please write us, using this form or a photocopy. Be sure to include the exact module number, page number, a detailed description, and the correction, if applicable. Your input will be brought to the attention of the Technical Review Committee. Thank you for your assistance.

*Instructors* – If you found that additional materials were necessary in order to teach this module effectively, please let us know so that we may include them in the Equipment/Materials list in the Annotated Instructor's Guide.


**Write:** Product Development and Revision
National Center for Construction Education and Research
3600 NW 43rd St., Bldg. G, Gainesville, FL 32606

**Fax:** 352-334-0932

**E-mail:** curriculum@nccer.org


Craft ________________ Module Name ________________

Copyright Date ________ Module Number ________ Page Number(s) ________

Description ________________

________________

________________

________________

(Optional) Correction ________________

________________

________________

(Optional) Your Name and Address ________________

________________

________________

Module 15311-08

# *Installing Belt and Chain Drives*

## NCCER STANDARDIZED CRAFT TRAINING PROGRAM

The National Center for Construction Education and Research (NCCER) provides a standardized national program of accredited craft training. Key features of the program include instructor certification, competency-based training, and performance testing. The program provides trainees, instructors, and companies with a standard form of recognition through a National Craft Training Registry. The program is described in full in the *Guidelines for Accreditation*, published by NCCER. For more information on standardized craft training, contact the NCCER by writing us at 3600 NW 43rd St., Bldg. G, Gainesville, FL 32606; calling 352-334-0911; or emailing info@nccer.org. More information may be found at our website, www.nccer.org.

## HOW TO USE THIS ANNOTATED INSTRUCTOR'S GUIDE

Each page presents two sections of information. The larger section displays each page exactly as it appears in the Trainee Module. The narrow column ties suggested trainee and instructor actions to each page and provides icons (detailed below) to call your attention to material, safety, audiovisual, or testing requirements. The bottom of each page includes space for your notes.

The **Audiovisual** icon indicates an appropriate time to show a transparency or other audiovisual aid.

The **Classroom** icon prompts you to define a term, stress a point, ask trainees to explain a concept, or give examples.

The **Demonstration** icon directs you to show trainees how to perform tasks.

The **Examination** icon tells you to administer the written module examination.

The **Homework** icon is placed where you may wish to assign reading for the next class, assign a project, or advise trainees to prepare for an examination.

The **Laboratory** icon is used when trainees are to practice performing tasks.

The **Materials** icon is a reminder for you to gather materials needed for classes, labs, and testing.

The **Performance Testing** icon tells you to administer a performance test or a portion thereof.

The **Safety** icon is used to emphasize safety issues. It is often keyed to *Caution* and *Warning!* statements in the Trainee Module.

The **Teaching Tip** icon indicates additional guidance is available, such as how to conduct an exercise, get the most educational value from a field trip, or encourage class participation. Teaching Tips may expand on a feature (*Think About It, Did You Know?*) or provide *Quick Quizzes* or similar exercises. You will be referred to the Teaching Tips section at the back of the module if there is additional material.

The **Combination** icon indicates that the laboratory listed corresponds with a performance task. If desired, you can note the proficiency of the trainees during the laboratory, and use it to satisfy performance testing requirements.

## PREPARATION

Before teaching this module, you should review the Objectives, Performance Tasks, Materials and Equipment List, and Module Outline. Be sure to allow ample time to prepare your own training or lesson plan and gather all required materials and equipment.

## MODULE OVERVIEW

This module explains how belts and chains are used to drive parallel shafts. It also provides information on different types of chains and belts.

## PREREQUISITES

Prior to training with this module, it is recommended that the trainee shall have successfully completed *Core Curriculum; Millwright Level One; Millwright Level Two;* and *Millwright Level Three,* Modules 15301-08 through 15310-08.

## OBJECTIVES

Upon completion of this module, the trainee will be able to do the following:

1. Identify and explain belt drive types.
2. Install belt drives.
3. Identify and explain chain drive types.
4. Install chain drives.

## PERFORMANCE TASKS

Under the supervision of the instructor, the trainee should be able to do the following:

1. Identify belt drive types.
2. Install belt drives.
3. Identify chain drive types.
4. Install chain drives.

## MATERIALS AND EQUIPMENT LIST

Overhead projector and screen
Transparencies
Blank acetate sheets
Transparency pens
Whiteboard/chalkboard
Markers/chalk
Pencils and scratch paper
Appropriate personal protective equipment
Basic trainee tools
*ANSI Standard B29.1, Transmission Roller Chains and Sprocket Teeth*
*ANSI Standard B29.2, Inverted-Tooth Chains and Sprocket Teeth*
Specialized tools, including laser alignment tools, simple sprocket alignment tools, and chain tools
Quick Quiz*
Module Examinations**
Performance Profile Sheets**

* Located in the back of this module
**Located in the Test Booklet

## SAFETY CONSIDERATIONS

Ensure that the trainees are equipped with appropriate personal protective equipment and know how to use it properly. This module may require that the trainees visit job sites. Ensure that trainees are briefed on site safety policies prior to any site visits.

## ADDITIONAL RESOURCES

This module is intended to present thorough resources for task training. The following reference works are suggested for both instructors and motivated trainees interested in further study. These are optional materials for continued education rather than for task training.

The Complete Guide to Chain, http://chain-guide.com/

Michigan Industrial Belting, Inc., http://www.mibelting.com/indbelt.htm?gclid=CL_d0aS0rZECFQIglgodrx6Aew

GlobalSpec: The Engineering Search Engine, http://mechanical-components.globalspec. com/Industrial-Directory/drive_belt

## TEACHING TIME FOR THIS MODULE

An outline for use in developing your lesson plan is presented below. Note that each Roman numeral in the outline equates to one session of instruction. Each session has a suggested time period of 2½ hours. This includes 10 minutes at the beginning of each session for administrative tasks and one 10-minute break during the session. Approximately 10 hours are suggested to cover *Installing Belt and Chain Drives*. You will need to adjust the time required for hands-on activity and testing based on your class size and resources. Because laboratories often correspond to Performance Tasks, the proficiency of the trainees may be noted during these exercises for Performance Testing purposes.

| Topic | Planned Time |
|---|---|
| **Session I. Introduction; Belt Drive Types** | |
| A. Introduction | ____________ |
| B. Belt Drive Types | ____________ |
| C. Laboratory<br>Trainees identify belt drive types. This laboratory corresponds to Performance Task 1. | ____________ |
| **Session II. Installing Belt Drives** | |
| A. Installing Belt Drives | ____________ |
| B. Laboratory<br>Have trainees install belt drives. This laboratory corresponds to Performance Task 2. | ____________ |
| **Session III. Chain Drive Types; Installing Chain Drives; Chain Tools** | |
| A. Chain Drive Types | ____________ |
| B. Laboratory<br>Have trainees identify chain drive types. Note the proficiency of each trainee. This laboratory corresponds to Performance Task 3. | ____________ |
| C. Installing Chain Drives | ____________ |
| D. Laboratory<br>Have trainees install chain drives. Note the proficiency of each trainee. This laboratory corresponds to Performance Task 4. | ____________ |

**Session IV. Review and Testing**

A. Module Review

B. Module Examination

1. Trainees must score 70% or higher to receive recognition from NCCER.
2. Record the testing results on Craft Training Report Form 200, and submit the results to the Training Program Sponsor.

C. Performance Testing

1. Trainees must perform each task to the satisfaction of the instructor to receive recognition from NCCER. If applicable, proficiency noted during laboratory exercises can be used to satisfy the Performance Testing requirements.
2. Record the testing results on Craft Training Report Form 200, and submit the results to the Training Program Sponsor.

15311-08

# Installing Belt and Chain Drives

Assign reading of Module 15311-08.

# 15311-08
# *Installing Belt and Chain Drives*

*Topics to be presented in this module include:*



## Overview

The alternative to coaxial shafts is parallel shafts. Couplings are used to connect and transfer force between coaxial shafts. In this module, you will learn how belts and chains are used to drive parallel shafts. Information on different types of chains and belts is also included.

Instructor's Notes:

## Objectives

When you have completed this module, you will be able to do the following:

1. Identify and explain belt drive types.
2. Install belt drives.
3. Identify and explain chain drive types.
4. Install chain drives.

## Trade Terms

| | |
|---|---|
| Datum width | Top width |
| Dress down | Vulcanized |

## Required Trainee Materials

1. Pencil and paper
2. Appropriate personal protective equipment

## Prerequisites

Before you begin this module, it is recommended that you successfully complete *Core Curriculum*; *Millwright Level One*; *Millwright Level Two*; and *Millwright Level Three*, Modules 15301-08 through 15310-08.

This course map shows all of the modules in the third level of the *Millwright* curriculum. The suggested training order begins at the bottom and proceeds up. Skill levels increase as you advance on the course map. The local Training Program Sponsor may adjust the training order.

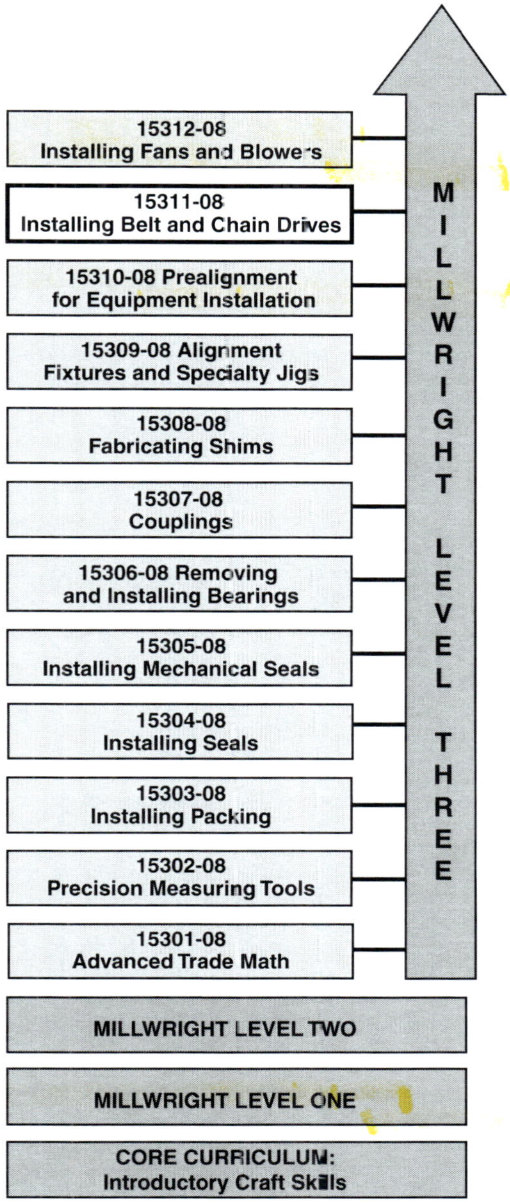


311CMAP.EPS



**Ensure that you have everything required to teach the course. Check the Materials and Equipment list at the front of this module.**

**See the general Teaching Tip at the end of this module.**

**Explain that terms shown in bold are defined in the Glossary at the back of this module.**

**Show Transparency 1, Objectives, and Transparency 2, Performance Tasks. Review the goals of the module, and explain what will be expected of the trainee.**

**Review the modules covered in Level Three and explain how this module fits in.**

MODULE 15311-08 ◆ INSTALLING BELT AND CHAIN DRIVES 11.1

**Discuss the purpose and function of belt and chain drives.**

**Explain the millwright's role in installing, adjusting, and maintaining belt and chain drives.**

**Stress that safety policies must be strictly followed when working with belt and chain drives.**

**Discuss the use of belt drives for power transmission.**

**Explain how belts are dimensioned.**

**Show Transparencies 3 and 4 (Figure 1 and Table 1).**

**Introduce V-belts and synchronous belts.**

**Describe the shape of a V-belt. Explain how V-belts work.**

**Show Transparency 5 (Figure 2).**

## 1.0.0 ◆ INTRODUCTION

Belt and chain drives are types of power transmission equipment. Practically all equipment that moves does so by means of power transmission equipment. The most common form of power transmission is the transmission of power from one shaft to another. Shaft-to-shaft transmission can be accomplished in two ways: shaft to shaft axially or shaft to parallel shaft. Couplings or clutches are used to transmit power from shaft to shaft axially. Belt and chain drives are used to transmit power from one shaft to a parallel shaft.

Installing, adjusting, and maintaining belt and chain drives are common parts of a millwright's job. The service life of belt and chain drives depends heavily on these procedures being performed properly.

> **WARNING!**
> Always lock out and tag equipment according to plant policy before any guard is removed or any work is performed. Ensure that all safety policies and procedures are strictly followed.

## 2.0.0 ◆ BELT DRIVE TYPES

Belt drives are a quiet, smooth, and economical form of power transmission. They are available in many forms and styles and are widely used in almost all industries. Belts are made of a combination of fabric, cord, and/or metal reinforcement **vulcanized** with natural rubber compounds.

Metric belts have different dimensions from the American or English belts, and should not be interchanged with American standard belts. Cross sections may vary, as well as length. In working with foreign machinery, be sure to use the belts specified. The normal V-belt is measured by the length and by one version or another of the width. Belts are dimensioned in the German (DIN) specification by their width at the top, called the **top width**, and in the ISO standard by the **datum width**, the width at an arbitrary point *(Figure 1)*. The sizes in use are named by a code similar to the American standard. A general category designation of XP denotes molded notch metric narrow belts, while SP identifies metric narrow belts.

Several types of belt designations may be used for belts of the same top width, especially when English designations are included. *Table 1* shows several designations and sizes of V-belts. Belts are also called by their top width alone, such as ³⁄₁₆, ¼, 5 mm, 8 mm, etc. There are also designations for a specific cross-section, as *Table 1* shows.

It is clear from this information that even single belts may not be interchangeable, even if the top width is identical. While interchange charts are available for metric and English or U.S. belts, the difference may still be sufficient to cause early belt failure or poor performance. In the case of the multiple joined belts, this problem becomes more problematic because the width of the notch in between, included angle between sidewalls, or length may be different for metric and inch belts. To prevent problems, use the belt recommended by the manufacturer for a particular machine.

A belt drive consists of a driver with one or more sheaves, a driven with a matching number of sheaves, and belts to match the sheaves. Belt drives can be divided into the two basic types: V-belts and synchronous belts.

### 2.1.0 V-Belts

The V-belt is the most common type of belt. It has a tapered shape that causes it to wedge firmly into the groove of the sheave when it is under load (see

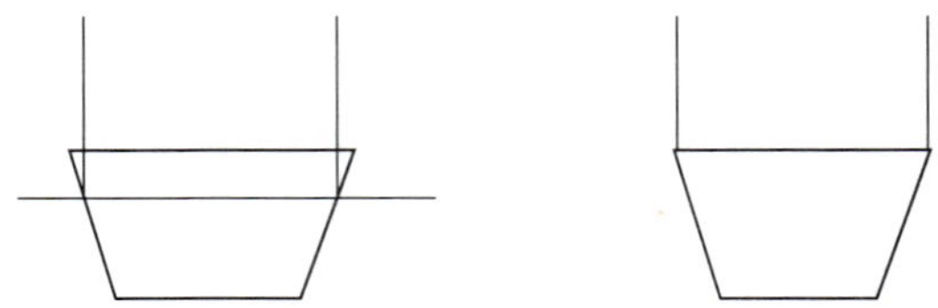


*Figure 1* ◆ Top and datum width.

**Table 1** Belt Designation Codes

| Belt Designation | Belt Size |
|---|---|
| O, 3L. Z | ⅜" |
| A, 4L | ½" × 5⁄16" |
| B, 5L | 21⁄32" × 13⁄32" |
| C | ⅞" × 17⁄32" |
| D | 1¼" × ¾" |
| E | 1½" × 29⁄32" |
| **Wedge belts** | |
| 3V | ⅜" × 5⁄16" |
| 4V | ½" × 13⁄32" |
| 5V | ⅝" × 17⁄32" |
| 8V | 1" × 29⁄32" |
| **Metric belts** | |
| SPA | 13mm × 8mm |
| SPB | 17mm × 11mm |
| SPC | 22mm × 18mm |
| SPZ | 10mm × 6mm |

311T01.EPS



Instructor's Notes:

11.2 MILLWRIGHT ◆ LEVEL THREE

*Figure 2*) A V-belt works through frictional contact between the sides of the belt and the tapered sheave groove.

The six main types of V-belts are the following:

- Fractional horsepower
- Standard multiple
- Wedge
- Double-angle
- Joined
- Notched

### 2.1.1 Fractional Horsepower Belts

Fractional horsepower (FHP) belts are light-duty belts that are used in appliances and small machines in industry and in the home. They are generally used singularly instead of in sets of two or more, as is the case with other types of V-belts.

The size of FHP V-belts is indicated by a code printed on the outside of the belt. The first number and letter in the code tell the width of the belt in eighths of an inch. The next three numbers in the code tell the length of the belt in inches, with the last number indicating tenths of an inch. For example, a belt that is coded 4L300 is 4/8- (1/2) inch wide and 30 inches long. FHP belts are measured on the outside surface of the belt. FHP belts come in the following standard widths and thicknesses (*Figure 3*):

- 2L – 2/8 (1/4) inch wide
- 3L – 3/8 inch wide
- 4L – 4/8 (1/2) inch wide
- 5L – 20/32 (5/8) inch wide

### 2.1.2 Standard Multiple Belts

Standard multiple belts are designed for the continuous service that is required in industrial applications. As the name implies, standard multiple belts are used in sets of two or more. They are used for industrial drives with normal loads, speeds, sheave diameters, center distances, and operating conditions. A higher grade belt is used for severe conditions.

The size of standard multiple belts (*Figure 4*) is indicated by a code printed on the belt. In the code for a standard belt, a letter indicates the width of the belt and a number indicates the length of the belt. A belt that is coded A42 is 1/2-inch wide and 42 inches long. The length of standard belts is measured on the inside surface of the belt. This is called the standard length designation. Standard multiple

**List the six main types of V-belts.**

**Discuss fractional horsepower (FHP) belts and their applications.**

**Show Transparency 6 (Figure 3). Explain the standard sizes of FHP belts.**

**Discuss standard multiple belts and their applications. Explain that they are used in sets of two or more.**

**Show Transparency 7 (Figure 4). Explain the standard sizes of multiple belts.**

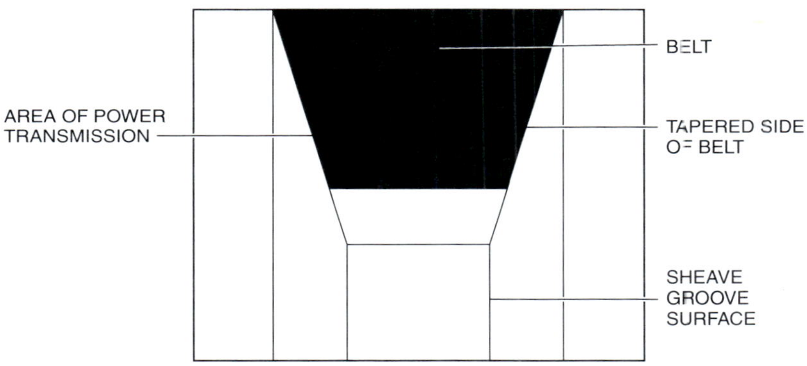


***Figure 2*** ◆ Area of power transmission of V-belt.

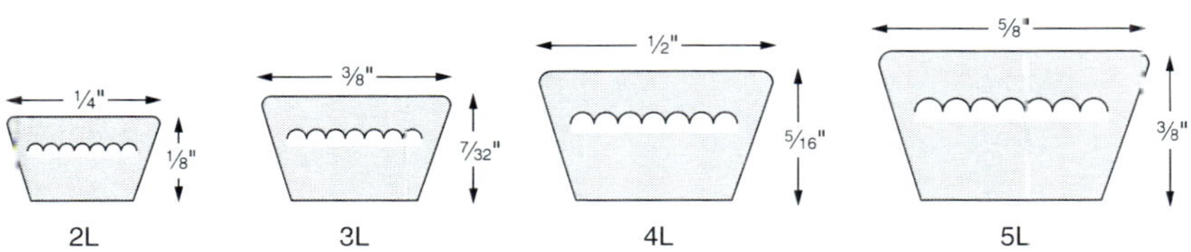


***Figure 3*** ◆ Standard FHP belt sizes.



MODULE 15311-08 ◆ INSTALLING BELT AND CHAIN DRIVES 11.3

**Describe wedge belts.**

**Show Transparency 8 (Figure 5). Explain the standard sizes for wedge belts.**

belts come in various lengths for each width size and are available in the following standard widths:

- A – ½ inch wide
- B – ⅝ inch wide
- C – ⅞ inch wide
- D – 1¼ inches wide
- E – 1½ inches wide

## *2.1.3 Wedge Belts*

The wedge belt is an improved design V-belt that allows a reduction in size, weight, and cost of V-belt drives. It is a type of multiple belt, but has a smaller cross section per horsepower rating than standard multiple V-belts. Also, it can be used on smaller diameter sheaves with shorter center distances than the standard belt. Wedge belts are not interchangeable with standard multiple belts and should not be run on sheaves for standard belts.

The code markings for wedge belts are similar to the markings for FHP belts. The first number and letter of the code indicate the width and cross section of the belt, and the last three numbers indicate the length of the belt. A 3V500 belt is defined as a 3V cross section that is 50 inches long. The length of a wedge belt is measured along the pitch line, which runs along the center of the belt thickness. Wedge belts come in the following widths:

- 3V – ⅜ inch wide
- 5V – ⅝ inch wide
- 8V – ⅞ (or 1 inch) wide

*Figure 5* shows the sizes and cross sections of wedge belts.

There is another code, called a match code, that is separate from the regular belt number and is used to match multiple belts. The match code includes the belt codes and the manufacturer's name. This code is used to ensure that replacement

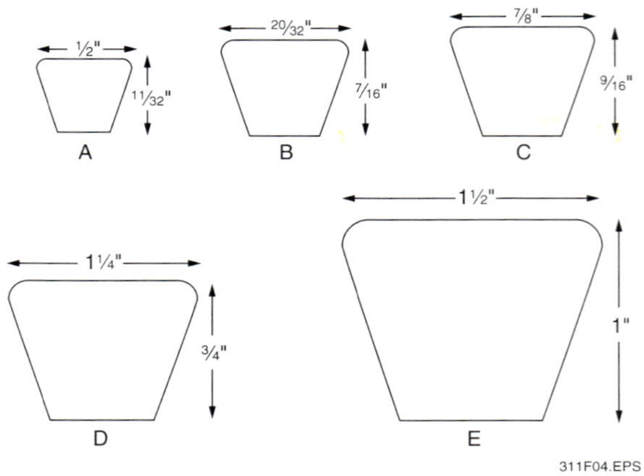


*Figure 4* ◆ Standard multiple belt sizes.

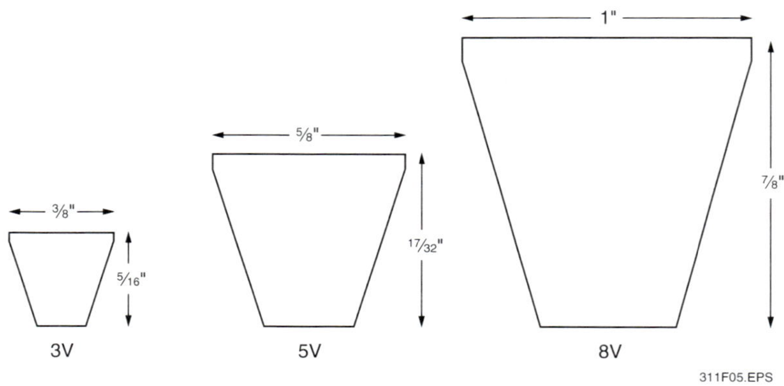


*Figure 5* ◆ Wedge belts.



Instructor's Notes:

11.4 MILLWRIGHT ◆ LEVEL THREE

belts in a multiple-belt application are all the same length. When selecting belts for a multiple-belt application, the belts must be made by the same manufacturer and have the same match code.

### 2.1.4 Double-Angle Belts

Double-angle belts (*Figure 6*) are used on multiple-sheave drives with reverse bends that would damage regular V-belts. Double-angle belts are V-shaped on both sides and can handle reverse bends and still transmit the required power.

### 2.1.5 Joined Belts

Joined belts (*Figure 7*) are standard or wedge V-belts that have a common back which joins them. They are used to provide extra stability on applications that experience severe shock loads by preventing the belts from turning over in the sheaves. The extra support of the back also helps keep all the belts in the multiple series the same length.

### 2.1.6 Notched Belts

Notched belts (*Figure 8*) are V-belts that have notches along the inner surface. The notches allow for more bend in the belt and relieve some of the bending stress. Notched belts are used on applications where the sheaves are very small.

## 2.2.0 Synchronous Belts

Synchronous, or timing, belts are used as a standard method of power transmission. They have teeth molded into them and are used to synchronize, or time, the action of pulleys and related devices. They have an advantage over V-belts because the action of the teeth and the toothed pulleys prevent the belt from slipping. Synchronous belts are available as single-sided or dual-sided.

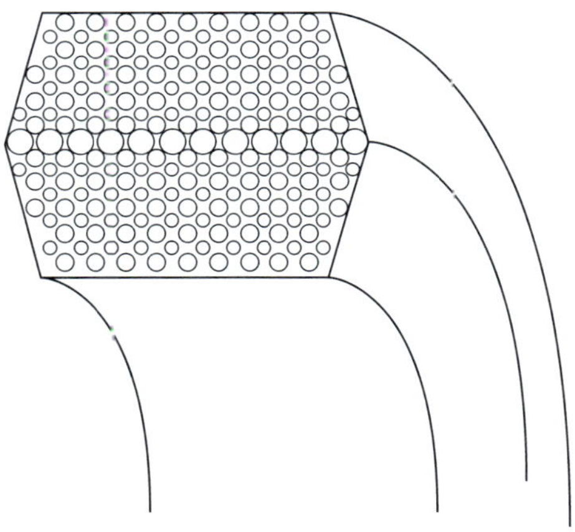


*Figure 6* ◆ Double-angle V-belt.

### 2.2.1 Single-Sided Synchronous Belts

The single-sided synchronous belt has teeth on only one side and runs on the outside of toothed pulleys (*Figure 9*). It is the most common type of timing belt.

### 2.2.2 Dual-Sided Synchronous Belt

The dual-sided synchronous belt may have teeth on both the inside and the outside. These are often used as timing belts for machines. It runs on four-point or serpentine drives where both sides of the

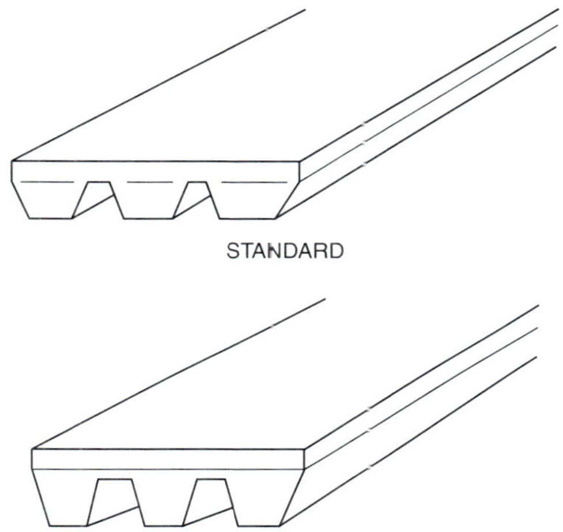


*Figure 7* ◆ Joined belts.

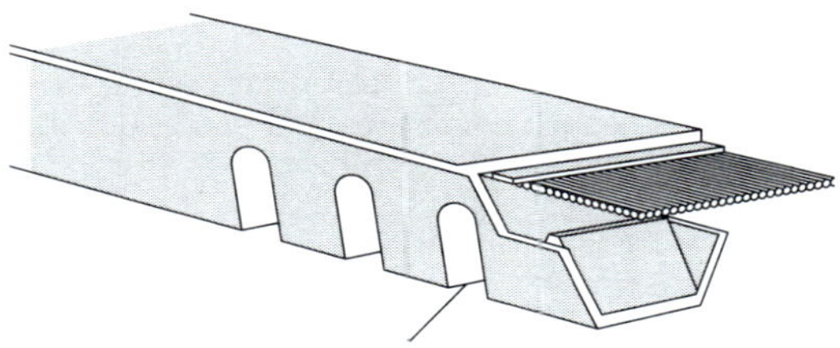


*Figure 8* ◆ Notched belt.

**Discuss double-angle belts, joined belts, and notched belts.**

**Show Transparencies 9-11 (Figures 6-8).**

**Explain that synchronous (timing) belts have teeth molded into them and are available as either single- or double-sided.**

**Show Transparency 12 (Figure 9). Describe the components of a synchronous belt.**



MODULE 15311-08 ◆ INSTALLING BELT AND CHAIN DRIVES 11.5

**Show Transparency 13 (Figure 10). Discuss drive configurations for dual-sided timing belts.**

**Discuss the standard tooth pitch codes for timing belts.**

**Have trainees identify belt drive types. Note the proficiency of each trainee. This laboratory corresponds to Performance Task 1.**

**Assign reading of Section 3.0.0.**

**Ensure that you have everything required to teach this session.**

**Explain that belt-drive failure can often be traced to improper installation procedures.**

**Show Transparency 14 (Figure 11). Discuss the three basic types of misalignment.**

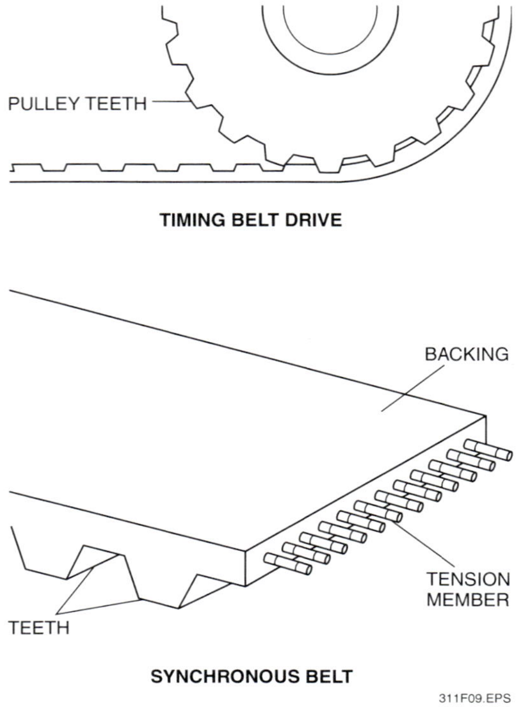


*Figure 9* ◆ Timing belt drive and synchronous belt components.

belt come in contact with the pulleys. *Figure 10* shows examples of drives that use dual-sided timing belts.

Matching, or sizing, codes for timing belts indicate the pitch of the teeth and the width of the belt. The five standard tooth pitch codes are the following:

- XL *(extra light)* – ⅕-inch pitch
- L *(light)* – ⅜-inch pitch
- H *(heavy)* – ½-inch pitch
- XH *(extra heavy)* – ⅞-inch pitch
- XXH *(double extra-heavy)* – 1¼-inch pitch

These are standard. Other parts of the codes vary between manufacturers. A typical matching code is 225 L 075. The number 225 is a particular manufacturer's code and indicates the length of the belt. To determine the actual length of the belt in inches, divide 225 by 10. The actual length of the belt is 22.5 inches. This is only an example; manufacturers' codes vary.

The letter L indicates that the belt is a light belt that has a ⅜-inch pitch. The last three numbers of this example indicate the width of the belt. To determine the actual width of the belt, place a decimal in front of the number 075 to read 0.075 and multiply by 10. (The actual width of the belt is 0.75, or ¾ inch.)

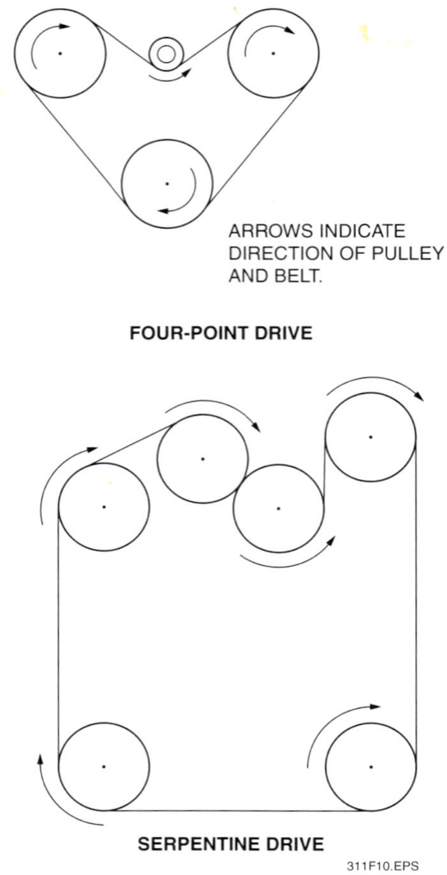


*Figure 10* ◆ Dual-sided belt drives.

## 3.0.0 ◆ INSTALLING BELT DRIVES

Belt drives are designed to give many hours of service in their particular application. The service life of belt drives depends on the quality of materials used to manufacture the belts and the proper installation and maintenance of the drives.

Belt-drive failure can often be traced to improper installation procedures. The most frequent cause of drive failure is excessive misalignment. There are basically three kinds of misalignment: angular, parallel, and sheave groove misalignment. See *Figure 11*.



Instructor's Notes:

11.6 MILLWRIGHT ◆ LEVEL THREE

Another cause of premature failure is improper tension on the belts. All belts should be tensioned according to the manufacturer's instructions using a tension gauge.

Belt drives are used on many types of equipment. This section explains how to install a belt drive on a pump and motor. Follow these steps to install a belt drive:

*Step 1* Clean the shafts thoroughly, using solvent and a rag.

*Step 2* Inspect the shafts, and **dress down** any burrs or nicks, using a hone or a fine file, or a light grinder.

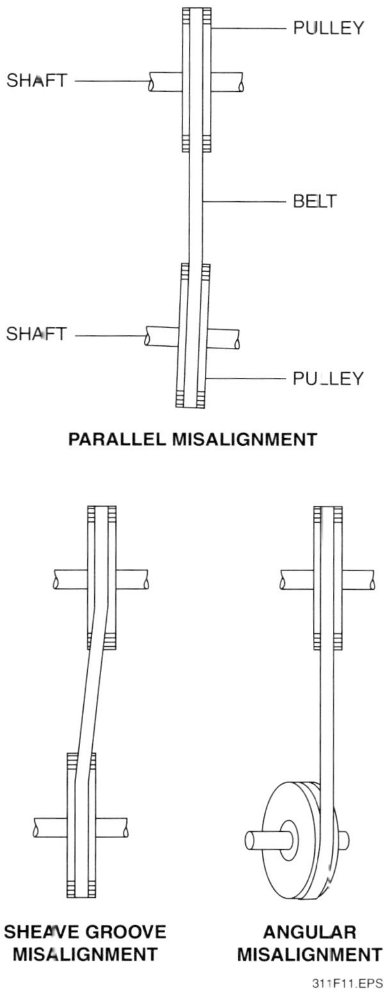


*Figure 11* ◆ Misalignment.

*Step 3* Measure the distance between the shafts to ensure that they are parallel with one another. To check parallelism, take one measurement between the shafts near the end of the shafts and one near the driven and driver (*Figure 12*). If the shafts are perfectly parallel, the measurements will be the same. If the measurements are not the same, the shafts need to be adjusted until they are parallel before the chain drive is installed.

> **NOTE**
> For the most precise measurements, optical and electronic equipment is available.

*Step 4* Slip the motor pulley onto the motor shaft, and place the pulley in the proper position. The proper pulley position depends on the application. The pulleys are usually positioned as near the motor as is practical.

*Step 5* Lock the motor pulley in position. The method for locking a pulley to a shaft depends on the pulley type. Some pulleys have a setscrew that is used to lock them to the shaft, some have taperlock hubs. Some multisheave and adjustable sheave pulleys come in several parts that must be assembled on the shaft. Follow the pulley manufacturer's instructions for mounting the pulley on the shaft.

*Step 6* Slip the pump pulley onto the shaft.

*Step 7* Position a straightedge across the face of both pulleys, and adjust the pump pulley until the pulleys are perfectly aligned. When the pulleys are aligned, the straightedge will be flat on the faces of both pulleys. An alternate method of checking pulley alignment is to pull a piano wire taut across the faces (*Figure 13*).

*Step 8* Lock the pump pulley in position on the shaft.

*Step 9* Move the motor toward the pump until the center-to-center distance of the pulleys has been reduced enough to allow the belts to be slipped onto the pulleys.

*Step 10* Slip the belts onto the pulleys one at a time. *Figure 14* shows the incorrect and correct methods for mounting belts.

**Explain that improper tension on the belts is also a cause of premature failure.**

**Walk through the steps for installing a belt drive on a pump and motor.**

**Show Transparency 15 (Figure 12). Explain how to measure for parallelism.**

**Show Transparency 16 (Figure 13). Discuss how to check pulley alignment.**

**Show Transparency 17 (Figure 14).**



MODULE 15311-08 ◆ INSTALLING BELT AND CHAIN DRIVES 11.7

**Stress the importance of safety guards.**

**Demonstrate how to install a belt drive.**

**Have trainees install belt drives. Note the proficiency of each trainee. This laboratory corresponds to Performance Task 2.**

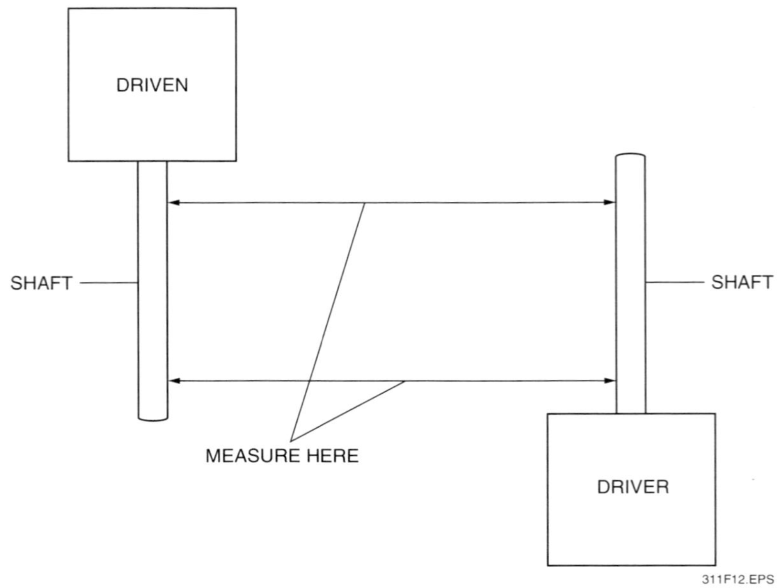


*Figure 12* ◆ Measuring for parallelism.

> **CAUTION**
> Do not try to run the belts on the pulleys. This will place excessive stress on the cords of the belt and cause it to flop under load and possibly turn over in the sheaves. The belts should be slipped loosely over the sheaves.

*Step 11* Position the belts so that the slack of all the belts is on top of the drive.

> **CAUTION**
> It is important that the slack of all belts be on the same side of the drive, preferably the top, to prevent any of the belts from being overstretched when they are tightened.

*Step 12* Move the driver away from the driven to tighten the belts to the proper tension. After the driven has been set, the driver can usually be moved by turning adjusting screws on the motor base. Refer to the belt manufacturer's installation procedures for the proper tension for the belts. Use a tensiometer to check the tension of the belts as they are being tightened.

*Step 13* Install all safety guards.

> **WARNING!**
> Do not run the equipment without the safety guards. Serious injury can result if your hand or clothing gets caught in the belts.

*Step 14* Start the driver, and allow the belts to seat in the sheaves.

*Step 15* Shut down the driver.

*Step 16* Lock out and tag the motor according to plant policies.

*Step 17* Remove the safety guards.

*Step 18* Check the belt tension, and adjust it if necessary.

*Step 19* Install all safety guards.

*Step 20* Start up the equipment, and allow it to run for 24 to 48 hours.

*Step 21* Repeat Steps 15 through 19.



Instructor's Notes:

11.8 MILLWRIGHT ◆ LEVEL THREE

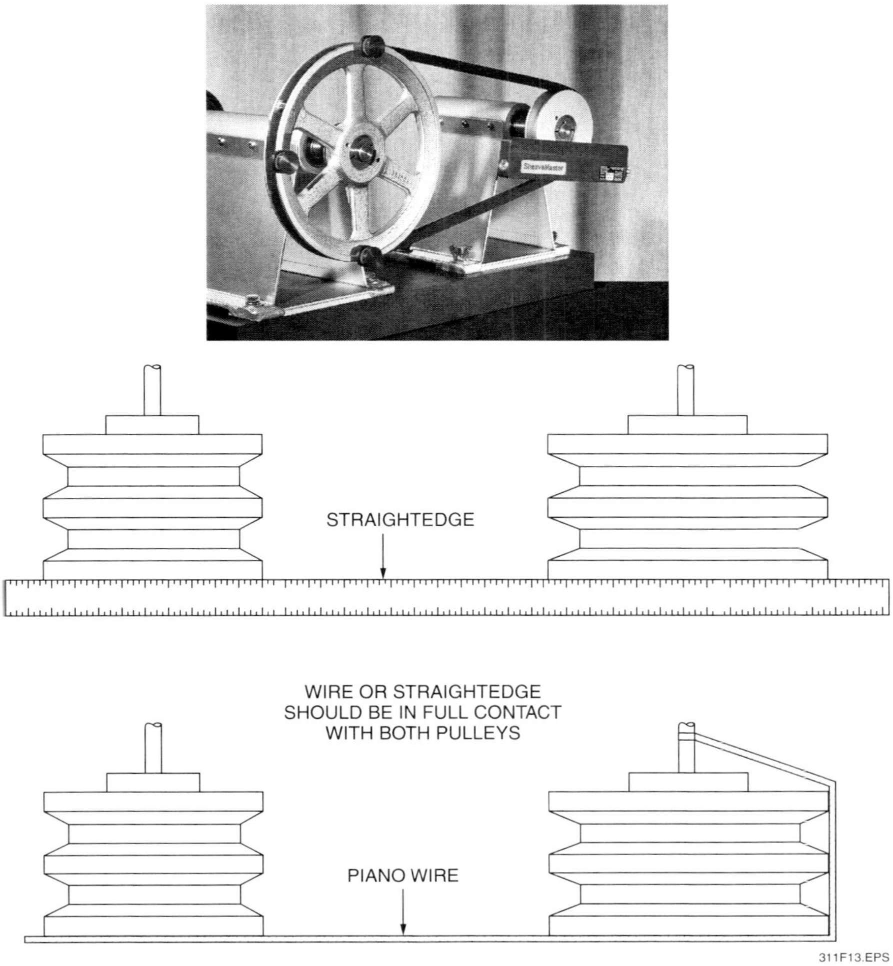


*Figure 13* ◆ Checking pulley alignment.

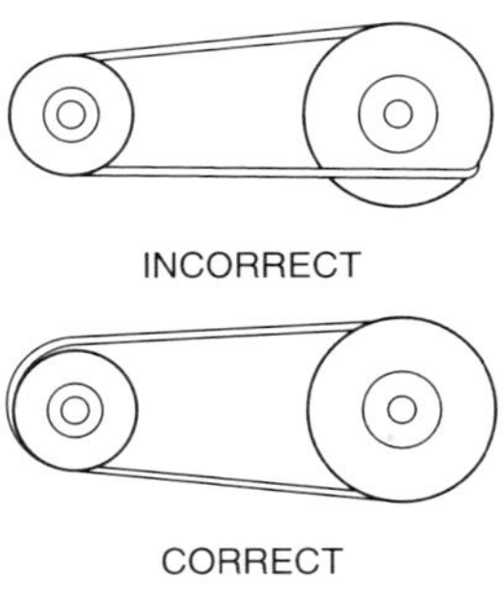


*Figure 14* ◆ Mounting belts.

## 4.0.0 ◆ CHAIN DRIVE TYPES

Chain drives are widely used for power transmission in industry. A chain drive consists of a driving sprocket, one or more driven sprockets, and an endless chain that runs around and meshes with the sprocket teeth. Chain drives maintain a positive speed ratio between the driving and the driven sprockets because they do not slip or creep. The biggest advantages that chain drives have over belt drives are their simplicity, economy, efficiency, and adaptability. The biggest disadvantage of chain drives is their need for adequate lubrication. The two most common types of power transmission chains are roller and silent chains.

**Assign reading of Sections 4.0.0–6.0.0.**

**Ensure that you have everything required to teach this session.**

**Explain the functions and components of chain drives.**

**Discuss the advantages and disadvantages of chain drives.**



MODULE 15311-08 ◆ INSTALLING BELT AND CHAIN DRIVES 11.9

**Discuss roller chain components.**

**Show Transparency 18 (Figure 15). Explain single- and multiple-strand roller chains.**

**Explain the standardization of roller chain.**

**Provide a copy of *ANSI Standard B29.1* for trainees to examine.**

**Explain that roller chain is sized according to pitch, chain width, and roller diameter.**

**Show Transparency 19 (Figure 16). Describe chain pitch.**

**Explain that the dimensions of chain parts are based on a ratio of the pitch.**

## 4.1.0 Roller Chains

Roller chains are the most commonly used type of power transmission chain. A roller chain consists of a series of connected links that are an assembly of pins, rollers, plates, and bushings. A roller chain is made by alternately connecting roller links and pin links and can be made to any length.

Roller chains can be assembled in single strands (simplex) or multiple strands (multiplex) for different applications (*Figure 15*). Multiple-strand chain assemblies are used on drives that have multiple sprockets. Multiple-strand chains are used to increase the horsepower of a drive.

Roller chains are made of case-hardened material. This means that they are hard on the outside and soft on the inside. The chain operates properly until the hardened outside material is worn away. Once this happens, the softer material wears rapidly until the chain fails. When a chain begins to wear, it elongates. When it has elongated to 2 percent over its original length, it is considered worn out and should be replaced.

Standard roller chain is manufactured to the specifications of *ANSI Standard B29.1, Transmission Roller Chains and Sprocket Teeth,* which standardizes the way roller chains are made, no matter which manufacturer makes them. Because of this standardization, chains and sprockets from different manufacturers are interchangeable. Identification is also standardized, and replacements can easily be selected from different manufacturer's stock.

Chains to be used in especially severe contexts such as oilfield drives are subject to the API 7F7 standard, in addition to *ANSI 29.1*. These chains are usually made with wider plate waists.

### 4.1.1 Roller Chain Sizing

Roller chain sizing is also standardized. The three principal dimensions used to size roller chains are pitch, chain width, and roller diameter. Pitch (*Figure 16*) is the center-to-center distance from one hinged point to the next.

Chain width, also called nominal width, is the minimum distance between the plates of a roller link. This dimension determines the width of the sprocket teeth that the chain will mesh with. The roller diameter is the outside diameter of the chain rollers. Roller diameter is usually ⅝ of the pitch. The form of the sprocket teeth is determined by this dimension. The dimensions of standard roller chain parts are based on a ratio to the pitch dimension. This proportion is standard for all manufacturers and is as follows:

- Chain width is approximately ⅝ of the pitch.
- Roller diameter is approximately ⅝ of the pitch.
- Pin diameter is approximately 5⁄16 of the pitch.
- Link plate thickness is approximately ⅛ of the pitch.

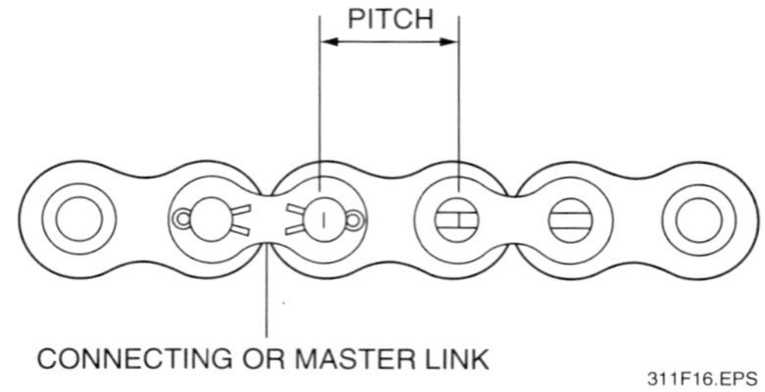


*Figure 16* ◆ Chain pitch.

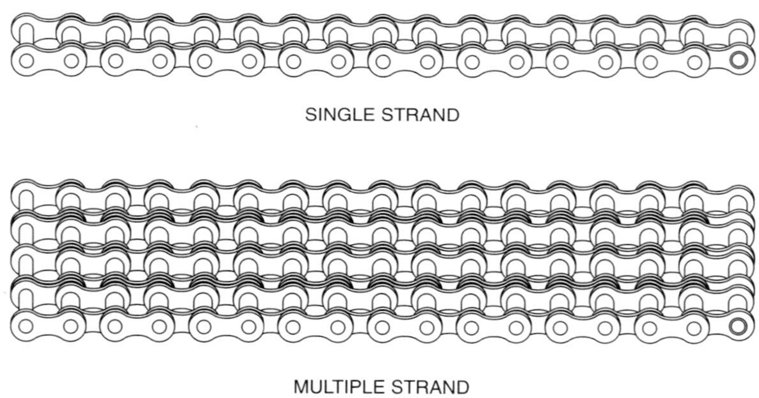


*Figure 15* ◆ Single- and multiple-strand chains.



Instructor's Notes:

11.10 MILLWRIGHT ◆ LEVEL THREE

### 4.1.2 Roller Chain Numbers

The numbering system for roller chains is also standard and provides a complete identification of the chains by number. The first one or two digits in the number denote the number of ⅛ inches in pitch. For example, a 2 indicates 2/8- (¼) inch pitch, and a 12 indicates 12/8- (1½) inch pitch. The right-hand digit of the number indicates the type of chain. For instance, a right-hand digit 0 indicates that the chain is of standard proportions. The digit 1 indicates a lightweight chain, and the digit 5 indicates a rollerless bushing chain. If the chain is a multiple-row chain, the number has a hyphenated digit suffixed to the end of the chain number. For example, -2 suffixed to the number denotes a 2-row chain, and -3 denotes a 3-row chain.

For example, the number 30 denotes a ⅜-inch chain of standard proportions. The number 25 denotes a rollerless link 2/8- (¼) inch chain. The number 120-4 denotes a 4-row 12/8- (1½) inch chain of standard proportions. The number 60H-3 denotes a 3-row, 6/8- (¾) inch heavy series chain. Size charts are used to identify and select chains. *Table 2* lists sizes of roller chain.

### 4.1.3 Metric Roller Chain

Metric chains are either according to European standard or International Chain Standard. The ISO 606 and DIN 8187 standards cover three versions:

- *Simplex* – Single line of chain
- *Duplex* – Doubled lines of side-by-side chain
- *Triplex* – Three lines of side-by-side chains.

The pitches available vary from 4 millimeter (about 3/16 inches) to 114.3 millimeters (4.5 inches) The ISO standard labels the chains by a system similar to the ANSI system, a three-part code. The first part uses the first two digits to give the pitch size in 16ths of an inch, so that 04 would have a pitch of 4/16 (¼) of an inch. The second part is a letter code for the particular relevant standard. The letter B stands for the European Standard. Finally, the number of strands is given. Thus a part number of 04B-3 would be a triplex chain by European Standard with a ¼-inch pitch.

## 4.2.0 Silent Chains

Silent chain, also called inverted-tooth chain, is constructed with inverted teeth that are designed to engage cut-tooth sprockets (see *Figure 17*). Silent chains have the flexibility and quietness of a belt and the positive action and efficiency of a chain. The links of silent chains are alternately assembled with either pins or a combination of joint components. Silent chains are often used as conveyor chains because of the durability and positive traction of the gears. Silent chains take a lot of abuse and can run at relatively high speeds.

There are two basic types of silent chains: the side-guide chain and the center-guide chain (*Figure 18*). They are classified according to the position of the guide plates. Silent chains smaller than ¾-inch pitch have side-guide plates; chains with ¾-inch pitch and larger have center-guide plates.

Silent chain is manufactured to the specifications of *ANSI Standard B29.2, Inverted-Tooth Chains and Sprocket Teeth*, which is primarily intended for

**Discuss the numbering system for roller chains.**

**Show Transparency 20 (Table 2).**

**Discuss sizing of metric roller chains.**

**Describe the configuration of silent chains.**

**Discuss side-guide chain and center-guide chain.**

**Explain that silent chain is manufactured to the specifications of ANSI *Standard B29.2*.**

**Provide a copy of ANSI *Standard B29.2* for trainees to examine.**

**Show Transparencies 21 and 22 (Figures 17 and 18).**

**Table 2** Roller Chain Sizes

| Chain Numbers | Pitch (inches) | Type of Roller Chain Link |
|---|---|---|
| 25 | 2/8 = ¼ | Rollerless link |
| 30 | ⅜ | Standard link |
| 35 | ⅜ | Rollerless link |
| 40 | 4/8 = ½ | Standard link |
| 41 | ½ | Light duty, narrow link |
| 50 | ⅝ | Standard link |
| 60 | 6/8 = ¾ | Standard link |
| 60H | ¾ | Heavy series link |
| 80 | 1 | Standard link |
| 100 | 10/8 = 1¼ | Standard link |
| 120 | 12/8 = 1½ | Standard link |
| 140 | 14/8 = 1¾ | Standard link |
| 160 | 16/8 = 2 | Standard link |
| 180 | 18/8 = 2¼ | Standard link |
| 200 | 20/8 = 2½ | Standard link |
| 240 | 24/8 | Standard link |

311T02.EPS



MODULE 15311-08 ◆ INSTALLING BELT AND CHAIN DRIVES 11.11

**Discuss the sizing code for silent chains.**

**Have trainees identify chain drive types. Note the proficiency of each trainee. This laboratory corresponds to Performance Task 3.**

**Explain that a misaligned or improperly tensioned chain drive will wear rapidly and fail prematurely.**

**Walk through the steps to install a chain drive.**

**Show Transparency 23 (Figure 19). Explain how to measure for parallelism.**

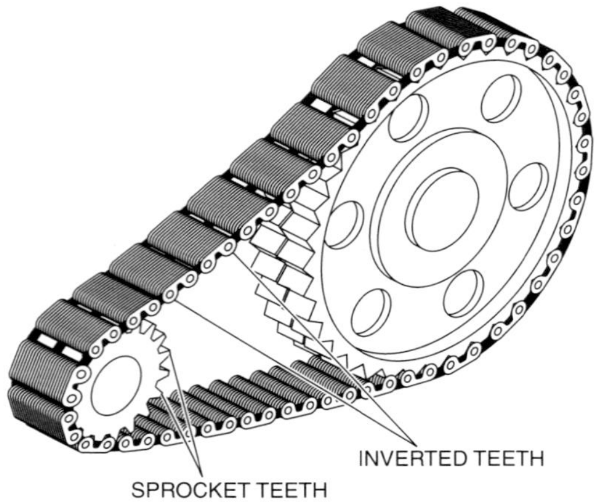


*Figure 17* ◆ Silent chain drive.

interchangeability between chains and sprockets of different manufacturers. It does not provide for standardization of joint components and link plate shape, which may be different for each manufacturer's design. However, all manufacturers' links are made to fit the standard sprocket tooth.

### 4.2.1 Silent Chain Numbers

Silent chains have a sizing code that consists of a two-letter symbol that indicates that it is a silent chain (SC), one or two digits that indicate the pitch in eighths of an inch, and two or three digits that indicate the width of the chain in quarter inches. For example, in the code SC102, the SC denotes silent chain; the 1 stands for $\frac{1}{8}$-inch pitch; and the 02 stands for $\frac{2}{4}$ ($\frac{1}{2}$) inch. The number SC1012 indicates a silent chain of $1\frac{1}{4}$-inch pitch and 3-inch width. Chains that are made by *ANSI Standard B29.2* have a number stamped on the link plate that indicates the pitch of the chain. For instance, the number SC3, or simply 3, stamped on the link plate of a chain indicates a $\frac{3}{8}$-inch pitch.

## 5.0.0 ◆ INSTALLING CHAIN DRIVES

Proper installation of chain drives requires that the shafts and the sprockets be accurately aligned and that the chain be accurately installed and set at the proper tension. A misaligned or improperly tensioned chain drive will wear rapidly and fail prematurely. Follow these steps to install a chain drive:

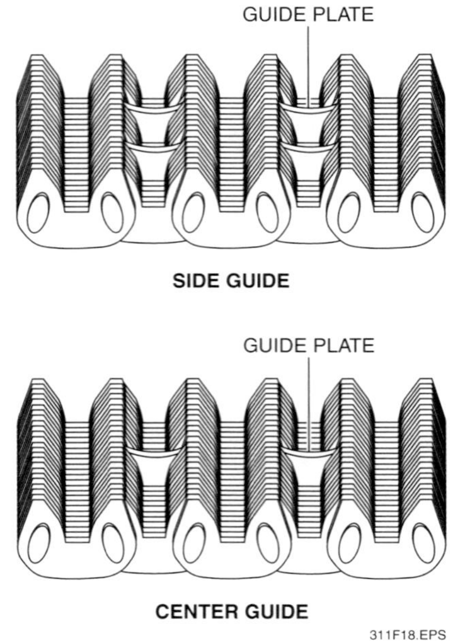


*Figure 18* ◆ Silent chain types.

*Step 1* Clean both shafts thoroughly.

*Step 2* Inspect the shafts, and dress any burrs or nicks that would interfere with the installation of the sprockets.

*Step 3* Polish the shafts lightly, using an emery cloth.

*Step 4* Measure the shafts in several places, using a micrometer, to ensure that the shaft diameter is within specifications for the sprockets being installed and to ensure that the shafts are not out of round.

*Step 5* Measure the distance between the shafts to ensure that they are parallel with one another. To check parallelism, take one measurement between the shafts near the end of the shafts and one near the driven and driver (*Figure 19*). If the shafts are perfectly parallel, the measurements will be the same. If the measurements are not the same, the shafts need to be adjusted until they are parallel before the chain drive is installed.

*Step 6* Oil the shaft lightly, if permitted by the specifications.



Instructor's Notes:

11.12 MILLWRIGHT ◆ LEVEL THREE

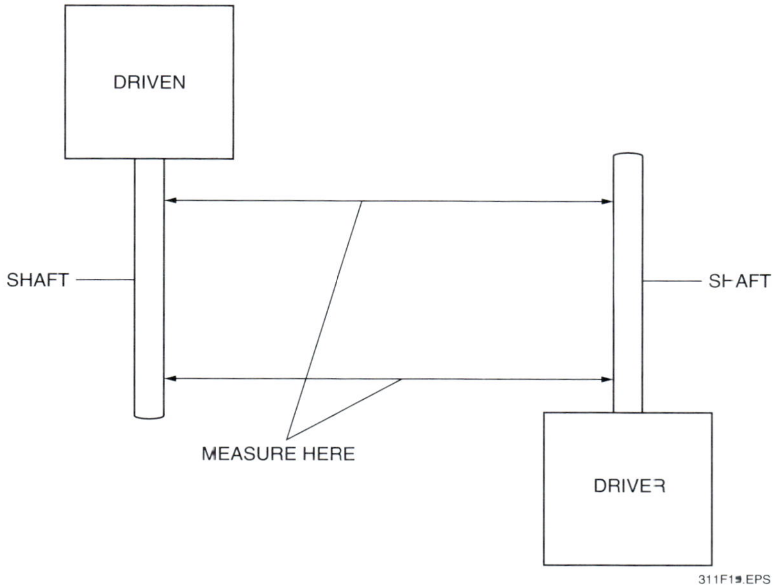


*Figure 19* ◆ Measuring for parallelism.

*Step 7* Slip the driven sprocket onto its shaft, and position the sprocket in the proper position. The proper sprocket position depends on the application.

*Step 8* Lock the driven sprocket to the shaft. Sprockets are usually locked to the shaft with setscrews and a key.

*Step 9* Position a straightedge across the face of both sprockets, and adjust the driver sprocket until the sprockets are perfectly aligned. When the sprockets are aligned, the straightedge will be flat on the faces of both sprockets. An alternate method of checking sprocket alignment is to pull a piano wire taut across the faces. *Figure 20* shows how to check sprocket alignment using a straightedge and a piano wire. This procedure can also be done with dial indicators or lasers.

*Step 10* Lock the driving sprocket to the shaft.

*Step 11* Install the chain (*Figure 21*) around the sprockets, with the ends coming together on the larger sprocket.

*Step 12* Connect the ends of the chain, using a connecting link (*Figure 22*), which is also known as a master link or a half link, to make it an endless chain.

*Step 13* Adjust the chain drive so that all of the chain slack is on top of the drive.

*Step 14* Place a straightedge on the chain from one sprocket to the other, and measure from the straightedge to the chain to check the chain tension (*Figure 23*). The sag should be measured midway between the sprockets. Chain tension should be such that the chain sags approximately 2 percent of the distance between the shaft centers. If the sprockets are too far apart to use a straightedge, piano wire can be used instead.

*Step 15* Adjust the chain tension until the required 2 percent sag is achieved.

*Step 16* Lock the adjusting screws in position. The adjusting screws are usually locked with a jam nut.

*Step 17* Lubricate the chain according to the manufacturer's recommendations.

*Step 18* Install all safety guards.

**Explain how to check sprocket alignment.**

**Explain how to measure the chain tension, using a straightedge to measure sag.**

**Show Transparencies 24-27 (Figures 20-23).**

**Demonstrate how to install a chain drive.**

**Have trainees install chain drives. Note the proficiency of each trainee. This laboratory corresponds to Performance Task 4.**



MODULE 15311-08 ◆ INSTALLING BELT AND CHAIN DRIVES 11.13

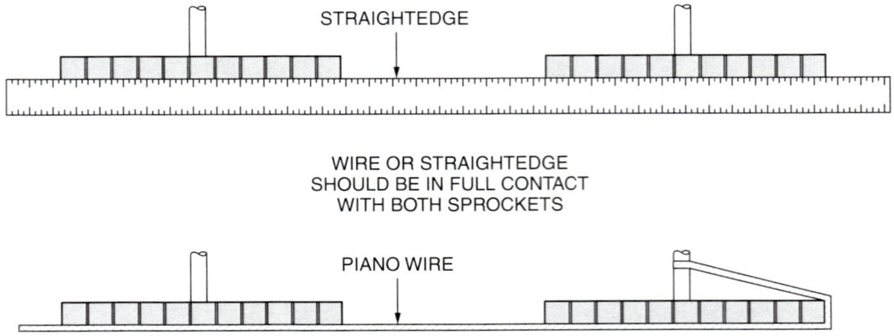


*Figure 20* ◆ Checking sprocket alignment.

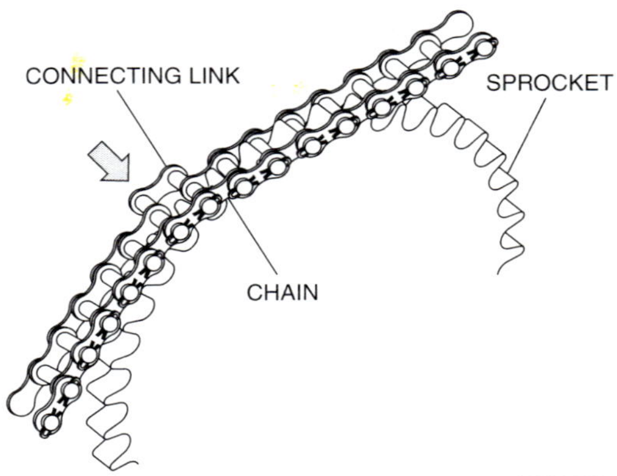


*Figure 21* ◆ Installing chain.

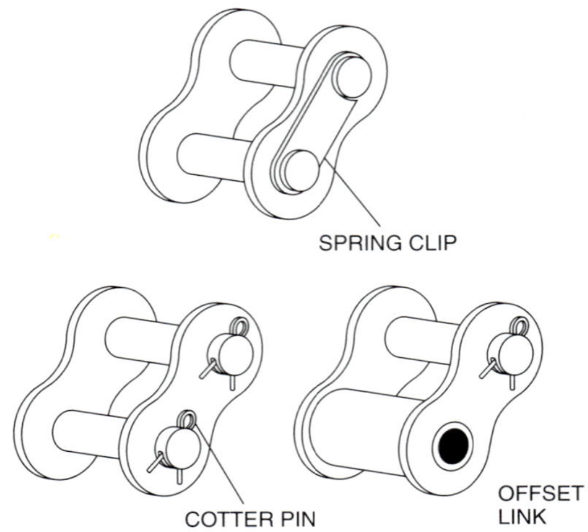


*Figure 22* ◆ Connecting links.

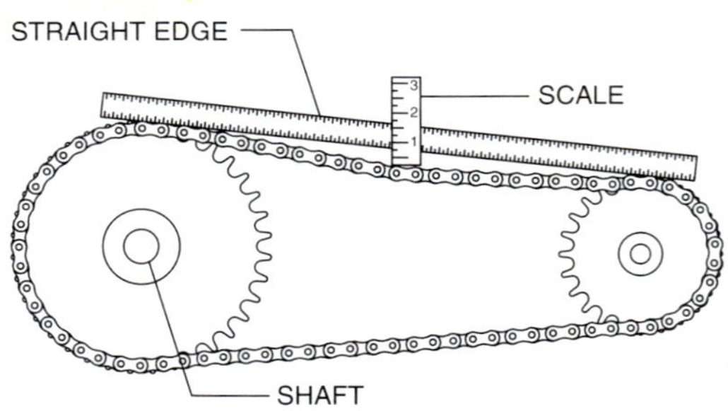


*Figure 23* ◆ Measuring chain tension.



Instructor's Notes:

11.14 MILLWRIGHT ◆ LEVEL THREE

## 6.0.0 ◆ CHAIN TOOLS

Specialized tools are used for working on roller chains and pulleys. Pulleys and sprockets can be checked for alignment with a magnetically attached laser such as the Dotline Laser® (*Figure 24*). This makes the alignment operation a one-person job. Chains can be checked visually for alignment with the sprockets with chain alignment tools such as the one shown in *Figure 25*. This model simply clamps onto the sprocket and holds the rod shown in line with the sprocket, allowing the worker to see whether chain is in line with the rod, and therefore with the sprocket.

The removal and replacement of chain links is accomplished with a chain breaker and riveter (*Figure 26*), which holds the chain in place while it pushes the pin out or rivets the end of the pin in place.

311F24.EPS

*Figure 24* ◆ Laser alignment tool.

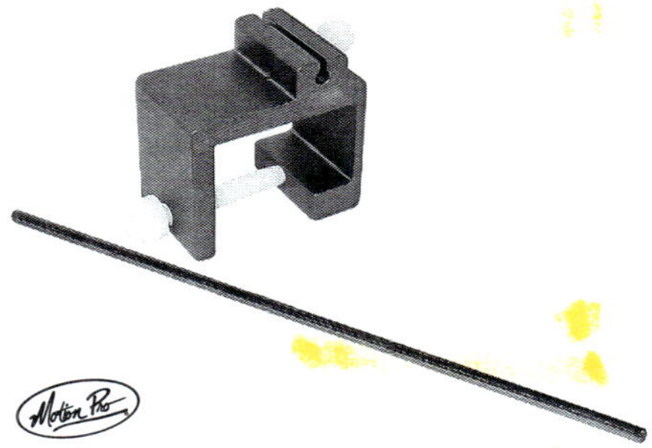

311F25.EPS

*Figure 25* ◆ Simple sprocket alignment tool.

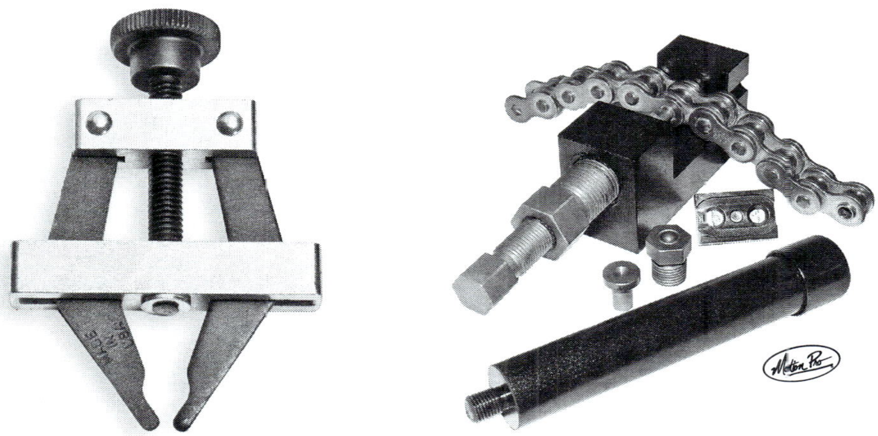

311F26.EPS

*Figure 26* ◆ Chain tool.



**Discuss some of the specialized tools designed for working on roller chains and pulleys.**

**Provide samples of laser alignment tools, simple sprocket alignment tools, and chain tools for trainees to examine.**

MODULE 15311-08 ◆ INSTALLING BELT AND CHAIN DRIVES 11.15

**Have trainees complete the Review Questions, and go over the answers prior to administering the Module Examination.**

## Review Questions

1. The most common form of power transmission is _____.
   a. the transformer
   b. gear drive
   c. from one shaft to another
   d. from pump to pump

2. Belt and chain drives are used to transmit power from one shaft to a(n) _____.
   a. parallel shaft
   b. angular shaft
   c. driver
   d. ground

3. A metric V-belt is specified in the ISO standard by the _____.
   a. top width
   b. dot width
   c. datum width
   d. length in meters

4. A V-belt labeled 3L has a top width of _____.
   a. 3 mm
   b. ½ inch
   c. 3 inches
   d. ⅜ inch

5. The length of FHP belts is measured on the _____ of the belt.
   a. inside
   b. ends
   c. outside
   d. datum width

6. The length of wedge belts is measured on the _____.
   a. bottom of the groove
   b. pitch line
   c. outside of the belt
   d. side of the belt

7. The code used to ensure that the replacement belts in a multiple-belt application are the same length is called a _____ code.
   a. batch
   b. match
   c. length
   d. line

8. The center-to-center distance between hinged points in a roller chain is called the _____.
   a. chain width
   b. pitch
   c. hinge length
   d. grade

9. A standard ¾-inch roller chain that was three rows wide would have the part number _____.
   a. 31-3
   b. 61-3
   c. 60-3
   d. 30-2

10. Chain tension should be such that a chain sags approximately _____.
    a. ¼ of the diameter of the sprocket
    b. 25 percent of the distance between the shaft centers
    c. 5 percent of the diameter of the sprockets
    d. 2 percent of the distance between the shaft centers



Instructor's Notes:

11.16 MILLWRIGHT ◆ LEVEL THREE

## Summary

Practically all equipment that moves does so by means of power transmission equipment. The most common form of power transmission is the transmission of power from one shaft to another. Belt and chain drives are used to transmit power from one shaft to a parallel shaft.

Installing, adjusting, and maintaining belt and chain drives are part of a millwright's job. The service life of belt and chain drives depends heavily on these procedures being performed properly.

## Notes

**Summarize the major concepts presented in the module.**

**Administer the Module Examination. Record the results on Craft Training Report Form 200, and submit the results to the Training Program Sponsor.**

**Administer the Performance Test, and fill out Performance Profile Sheets for each trainee. If desired, trainee proficiency noted during laboratory sessions may be used to complete the Performance Test. Record the results on Craft Training Report Form 200, and submit the results to the Training Program Sponsor.**



MODULE 15311-08 ◆ INSTALLING BELT AND CHAIN DRIVES 11.17

## Trade Terms Introduced in This Module

***Datum width:*** The width of a V-belt at a particular point, as used in ISO standard codes.

***Dress down:*** To smooth and remove all extrusions from a belt.

***Top width:*** The width across the top of a V-belt, used in DIN and US codes.

***Vulcanized:*** Combined with natural rubber additives to improve strength and resiliency.



Instructor's Notes:

11.18 MILLWRIGHT ◆ LEVEL THREE

# Resources & Acknowledgments

## Additional Resources

This module is intended to present thorough resources for task training. The following reference works are suggested for further study. These are optional materials for continued education rather than for task training.

The Complete Guide to Chain, http://chain-guide.com/

Michigan Industrial Belting, Inc., http://www.mibelting.com/indbelt.htm?gclid=CL_d0aS0rZECFQIglgodrx6Aew

GlobalSpec: The Engineering Search Engine, http://mechanical-components.globalspec.com/Industrial-Directory/drive_belt

## Figure Credits

Photo courtesy of Ludeca, Inc., 311F13 (photo), 311F24, www.ludeca.com

Motion Pro, 311F25, 311F26 (right)

W.W. Grainger, Inc., 311F26 (left)



MODULE 15311-08 ◆ INSTALLING BELT AND CHAIN DRIVES 11.19

# MODULE 15311-08 — TEACHING TIPS

The following are suggested activities or instructional methods to help you teach the material in this module.

## General

When you call on someone to answer a question, the rest of the class relaxes or even tunes out because they expect that the question and answer will take place only between you and the trainee you called on. Instead, use this technique to involve more trainees in answering questions and to keep them on their toes.

1. Ask trainees to define a term or explain a concept.
2. After one trainee has answered, ask a trainee seated nearby if the answer is right. Then ask whether a trainee in the back of the room agrees.
3. Ask trainees to explain why they think an answer is right or wrong.
4. Use the session to clear up incorrect ideas and encourage trainees to learn from their mistakes.

## Section 2.0.0 *Belt Drive Types*

This exercise will help familiarize trainees with various types of belt drives. Trainees will need pencils and paper. You will need to arrange workstations with a different type of belt drive at each station. Allow 20 to 30 minutes for this exercise. This exercise corresponds to Performance Task 1.

1. Discuss various types of belt drives with trainees.
2. Have trainees rotate between the workstations and identify the type of belt drive at each station.
3. Answer any questions trainees may have.

## Section 4.0.0 *Chain Drive Types*

This exercise will help familiarize trainees with various types of chain drives. Trainees will need pencils and paper. You will need to arrange workstations with a different type of chain drive at each station. Allow 20 to 30 minutes for this exercise. This exercise corresponds to Performance Task 3.

1. Discuss various types of chain drives with trainees.
2. Have trainees rotate between the workstations and identify the type of chain drive at each station.
3. Answer any questions trainees may have.

## MODULE 15311-08 — ANSWERS TO REVIEW QUESTIONS

| Answer | Section |
|---|---|
| 1. c | 1.0.0 |
| 2. a | 1.0.0 |
| 3. c | 2.0.0 |
| 4. d | 2.1.1 |
| 5. c | 2.1.1 |
| 6. b | 2.1.3 |
| 7. b | 2.1.3 |
| 8. b | 4.1.1 |
| 9. c | 4.1.2 |
| 10. d | 5.0.0 |

## CONTREN® LEARNING SERIES — USER UPDATE

The NCCER makes every effort to keep these textbooks up-to-date and free of technical errors. We appreciate your help in this process. If you have an idea for improving this textbook, or if you find an error, a typographical mistake, or an inaccuracy in NCCER's Contren® textbooks, please write us, using this form or a photocopy. Be sure to include the exact module number, page number, a detailed description, and the correction, if applicable. Your input will be brought to the attention of the Technical Review Committee. Thank you for your assistance.

*Instructors* – If you found that additional materials were necessary in order to teach this module effectively, please let us know so that we may include them in the Equipment/Materials list in the Annotated Instructor's Guide.


**Write:** Product Development and Revision
National Center for Construction Education and Research
3600 NW 43rd St., Bldg. G, Gainesville, FL 32606

**Fax:** 352-334-0932

**E-mail:** curriculum@nccer.org


Craft ______________________ Module Name ______________________

Copyright Date ____________ Module Number ____________ Page Number(s) ____________

Description ______________________

______________________

______________________

______________________

(Optional) Correction ______________________

______________________

______________________

(Optional) Your Name and Address ______________________

______________________

______________________

Module 15312-08

# *Installing Fans and Blowers*

## NCCER STANDARDIZED CRAFT TRAINING PROGRAM

The National Center for Construction Education and Research (NCCER) provides a standardized national program of accredited craft training. Key features of the program include instructor certification, competency-based training, and performance testing. The program provides trainees, instructors, and companies with a standard form of recognition through a National Craft Training Registry. The program is described in full in the *Guidelines for Accreditation*, published by NCCER. For more information on standardized craft training, contact the NCCER by writing us at 3600 NW 43rd St., Bldg. G, Gainesville, FL 32606; calling 352-334-0911; or emailing info@nccer.org. More information may be found at our website, www.nccer.org.

## HOW TO USE THIS ANNOTATED INSTRUCTOR'S GUIDE

Each page presents two sections of information. The larger section displays each page exactly as it appears in the Trainee Module. The narrow column ties suggested trainee and instructor actions to each page and provides icons (detailed below) to call your attention to material, safety, audiovisual, or testing requirements. The bottom of each page includes space for your notes.

The **Audiovisual** icon indicates an appropriate time to show a transparency or other audiovisual aid.

The **Classroom** icon prompts you to define a term, stress a point, ask trainees to explain a concept, or give examples.

The **Demonstration** icon directs you to show trainees how to perform tasks.

The **Examination** icon tells you to administer the written module examination.

The **Homework** icon is placed where you may wish to assign reading for the next class, assign a project, or advise trainees to prepare for an examination.

The **Laboratory** icon is used when trainees are to practice performing tasks.

The **Materials** icon is a reminder for you to gather materials needed for classes, labs, and testing.

The **Performance Testing** icon tells you to administer a performance test or a portion thereof.

The **Safety** icon is used to emphasize safety issues. It is often keyed to *Caution* and *Warning!* statements in the Trainee Module.

The **Teaching Tip** icon indicates additional guidance is available, such as how to conduct an exercise, get the most educational value from a field trip, or encourage class participation. Teaching Tips may expand on a feature (*Think About It*, *Did You Know?*) or provide *Quick Quizzes* or similar exercises. You will be referred to the Teaching Tips section at the back of the module if there is additional material.

The **Combination** icon indicates that the laboratory listed corresponds with a performance task. If desired, you can note the proficiency of the trainees during the laboratory, and use it to satisfy performance testing requirements.

## PREPARATION

Before teaching this module, you should review the Objectives, Performance Tasks, Materials and Equipment List, and Module Outline. Be sure to allow ample time to prepare your own training or lesson plan and gather all required materials and equipment.

## MODULE OVERVIEW

This module explains the principles of operation of various types of fans and blowers, and explains how they are used. It also covers basic maintenance procedures.

## PREREQUISITES

Prior to training with this module, it is recommended that the trainee shall have successfully completed *Core Curriculum; Millwright Level One; Millwright Level Two;* and *Millwright Level Three,* Modules 15301-08 through 15311-08.

## OBJECTIVES

Upon completion of this module, the trainee will be able to do the following:

1. Identify and explain types of fans.
2. Explain how to install fans.
3. Identify and explain types of blowers.
4. Explain how to install blowers.

## PERFORMANCE TASKS

Under the supervision of the instructor, the trainee should be able to do the following:

1. Identify types of fans.
2. Identify centrifugal fan wheel types.
3. Identify types of blowers.

## MATERIALS AND EQUIPMENT LIST

Overhead projector and screen
Transparencies
Blank acetate sheets
Transparency pens
Whiteboard/chalkboard
Markers/chalk
Pencils and scratch paper
Appropriate personal protective equipment
Basic trainee tools
Various types of fans
Various types of blowers
Module Examinations*
Performance Profile Sheets*

* Located in the Test Booklet

## SAFETY CONSIDERATIONS

Ensure that the trainees are equipped with appropriate personal protective equipment and know how to use it properly. This module may require that the trainees visit job sites. Ensure that trainees are briefed on site safety policies prior to any site visits.

## ADDITIONAL RESOURCES

This module is intended to present thorough resources for task training. The following reference works are suggested for both instructors and motivated trainees interested in further study. These are optional materials for continued education rather than for task training.

Greenheck Fan Corporation, http://www.greenheck.com/

Gardner Denver, Inc., http:// www.gardnerdenver.com/

## TEACHING TIME FOR THIS MODULE

An outline for use in developing your lesson plan is presented below. Note that each Roman numeral in the outline equates to one session of instruction. Each session has a suggested time period of 2½ hours. This includes 10 minutes at the beginning of each session for administrative tasks and one 10-minute break during the session. Approximately 10 hours are suggested to cover *Installing Fans and Blowers*. You will need to adjust the time required for hands-on activity and testing based on your class size and resources. Because laboratories often correspond to Performance Tasks, the proficiency of the trainees may be noted during these exercises for Performance Testing purposes.

| Topic | Planned Time |
|---|---|
| **Session I. Introduction; Types of Fans** | |
| A. Introduction | __________ |
| B. Types of Fans | __________ |
| C. Laboratory<br>Trainees practice identifying types of fans. This laboratory corresponds to Performance Task 1. | __________ |
| D. Laboratory<br>Trainees practice identifying centrifugal fan wheel types. This laboratory corresponds to Performance Task 2. | __________ |
| **Session II. Installing Fans** | |
| A. Installing Fans | __________ |
| B. Installing Preassembled Fans | __________ |
| C. Installing Disassembled Fans | __________ |
| **Session III. Types of Blowers; Installing Blowers** | |
| A. Types of Blowers | __________ |
| B. Laboratory<br>Trainees practice identifying types of blowers. This laboratory corresponds to Performance Task 3. | __________ |
| C. Installing Blowers | __________ |
| **Session IV. Review and Testing** | |
| A. Module Review | __________ |
| B. Module Examination<br>1. Trainees must score 70% or higher to receive recognition from NCCER.<br>2. Record the testing results on Craft Training Report Form 200, and submit the results to the Training Program Sponsor. | __________ |
| C. Performance Testing<br>1. Trainees must perform each task to the satisfaction of the instructor to to receive recognition from NCCER. If applicable, proficiency noted during laboratory exercises can be used to satisfy the Performance Testing requirements.<br>2. Record the testing results on Craft Training Report Form 200, and submit the results to the Training Program Sponsor. | __________ |

15312-08

# Installing Fans and Blowers

Assign reading of Module 15312-08.

# 15312-08
# *Installing Fans and Blowers*

*Topics to be presented in this module include:*



## Overview

In this module, you will learn about the various types of fans and blowers. These are the air transfer systems, and some fans and blowers are used to move solid materials. This module covers the principles of operation of various types, and how they are used. The pressure and capacity of different types are described. Basic maintenance and installation procedures are also presented.

*Instructor's Notes:*

## Objectives

When you have completed this module, you will be able to do the following:

1. Identify and explain types of fans.
2. Explain how to install fans.
3. Identify and explain types of blowers.
4. Explain how to install blowers.

## Trade Terms

Airfoil
Axial flow
Centrifugal

## Required Trainee Materials

1. Pencil and paper
2. Appropriate personal protective equipment

## Prerequisites

Before you begin this module, it is recommended that you successfully complete *Core Curriculum; Millwright Level One; Millwright Level Two;* and *Millwright Level Three,* Modules 15301-08 through 15311-08.

This course map shows all of the modules in the third level of the *Millwright* curriculum. The suggested training order begins at the bottom and proceeds up. Skill levels increase as you advance on the course map. The local Training Program Sponsor may adjust the training order.

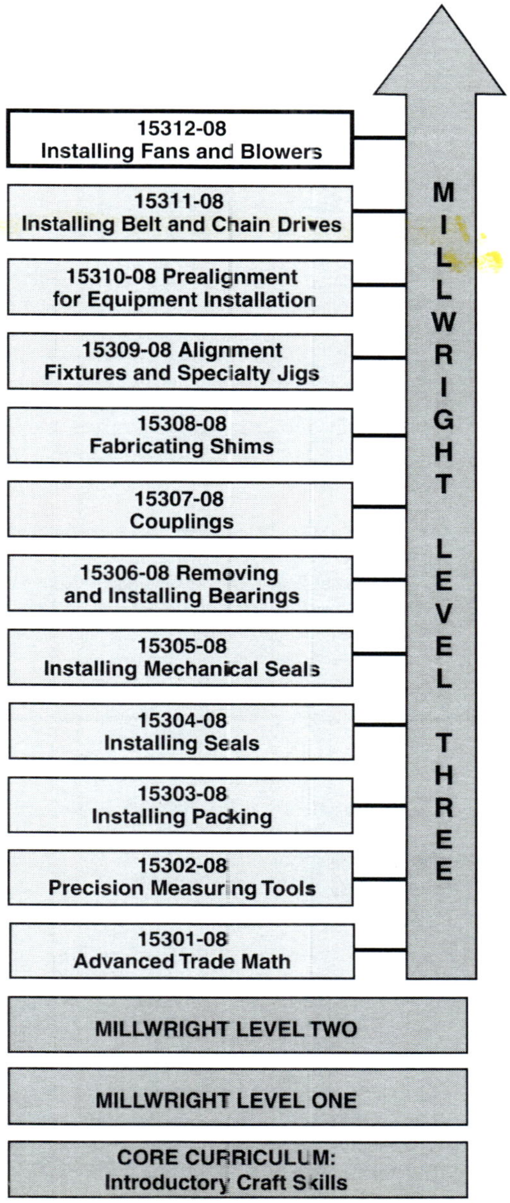


312CMAP.EPS

**Ensure that you have everything required to teach the course. Check the Materials and Equipment list at the front of this module.**

**See the general Teaching Tip at the end of this module.**

**Explain that terms shown in bold are defined in the Glossary at the back of this module.**

**Show Transparency 1, Objectives, and Transparency 2, Performance Tasks. Review the goals of the module, and explain what will be expected of the trainee.**

**Review the modules covered in Level Three and explain how this module fits in.**

**Discuss the many ways that fans and blowers are used.**

**Explain the millwight's role in installing and maintaining fans and blowers.**

**Stress the importance of following safety policies, including lockout/tagout procedures.**

**Introduce axial-flow and centrifugal fans.**

**Discuss the categories of axial-flow fans.**

**Discuss tube-axial and vane-axial fans.**

**Show Transparencies 3 and 4 (Figures 1 and 2).**

**See the Teaching Tip for Section 2.0.0 at the end of this module.**

## 1.0.0 ◆ INTRODUCTION

Fans and blowers have a wide range of applications in industry, the office, and the home. They are used to exhaust or introduce air and other gases into process reactors, dryers, and cooling towers. They are also used to assist combustion in furnaces, to convey material pneumatically, and to ventilate for comfort and safety.

Fans have pressure rises up to approximately 2 psig. Blowers have pressure rises between 2 and 10 psig. For discharge pressures over 10 psig, the machine is called a compressor.

For fans and blowers to operate safely and efficiently, they must be installed and maintained properly. Since fans and blowers are widely used in industry, installing them is a major part of a millwright's job. This module explains the most common types of fans and blowers and the methods for installing each.

> **WARNING!**
> All equipment should be locked and tagged out to protect against accidental operation when such operation could cause injury to personnel. Ensure that all machinery has been properly locked or tagged out prior to operating.

## 2.0.0 ◆ TYPES OF FANS

Although there are many variations in the styles of fans for different applications, most are either **axial-flow** fans or **centrifugal** fans.

### 2.1.0 Axial-Flow Fans

Axial-flow (AF) fans are capable of moving large amounts of air at low pressures and are usually considered for low-resistance applications. AF fans are classified as either tube-axial or vane-axial fans. AF fans are divided into the following categories:

- *Free fans* – Circulate air by rotating in an unrestricted air space. Ceiling fans and table-top fans are free fans.
- *Diaphragm-mounted fans* – Transfer air from one space to another. They are usually mounted in a wall.
- *Ducted fans* – Constrain air by a duct to enter and leave the fan blades in an axial direction. A fan is considered ducted if the duct length is more than the distance between the inlet to and the outlet from the fan blades.

#### *2.1.1 Tube-Axial Fans*

Tube-axial fans are designed for a wide range of volumes at medium pressures. They consist of a propeller enclosed in a cylinder that collects and directs air flow. A tube-axial fan discharges air in a helical, or screw-like, motion (*Figure 1*).

#### *2.1.2 Vane-Axial Fans*

Unlike a tube-axial fan, a vane-axial fan has vanes on the discharge side of the propeller that cause the air to discharge in a straight line (*Figure 2*). Therefore, turbulence is reduced. This reduction improves fan efficiency and pressure capabilities.

Vane-axial fans can develop high pressures and move large volumes of air. The driver for vane-axial fans can be sized to meet most horsepower

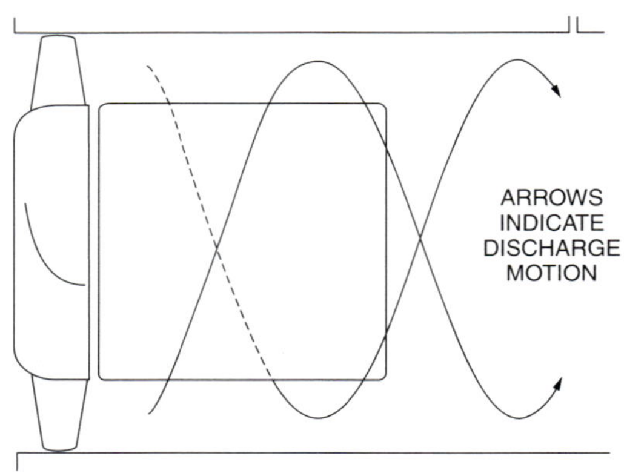


*Figure 1* ◆ Tube-axial fan and discharge motion.



Instructor's Notes:

12.2 MILLWRIGHT ◆ LEVEL THREE

requirements. These fans are available with adjustable-pitch propellers for performance variation.

## 2.2.0 Centrifugal Fans

Centrifugal fans are used in applications in which higher head pressures are needed, such as when moving air encounters high resistance. They are generally easier to control, stronger, and quieter than AF fans. *Figure 3* shows two models of centrifugal fans.

A centrifugal fan is identified by its wheel type. Centrifugal fans are classified into the following basic types:

- Radial
- Forward-curved
- Backward-curved
- Airfoil

### *2.2.1 Radial Fans*

A radial fan wheel can be used in many applications ranging from pneumatic conveying to exhausting gas or air in high-resistance systems. This fan can be made to work at various widths and capacities, which allows it to achieve high static pressure at relatively low capacity.

A radial fan can develop high pressures at high speeds. The following radial fan wheels are used for different applications (see *Figure 4*):

- *Open-type* – Used for general-purpose duties, self-cleaning
- *One side closed* – Used with stringy materials
- *Rim-type* – Used for severe duties
- *Backplate-type* – Creates steady draft but is not suitable for fibrous or chunky materials

### *2.2.2 Forward-Curved Fans*

A forward-curved fan discharges higher velocity air from the blade tip than a backward-curved fan running at the same tip speed. Forward-curved fans are more suitable for equipment with long shafts because the fans can produce high-velocity air while running at slower speeds than other types of fans. Forward-curved fans operate quietly and require little space.

### *2.2.3 Backward-Curved Fans*

A backward-curved fan (*Figure 5*) has blades that are tilted backward to the optimum angle, and the fan develops most of its energy directly as pressure.

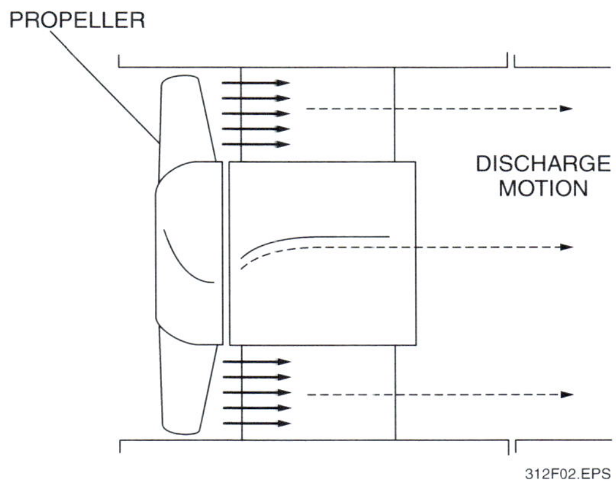


*Figure 2* ◆ Vane-axial fan and discharge motion.

*Figure 3* ◆ Centrifugal fans.

**Explain the classifications of centrifugal fans.**

**Discuss the types of radial fans and their applications.**

**Show Transparency 5 (Figure 4).**

**Describe forward-curved and backward-curved fans.**

**Show Transparency 6 (Figure 5).**

Describe airfoil fans.

Show Transparency 7 (Figure 6).

Have trainees practice identifying types of fans. Note the proficiency of each trainee. This laboratory corresponds to Performance Task 1.

Have trainees practice identifying centrifugal fan wheel types. Note the proficiency of each trainee. This laboratory corresponds to Performance Task 2.

Have trainees review Section 3.0.0–3.2.0.

Ensure that you have everything required to teach this session.

Introduce installation procedures for preassembled and disassembled fans.

Stress the importance of following lockout/tagout procedures.

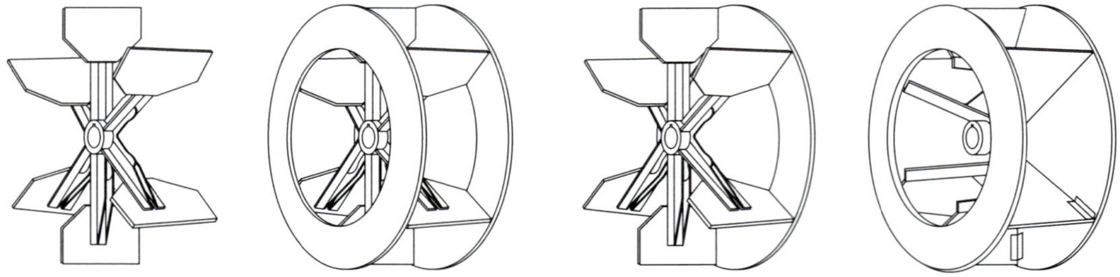


*Figure 4* ◆ Radial fan wheel types.

Backward-curved fans are extensively used as ventilators. They operate at medium speed and have broad volume and pressure capabilities. They develop less velocity head than forward-curved fans of the same size. Backward-curved fans are easy to control because small changes in system volume result in small variations of air pressure.

### 2.2.4 Airfoil Fans

**Airfoil** fans (*Figure 6*) are backward-curved fans that have an airfoil cross section which increases their stability, efficiency, and performance. These fans generally run more quietly than other types of centrifugal fans, and they do not pulsate within their operating range because air flows through the wheel with less turbulence.

## 3.0.0 ◆ INSTALLING FANS

Fans can be mounted horizontally or vertically, whether positive or negative flow. Installation procedures differ for each fan type and model. Some fans are preassembled by the manufacturer and shipped as a unit. Installation of a preassembled fan consists of properly mounting the fan on its base and tying it into the system. Larger fans are sometimes shipped disassembled and require site assembly. Installation of disassembled fans requires more attention to detail and precise performance of procedures. Usually, each fan comes with a complete set of installation, start-up, and operating procedures. Regardless of the type of fan being installed, the manufacturer's installation procedures should always be followed precisely. The following sections provide common installation procedures for both preassembled and disassembled fans.

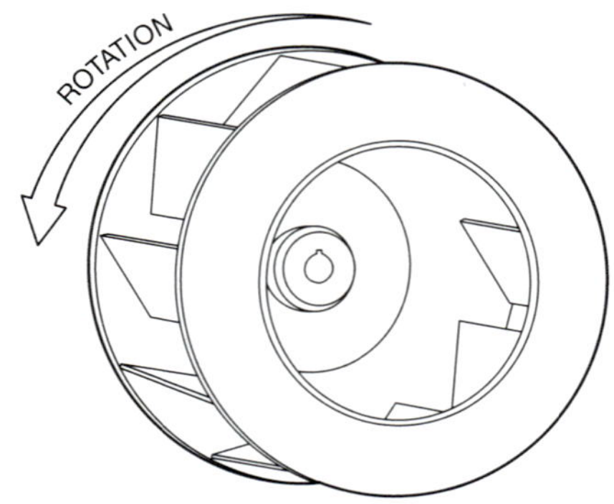


*Figure 5* ◆ Backward-curved fan wheel.

It is important to know whether the fan or blower will be heated or exposed to significant heat during operation. A fan or blower that is to be used to move hot air from some aspect of process is at some risk of alignment change due to thermal expansion of parts. Since the fan or blower will be run most often at temperature, thermal expansion must be allowed for during the alignment process. This information would be supplied by the manufacturer or engineer.

> **WARNING!**
> All equipment should be locked and tagged out to protect against accidental operation when such operation could cause injury to personnel. Ensure that all machinery has been properly locked or tagged out prior to operating.



Instructor's Notes:

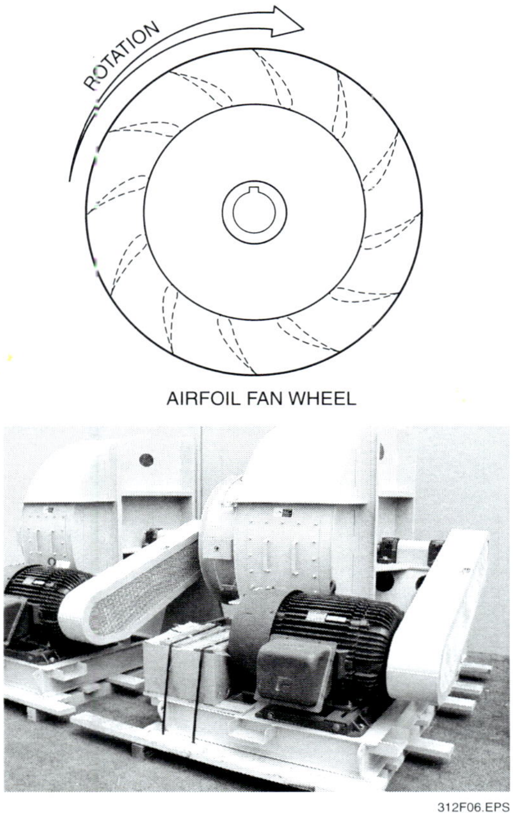


*Figure 6* ◆ Airfoil fan wheel.

### 3.1.0 Installing Preassembled Fans

Preassembled fans are completely assembled and balanced at the factory. They are packaged in crates and protective coverings to prevent damage during shipping. The fan should be properly installed and checked to ensure that it is still properly balanced and adjusted after shipping. Follow these steps to install a preassembled fan:

*Step 1* Move the fan near the intended fan placement location.

*Step 2* Carefully remove the crate, skids, and protective coverings.

*Step 3* Inspect the fan thoroughly to ensure that it was not damaged in shipping.

**NOTE**
Promptly report any damage to your supervisor so that a determination can be made as to who is responsible for the damage and needed repairs.

*Step 4* Measure the fan anchor bolt hole pattern, and compare it to the anchor bolt pattern to ensure that they match.

*Step 5* From the engineer's drawing, determine if the fan requires vibration isolators. Install vibration isolators if they are required.

*Step 6* Rig the fan for lifting. Lift the fan using wire rope or nylon slings and a crane, chain, or come-along. Fans usually have lifting lugs for rigging. Do not lift fans by the shaft with a forklift. The shaft can be warped or otherwise damaged if lifted with a forklift.

*Step 7* Move the fan into position over the anchor bolts, and set it in place.

**CAUTION**
Be careful not to damage the anchor bolts when setting the fan in place.

*Step 8* Install nuts on the anchor bolts.

*Step 9* Level the fan on its base, using a level and adding shims where necessary.

*Step 10* Tighten the nuts on the anchor bolts to the specified torque.

*Step 11* Check the fan to ensure that it is still level.

*Step 12* Turn the fan by hand to ensure that it turns freely.

*Step 13* Install all safety guards.

*Step 14* Start up the fan.

*Step 15* Check the rotation of the fan. If the motor can be uncoupled, check by bumping for rotation to determine if the direction is proper.

*Step 16* Monitor the fan while it is operating for any unusual noises or vibration.

**Discuss preassembled fans and review the steps for installation.**

**Discuss disassembled fans and review the steps for installation.**

**Emphasize that assembly drawings and instructions must be strictly followed.**

**Stress the importance of following lockout/tagout procedures.**

**Show Transparency 8 (Figure 7).**

## 3.2.0 Installing Disassembled Fans

Some specially built fans and very large fans are shipped disassembled. Assembly drawings and instructions are provided with disassembled fans. These drawings and instructions must be strictly followed. *Figure 7* shows a disassembled axial fan.

> **WARNING!**
> All equipment should be locked or tagged out to protect against accidental operation when such operation could cause injury to personnel. Ensure that all machinery has been properly locked or tagged out prior to operating.

The assembly procedures vary for each type of fan. The following is a sample procedure for installing a heavy-duty, double-inlet fan with inlet boxes and sleeve bearings on pedestals and soleplates. Follow these steps to install a disassembled fan:

*Step 1* Carefully uncrate the fan parts and accessories.

*Step 2* Identify all of the fan parts, using the parts list in the assembly drawings to familiarize yourself with the parts and to ensure that you have all of the parts.

*Step 3* Check the foundation bolts to ensure that they are located as shown on the assembly drawing and that they match the fan mounting holes.

*Step 4* Lay out and mark a center line on the foundation. This center line will be used as a reference throughout the assembly of the fan.

*Step 5* Determine from the assembly drawings the fan shaft center line height from the foundation.

*Step 6* Rig the lower half of the fan housing, and set it in place over the anchor bolts.

> **CAUTION**
> Do not lift the fan with a forklift. Use wire rope or nylon slings and a crane or chain fall to prevent damaging the fan.

*Step 7* Align the fan housing with the marked center line.

*Step 8* Shim the fan housing to the approximate finished elevation.

*Step 9* Attach the inlet boxes to the fan housing, following the assembly instructions.

*Step 10* Set the bearing pedestals in place.

*Step 11* Shim the pedestals to bring them to the proper elevation and level. The pedestals should be set so that the bearing center will be at the proper elevation.

*Step 12* Rig and lift the shaft out of its crate, using a nylon sling.

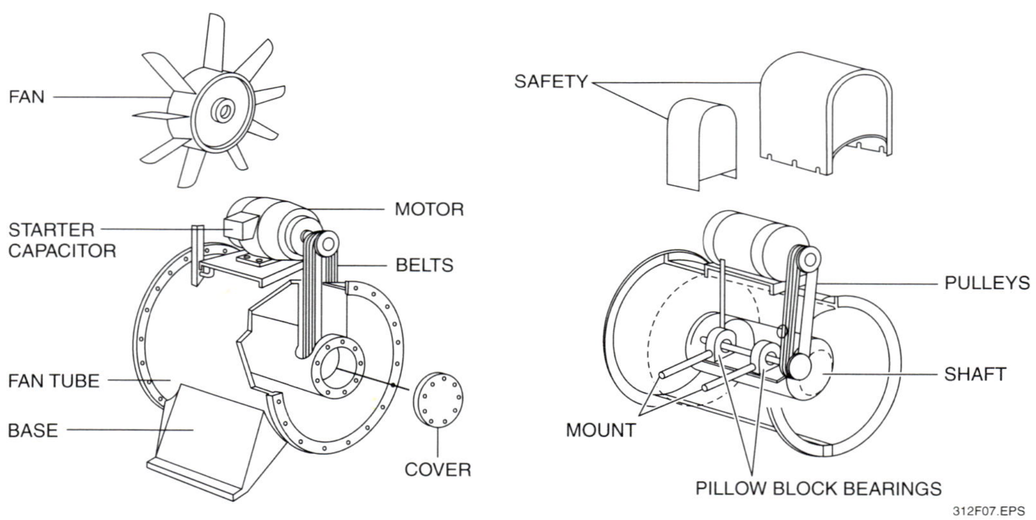


*Figure 7* ◆ Disassembled axial fan.



Instructor's Notes:

12.6 MILLWRIGHT ◆ LEVEL THREE

**CAUTION**

Do not lift the shaft by the journal surfaces where the bearing will fit. This may scratch or damage the journal surfaces.

*Step 13* Clean the bore of the wheels, using solvent.

*Step 14* Coat the bore of the wheels, using clean oil.

*Step 15* Slip the wheels onto the shaft so that the assembly can be set onto its supports.

*Step 16* Set the assembly onto its supports in the housing. Be sure that the rotation arrow on the fan corresponds to the rotation arrow on the wheel and that the shaft thrust collars are on the drive side of the housing.

*Step 17* Turn the shaft until the keys are on top.

*Step 18* Coat the wheel keys with an anti-seize compound, if that is specified, and insert them into the keyways.

*Step 19* Align the wheels with the keys, and slide the wheels into place.

*Step 20* Tighten the wheel setscrews to lock the wheels in place.

*Step 21* Slip the fan inlet bell housing and variable vanes over the shaft. Ensure that the rotation marks on the bell housing and vanes correspond with the housing and wheel arrows. At this point, the bell housing and vanes are only slipped over the shaft and will be attached later.

*Step 22* Prepare and install the bearings.

*Step 23* Clean the shaft, using solvent to remove protective coatings.

**NOTE**

Bearing preparation and installation procedures depend on the type of bearing used. Follow the manufacturer's installation procedures and good bearing installations practices.

*Step 24* Coat the journal area of the shaft, using clean oil, as specified by the manufacturer.

**CAUTION**

Do not touch the cleaned journal surfaces with your bare hand since perspiration can cause discoloration and pitting of the surface.

*Step 25* Set the wheels and shaft in the bearings, following the manufacturer's assembly instructions.

**CAUTION**

The procedure for setting the shaft and wheels in the bearings depends on the type of bearings used. Use extreme caution and follow instructions precisely because using incorrect procedures can permanently damage bearings.

*Step 26* Install gaskets for the top section of the fan housing.

*Step 27* Rig the top half of the fan housing, and set it in place.

*Step 28* Install the connecting bolts for the top half. A drift pin may be needed to align the holes for bolt installation.

*Step 29* Tighten the connecting bolts sequentially.

*Step 30* Attach the inlet bells to the fan housing, bolting from the outside and leaving the bolts loose.

*Step 31* Adjust the wheels and inlet bells for proper alignment according to the manufacturer's assembly instructions.

*Step 32* Tighten the inlet bells bolts.

*Step 33* Install the fan and shaft driver couplings.

*Step 34* Align the driver to the fan. At this point, you will have learned only rough prealignment. For now, precision alignment must be done by an experienced millwright or by your instructor.

*Step 35* Grout the housing base and bearing soleplates.

*Step 36* Install inlet boxes, outlet boxes, and variable vane-control mechanisms according to the manufacturer's assembly instructions.

Have trainees review Section 4.0.0–5.0.0.

Ensure that you have everything required to teach this session.

Discuss various types of blowers. Describe positive-displacement rotary blowers.

Discuss the function and configuration of roots-type blowers.

Provide a roots-type blower for trainees to examine.

Discuss screw-type blowers.

Show Transparency 9 (Figure 9).

See the Teaching Tip for Section 4.0.0 at the end of this module.

Introduce the topic of installing blowers.

Stress the importance of following lockout/tagout procedures.

*Step 37* Turn the fan by hand to ensure that it turns freely. Uncouple the motor and fan and check for direction, then recouple to the power transmission.

*Step 38* Install all safety guards.

*Step 39* Start up the fan.

*Step 40* Check the rotation of the fan.

**CAUTION**

Never try to operate any switch, valve, or energy isolating device when it is locked or tagged out.

*Step 41* Monitor the fan while it is operating for any unusual noises or vibration.

## 4.0.0 ◆ TYPES OF BLOWERS

Fans that have pressure rises between 2 and 10 psig are called blowers. Other types of blowers, called positive-displacement rotary blowers, work like compressors except that they are used to move large volumes of air at relatively low pressure. Unlike reciprocating compressors, blowers produce a smooth flow of air that is compressed in the discharge piping, rather than in the blower. Roots-type and screw-type are two types of positive-displacement rotary blowers.

### 4.1.0 Roots-Type Blowers

Roots-type blowers (*Figure 8*), also called lobe-type blowers, supply low-pressure, oil-free air. The pumping capacity of the blower is determined by its size, operating speed, and pressure conditions. A roots-type blower has two double-lobe impellers mounted on parallel shafts. The impellers rotate in opposite directions within a cylinder that is closed at both ends by head plates. As the impellers rotate, air is drawn into one side of the cylinder and forced out the opposite side against the existing pressure of the connected system. The differential pressure developed depends on the amount of resistance of the connected system.

Sealing between the blower inlet area and outlet area is accomplished by having very small clearances between the lobes and the cylinder. Clearances between the lobes are maintained by two accurately machined timing gears mounted on the shafts that extend to the outside of the cylinder. One gear is driven by the blower driver, which in turn drives the other gear.

### 4.2.0 Screw-Type Blowers

Screw-type blowers (*Figure 9*), also called axial-flow, cycloidal, or helical-rotor blowers, have two screw-shaped rotors. One rotor is called the male rotor, and the other is called the female rotor. In the screw-type blower, a motor drives one rotor, which in turn drives the other rotor through a set of gears. As the rotors turn, air is trapped in pockets between the threads or lobes of the two rotors at the blower inlet and is compressed by the rotors as it is moved along to the outlet at the opposite end of the blower. Oil is injected into the blower by a built-in oil pump to cool the air and to seal and lubricate between the rotors. Screw-type blowers do not produce oil-free air, but they are capable of higher pressures than roots-type blowers. Also, the discharge air temperature of a screw-type blower is lower than that of most other types of compressors.

## 5.0.0 ◆ INSTALLING BLOWERS

**WARNING!**

All equipment should be locked or tagged out to protect against accidental operation when such operation could cause injury to personnel. Ensure that all machinery has been properly locked or tagged out prior to operating.

312F08.EPS

*Figure 8* ◆ Roots-type blower.



Instructor's Notes:

12.8 MILLWRIGHT ◆ LEVEL THREE

Installation procedures vary with each blower type, size, and model. Always follow the manufacturer's installation procedures when installing blowers. The following basic steps apply to almost any of these blowers. Follow these steps to install a blower.

*Step 1* Remove all plugs, covers, and seals from the inlet and outlet connections.

*Step 2* Inspect the interior of the blower to ensure that there is no dirt or foreign material, and clean the blower if necessary.

*Step 3* Replace the plugs, covers, and seals to prevent dirt and other materials from entering the blower during installation.

*Step 4* Check the baseplate anchor bolt layout to ensure that it matches the layout of the mounting holes in the blower.

*Step 5* Rig the blower, and set it in place over the anchor bolts.

*Step 6* Check the blower baseplate to ensure that it is level, using a precision level.

*Step 7* Ensure that grout is properly placed.

*Step 8* Ensure that all of the feet of the blower are in solid contact with the baseplate.

*Step 9* Install the anchor bolt nuts.

*Step 10* Tighten the anchor bolt nuts.

*Step 11* Rotate the drive shaft by hand to ensure that the impellers turn freely.

**CAUTION**

If the impellers do not turn freely, the blower may be in a bind. Loosen the anchor bolts and check the blower again for soft foot.

*Step 12* Install coupling halves on the blower and the motor shafts.

**NOTE**

If installing a motor that is directly coupled to the blower, perform Steps 13 through 17 only. If the blower is belt-driven, follow the procedure for installing belt drives.

*Step 13* Set the motor over the anchor bolts on the baseplate.

*Step 14* Install the motor anchor bolt nuts, but leave them loose.

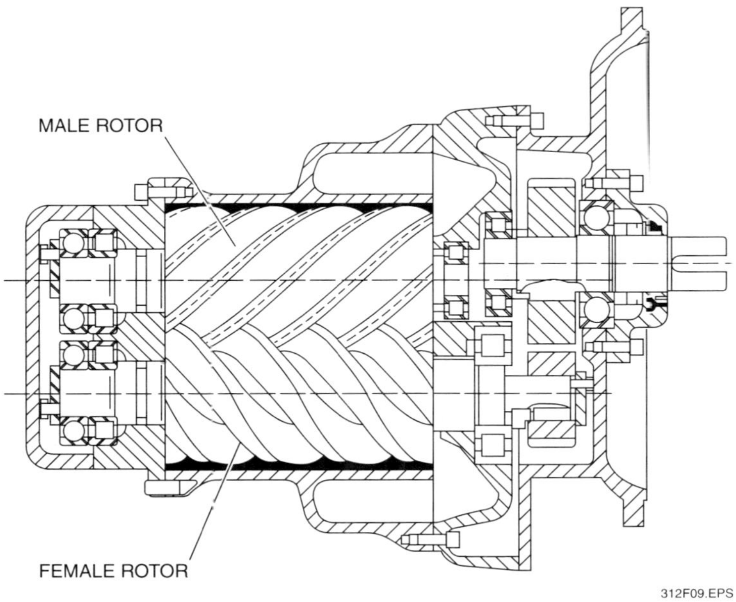


*Figure 9* ◆ Screw-type blower.



**Emphasize that the manufacturer's installation procedures must be followed when installing blowers.**

**Walk through the steps for installing a blower.**

Have trainees practice identifying types of blowers. Note the proficiency of each trainee. This laboratory corresponds to Performance Task 3.

*Step 15* Add and adjust shims under the motor to align the motor with the blower.

> **CAUTION**
> The motor must be aligned with the blower as precisely as possible to prevent operating problems and premature failure of the blower.

*Step 16* Tighten the motor anchor bolt nuts.

*Step 17* Test-run the motor to ensure that its rotation matches that of the blower.

*Step 18* Connect the couplings.

*Step 19* Install any blower accessories, such as filters and silencers, according to the manufacturer's installation instructions.

*Step 20* Connect air piping and eliminate pipe stress according to the engineer's design drawings.

> **CAUTION**
> All accessories and piping must be properly installed without stress to prevent binding that could damage the blower and motor during operation.

*Step 21* Turn the drive shaft by hand to ensure that the blower and motor turn freely.

*Step 22* Lubricate the blower according to the manufacturer's recommendations.

*Step 23* Install all safety guards.

*Step 24* Start up the blower.

*Step 25* Monitor the blower for any unusual noises or vibration while it is operating.



Instructor's Notes:

12.10 MILLWRIGHT ◆ LEVEL THREE

## Review Questions

1. The maximum pressure rise for a fan is _____ psig.
   a. 2
   b. 10
   c. 20
   d. 30

2. Diaphragm-mounted fans are usually mounted in a(n) _____.
   a. tube
   b. wall
   c. boiler
   d. heat exchanger

3. The _____ type of AF fan has reduced turbulence.
   a. tube-axial
   b. centrifugal
   c. vane-axial
   d. axial-frame

4. The _____ type of radial fan wheel is *not* used for fibrous material.
   a. one side closed
   b. rim-type
   c. open-type
   d. backplate-type

5. The quietest type of centrifugal fan is the _____ fan.
   a. tube-axial
   b. rim-type
   c. open-type
   d. airfoil

6. You should use a _____ sling to rig a fan.
   a. nylon
   b. chain
   c. bosun chair
   d. steel mesh

7. It is alright to use a forklift to lift a fan by its shaft.
   a. True
   b. False

8. The shaft thrust collars should be installed on the _____.
   a. left side of the fan
   b. front side of the fan
   c. drive side of the housing
   d. outboard side of the housing

9. The _____ blower has timing gears.
   a. axial-fan
   b. roots-type
   c. rotary-radial
   d. open fan

10. Oil is injected into a screw-type blower to _____.
    a. spray into the burner
    b. lubricate the bearings
    c. cool the air and seal and lubricate the rotors
    d. warm the air



Have trainees complete the Review Questions, and go over the answers prior to administering the Module Examination.

Summarize the major concepts presented in the module.

Administer the Module Examination. Record the results on Craft Training Report Form 200, and submit the results to the Training Program Sponsor.

Administer the Performance Test, and fill out Performance Profile Sheets for each trainee. If desired, trainee proficiency noted during laboratory sessions may be used to complete the Performance Test. Record the results on Craft Training Report Form 200, and submit the results to the Training Program Sponsor.

## Summary

Fans are specifically used for low-pressure air moving. They are used for convective cooling in relatively low-demand circumstances. Fans are also used for directional flow control, for cooling towers, and for comfort.

Blowers are used largely for furnace air supplies, for medium-demand air handling, and for confined space ventilation. They are also used for pneumatic bulk powder transport, for process gas direction, and for light powder fluidized bed work.

Compressors are used to pressurize air or gases, and for high-demand air supplies. Compressors operate above 10 psig.

Millwrights need to know how to install and balance fans and blowers correctly. The constant stresses and occasional contaminants can unbalance a blower, possibly ruining the bearings. Blowers, in particular, should be monitored for excessive vibration.

## Notes



Instructor's Notes:

12.12 MILLWRIGHT ◆ LEVEL THREE

## Trade Terms Introduced in This Module

***Airfoil:*** A fan blade that has a special shape or orientation that controls stability or direction of air.

***Axial flow:*** The flow of air in line with the axis of the fan shaft.

***Centrifugal:*** Moving or directed away from the center of the axis.

# Resources & Acknowledgments

## Additional Resources

This module is intended to present thorough resources for task training. The following reference works are suggested for further study. These are optional materials for continued education rather than for task training.

Greenheck Fan Corporation,
http://www.greenheck.com

Gardner Denver, Inc.,
http://www.gardnerdenver.com/

## Figure Credits

The New York Blower Company, 312F01 (photo)

Greenheck Fan Corporation 312F03 (top)

Topaz Publications, Inc., 312F03 (bottom)

FanAir Company, 312F06 (photo)

Used with permission. ©2008 Gardner Denver, Inc., 312F08



Instructor's Notes:

12.14 MILLWRIGHT ◆ LEVEL THREE

# MODULE 15311-08 — TEACHING TIPS

The following are suggested activities or instructional methods to help you teach the material in this module.

## General

When you call on someone to answer a question, the rest of the class relaxes or even tunes out because they expect that the question and answer will take place only between you and the trainee you called on. Instead, use this technique to involve more trainees in answering questions and to keep them on their toes.

1. Ask trainees to define a term or explain a concept.
2. After one trainee has answered, ask a trainee seated nearby if the answer is right. Then ask whether a trainee in the back of the room agrees.
3. Ask trainees to explain why they think an answer is right or wrong.
4. Use the session to clear up incorrect ideas and encourage trainees to learn from their mistakes.

## Section 2.0.0 *Types of Fans*

This exercise will help familiarize trainees with various types of fans. Trainees will need pencils and paper. You will need to arrange workstations with a different type of fan at each station. Allow 20 to 30 minutes for this exercise. This exercise corresponds to Performance Task 1.

1. Discuss various types of fans with trainees.
2. Have trainees rotate between the workstations and identify the type of fan at each station.
3. Answer any questions trainees may have.

## Section 4.0.0 *Types of Blowers*

This exercise will help familiarize trainees with various types of blowers. Trainees will need pencils and paper. You will need to arrange workstations with a different type of blower at each station. Allow 20 to 30 minutes for this exercise. This exercise corresponds to Performance Task 3.

1. Discuss various types of blowers with trainees.
2. Have trainees rotate between the workstations and identify the type of blower at each station.
3. Answer any questions trainees may have.

## MODULE 15312-08 — ANSWERS TO REVIEW QUESTIONS

| Answer | Section |
|---|---|
| 1. a | 1.0.0 |
| 2. b | 1.0.0 |
| 3. c | 2.1.2 |
| 4. d | 2.2.1 |
| 5. d | 2.2.4 |
| 6. a | 3.1.0 |
| 7. b | 3.1.0 |
| 8. c | 3.2.0 |
| 9. b | 4.1.0 |
| 10. c | 4.2.0 |

## CONTREN® LEARNING SERIES — USER UPDATE

The NCCER makes every effort to keep these textbooks up-to-date and free of technical errors. We appreciate your help in this process. If you have an idea for improving this textbook, or if you find an error, a typographical mistake, or an inaccuracy in NCCER's Contren® textbooks, please write us, using this form or a photocopy. Be sure to include the exact module number, page number, a detailed description, and the correction, if applicable. Your input will be brought to the attention of the Technical Review Committee. Thank you for your assistance.

*Instructors* – If you found that additional materials were necessary in order to teach this module effectively, please let us know so that we may include them in the Equipment/Materials list in the Annotated Instructor's Guide.


**Write:** Product Development and Revision
National Center for Construction Education and Research
3600 NW 43rd St., Bldg. G, Gainesville, FL 32606

**Fax:** 352-334-0932

**E-mail:** curriculum@nccer.org


Craft ______________________ Module Name ______________________

Copyright Date ____________ Module Number ____________ Page Number(s) ____________

Description ______________________________________________

______________________________________________

______________________________________________

______________________________________________

(Optional) Correction ______________________________________________

______________________________________________

______________________________________________

(Optional) Your Name and Address ______________________________________________

______________________________________________

______________________________________________

# Glossary of Trade Terms

***Adjacent side***: The side of a right triangle that is next to the reference angle.

***Airfoil:*** A fan blade that has a special shape or orientation that controls stability or direction of air.

***Alkalies:*** Various soluble mineral salts found in water.

***Angular misalignment:*** The condition that occurs when two shafts are at an angle to each other.

***Aqueous:*** Watery; containing or dissolved in water.

***Axial flow:*** The flow of air in line with the axis of the fan shaft.

***Axial movement:*** Movement in the direction of a shaft axis.

***Bore:*** The interior diameter of a shaft.

***Braiding:*** Twisting or interweaving three or more strands of fiber to form one rope-like strand.

***Brine:*** Salty water.

***Burr:*** A small, raised, uneven surface.

***Cam:*** A moving piece of machinery used to secure a bearing to a shaft.

***Carcinogenic:*** Cancer-causing.

***Cartridge mount:*** A complete seal assembly that is installed in one piece.

***Caustic:*** Capable of burning, corroding, dissolving, or eating away by chemical action.

***Centrifugal:*** Moving or directed away from the center of the axis.

***Clutch:*** A device used to engage or disengage a load from a driver.

***Corrosive:*** Inclined to produce corrosion.

***Cosine:*** Trigonometric ratio between the adjacent side and the hypotenuse, written as adjacent divided by the hypotenuse.

***Coupling gap:*** The space between the shaft faces within the couplings.

***Crimped:*** Pressed or pinched into small, regular folds.

***Datum width:*** The width of a V-belt at a particular point, as used in ISO standard codes.

***Deflection:*** The deviation from straight and true as shown by a measuring device, such as a dial indicator.

***Dress down:*** To smooth and remove all extrusions from a belt.

***Driven:*** The device being driven. The driven may be a gear case, pump, or generator.

***Driver:*** The prime mover of a system. The driver is usually a motor.

***Elastomer:*** A synthetic material with the elastic qualities of natural rubber.

***Ester:*** A compound formed by eliminating water and bonding an alcohol and an organic acid.

***Filament:*** A fine or thinly spun thread, fiber, or wire.

***Flaking:*** Cracks in the bearing housing.

***Flashing:*** The almost instantaneous vaporization of a liquid to a vapor.

***Fluting:*** Long, rounded grooves in the raceway of a bearing caused by electric arcing.

***Fretting:*** Losing material due to excessive vibration and rubbing.

***Gland:*** A part used to compress packing in a stuffing box.

***Graduated:*** Marked with degrees of measurement.

***Hypotenuse***: The longest side of a right triangle. It is always located opposite the right angle.

***Increment:*** One of a set of regular, consecutive additions.

***Indicator sag:*** The displacement reading on the dial indicator created by the weight of the dial indicator and the jig acting against the stiffness of the material from which the jig is made.

***Key:*** A device that fits between a coupling and a shaft to prevent slippage.

***Laminated shims:*** Layers of shims that are bonded together.

***Leaching:*** Causing a liquid to filter through a material.

***Migrating:*** Moving from one place to another.

***Nonpiloting:*** Not being automatically centered.

***Opposite side***: The side of a right triangle that is located directly across from the reference angle.

***Oxidizer:*** A substance that supports the combustion of a fuel or propellant.

***Parallel misalignment:*** The condition that occurs when two shafts are misaligned so that their axes never intersect.

***Perimeter:*** The outer boundary of an object or an area.

# Glossary of Trade Terms

***Piano wire:*** A high-strength wire that can be drawn very tight to make a straight line.

***Plumb:*** Vertically aligned.

***Ratio***: A comparison of one value to another value.

***Reciprocal:*** The inverse of any fraction. Since a whole number is equal to the number divided by one, the reciprocal of the number is one divided by the number.

***Reciprocating:*** Moving back and forth.

***Reference angle:*** The angle to which the sides are related as adjacent and opposite.

***Rigid:*** Stiff; not able to bend or flex.

***Rotating face:*** The sealing face of a mechanical seal that rotates with the pump shaft and presses against the stationary face.

***Seal face loading:*** The amount of pressure or force that is applied to the seal face and that acts to close it.

***Shear pin:*** A metal key installed between a drive shaft and a coupling or gear. It is designed to break in the event of a mechanical overload, preventing other, more expensive parts of the drive train from being damaged.

***Sine:*** Ratio between the opposite side and the hypotenuse, written as the opposite divided by the hypotenuse.

***Skew:*** A nonparallel and noncoaxial condition; at an angle.

***Skive:*** A cut made at an angle.

***Spalling:*** The chipping away or breaking of a bearing race.

***Stationary face:*** The nonmoving sealing face of a mechanical seal.

***Stuffing box:*** The housing that holds the packing in a pump, valve, or piece of equipment.

***Stuffing box:*** The housing that holds the packing that controls leakage along a shaft or rod.

***Tangent:*** Ratio between the opposite side and the adjacent side, written as the opposite divided by the adjacent.

***Thermal spike:*** A very rapid increase in temperature, which looks like a spike on a temperature chart graph.

***Thrust:*** The force applied to the sides of a bearing.

***Top width:*** The width across the top of a V-belt, used in DIN and US codes.

***Torque***: A turning or twisting force measured in foot-pounds (ft-lb), inch pounds (in-lb), or kilogram-meters (kgf-m).

***Torsional movement:*** The rotating movement of a shaft.

***Toxic:*** Harmful, destructive, or deadly.

***Trace element:*** A very minute amount of a chemical or particle in a fluid, which can only be detected by lab instruments.

***Transducer:*** A device that converts an input signal to a different type of output signal.

***Volatile:*** Able to rapidly change to vapor.

***Vortex:*** Fluid flow involving rotation about an axis.

***Vulcanized:*** Combined with natural rubber additives to improve strength and resiliency.

# Index

# Index